Scientific

VISIONS

MATHEMATICS

Secondary
Cycle Two, Year Two

Claude Boivin
Dominique Boivin
Antoine Ledoux
Étienne Meyer
Nathalie Ricard
Vincent Roy

9001, boul. Louis-H.-La Fontaine, Anjou (Québec) Canada H1J 2C5
Telephone: 514 351-6010 • Fax: 514 351-3534

ORIGINAL VERSION

Publishing Managers
Julie Duchesne
Véronique Lacroix

Production Manager
Danielle Latendresse

Coordination Manager
Rodolphe Courcy

Project Manager
Marguerite Champagne-Desbiens
Diane Karneyeff

Linguistic Review
Marguerite Champagne-Desbiens

Proofreader
Viviane Deraspe

Graphic Design
Dessine-moi un mouton

Technical Illustrations
Stéphan Vallières

General Illustrations
Yves Boudreau

Geographical Maps
Les Studios Artifisme

Iconographic Research
Jean-François Beaudette

These programs are funded by Quebec's Ministère de l'Éducation, du Loisir et du Sport, through contributions from the Canada-Québec Agreement on Minoroty-Language Education and Second-Language Instruction.

Visions, Technical and Scientific, Student Book, Volume 1,
Secondary Cycle Two, Year Two
© 2009, Les Éditions CEC inc.
9001, boul. Louis-H.-La Fontaine
Anjou, Québec H1J 2C5

Translation of *Visions, Technico-sciences, manuel de l'élève, volume 1*,
(ISBN 978-2-7617-2738-9) © 2009, Les Éditions CEC inc.

Legal Deposit: 2009
Bibliothèque et Archives nationales du Québec
Library and Archives Canada

ISBN 978-2-7617-2814-0

Imprimé au Canada
2 3 4 5 6 15 14 13 12 11

The authors and publisher wish to thank the following people for their collaboration in the evolution of this project.

Collaboration
Jocelyn Dagenais, Teacher, École secondaire André-Laurendeau, c.s. Marie-Victorin

Scientific Consultants
Driss Boukhssmi,Professor, Université du Québec en Abitibi-Témiscamingue

Pedagogical Consultants
Richard Cadieux, Teacher, École secondaire Jean-Baptiste-Meilleur, c.s. des Affluents
Sandra Gallup, Teacher, École secondaire Camille-Lavoie, c.s. du Lac-Saint-Jean
Sébastien Simard, Teacher, École Les Compagnons-de-Cartier, c.s. des Découvreurs
Ray Venables, Teacher, Evergreen High School, c.s. Eastern Shores

ENGLISH VERSION

Translators and Linguistic Review
Donna Aziz
Shona French
Alain Groven
Diane Lewis
Jennifer McCann

Project Managers
Patrick Bérubé
Rita De Marco
Stephanie Vucko
Valerie Vucko

Pedagogical Consultant
Joanne Malowany

Pedagogical Review
Vilma Scattolin

A special thank you to the following people for their collaboration in the evolution of this project.

Michael J. Canuel
Robert Costain
Margaret Dupuis
Rosie Himo
Doris Kerec
Louis-Gilles Lalonde
Denis Montpetit
Bev White

TABLE OF CONTENTS

volume 1

VISION 1

From functions to models 2

Arithmetic and algebra

REVISION 4

- Relations, independent variables and dependent variables
- Inverse
- Function
- Modes of representation
- Properties of functions

SECTION 1.1
From real life to a model 10

- Families of functions
- Choosing a model

SECTION 1.2
Multiplicative parameters 22

- The role of multiplicative parameters: multiplicative parameters a and b

SECTION 1.3
Graphical modelling 34

- Periodic function
- Piecewise function
- Step function

CHRONICLE OF THE PAST
The history of electricity 46

IN THE WORKPLACE
Investment advisors 48

overview 50

VISION 2

Congruent figures and similar figures 60

Arithmetic and algebra • Geometry

REVISION 62

- Classification of triangles
- Properties of quadrilaterals
- Regular polygon
- Area: triangle, quadrilateral, regular polygon and circle
- Angles created by a transversal intersecting two parallel lines
- Factoring: removing a common factor

SECTION 2.1
Congruent triangles 68

- Minimum conditions for congruent triangles
- Deductive reasoning

SECTION 2.2
Similar triangles 80

- Minimum conditions for similar triangles

SECTION 2.3
Manipulating algebraic expressions 89

- Factoring: grouping, difference of squares, perfect square trinomials
- Manipulating algebraic expressions: dividing a polynomial by a binomial, rational expressions and operations on rational expressions

SECTION 2.4
Optimizing a distance 102

- Optimizing a distance

SECTION 2.5
Metric relations 111

- Metric relations in a right triangle

CHRONICLE OF THE PAST
The Ionian School 120

IN THE WORKPLACE
Civil engineering technicians 122

overview 124

VISI3N

From lines to systems of equations . . 134

Arithmetic and algebra • Geometry

REVISION 136

- Solving systems of equations graphically, with a table of values and with the comparison method
- First-degree inequalities in one variable

SECTION 3.1
Points and segments in the Cartesian plane 143

- Change
- Slope of a segment
- Distance between two points
- Point of division

SECTION 3.2
Lines in the Cartesian plane 156

- Equation of a line
- Parallel lines
- Perpendicular lines
- Perpendicular bisector of a segment
- Distance from a point to a line

SECTION 3.3
System of equations 169

- Solving systems of equations: substitution and elimination methods
- Special systems of equations: coinciding and non-coinciding parallel lines

SECTION 3.4
Half-planes in the Cartesian plane 180

- First-degree inequality in two variables

CHRONICLE OF THE PAST
René Descartes 190

IN THE WORKPLACE
Lineperson 192

overview 194

LEARNING AND EVALUATION SITUATIONS 203

REFERENCE 215

Technology 216

Knowledge 222

PRESENTATION OF STUDENT BOOK

This *Student Book* contains three chapters each called "Vision." Each "Vision" presents various "Learning and evaluation situations (LES)" sections and special features "Chronicle of the past," "In the workplace" and "Overview." At the end of the *Student Book*, there is a "Reference" section.

Revision

The "Revision" section helps to reactivate prior knowledge and strategies that will be useful in each "Vision" chapter. This feature contains one or two activities designed to review prior learning, a "Knowledge summary" which provides a summary of the theoretical elements being reviewed and a "Knowledge in action" section consisting of reinforcement exercises on the concepts involved.

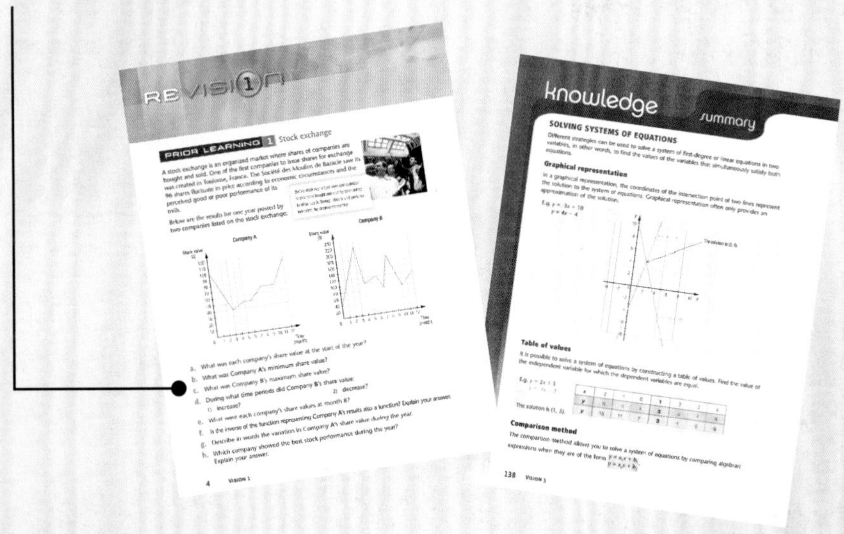

The sections

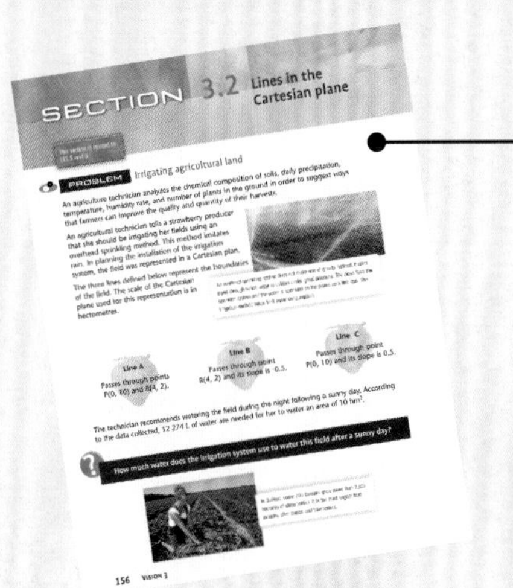

A "Vision" chapter is divided into sections, each starting with a problem and a few activities, followed by the "Technomath," "Knowledge" and "Practice" features. Each section is related to a LES that contributes to the development of subject-specific and cross-curricular competencies, as well as to the integration of mathematical concepts that underscore the development of these competencies.

Problem

The first page of a section presents a problem that serves as a launching point and is made up of a single question. Solving the problem engages several competencies and various strategies while calling upon the mobilization of prior knowledge.

Activity

The activities contribute to the development of subject-specific and cross-curricular competencies, require the use of various strategies, mobilize knowledge and further the understanding of mathematical notions. These activities can take on several forms: questionnaires, material manipulation, construction, games, stories, simulations, historical texts, etc.

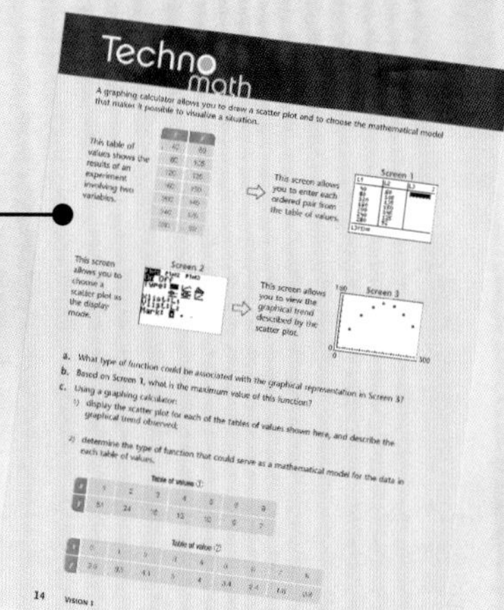

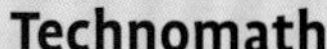

Technomath

The "Technomath" section allows students to use technological tools such as a graphing calculator, dynamic geometry software or a spreadsheet program. In addition, the section shows how to use these tools and offers several questions in direct relation to the mathematical concepts associated with the content of the chapter.

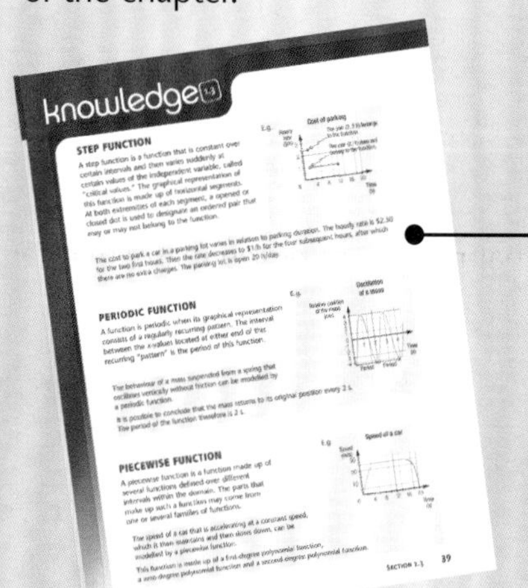

Knowledge

The "Knowledge" section presents a summary of the theoretical elements encountered in the section. Theoretical statements are supported with examples in order to foster students' understanding of the various concepts.

Practice
The "Practice" section presents a series of contextualized exercises and problems that foster the development of the competencies and the consolidation of what has been learned throughout the section.

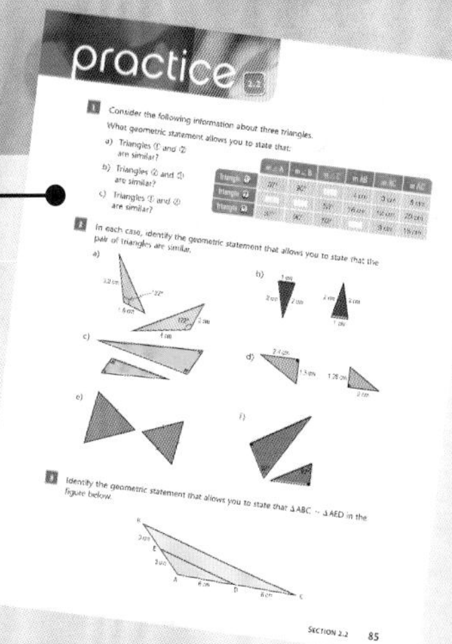

Special Features

Chronicle of the past
The "Chronicle of the past" feature recalls the history of mathematics and the lives of certain mathematicians who have contributed to the development of mathematical concepts that are directly related to the content of the "Vision" chapter being studied. This feature includes a series of questions that deepen students' understanding of the subject.

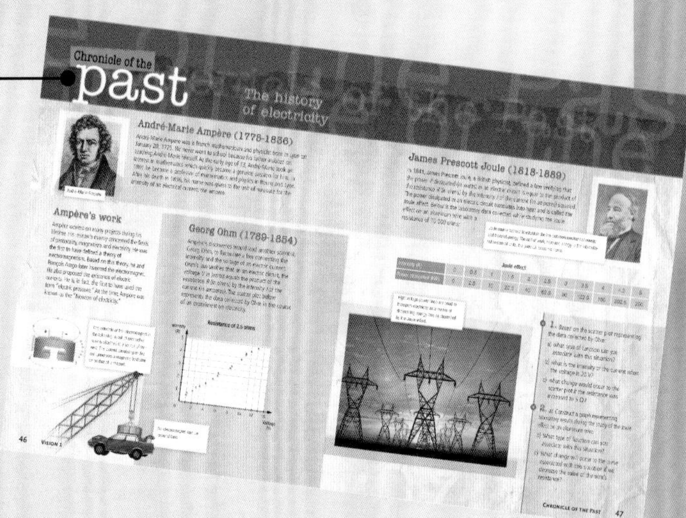

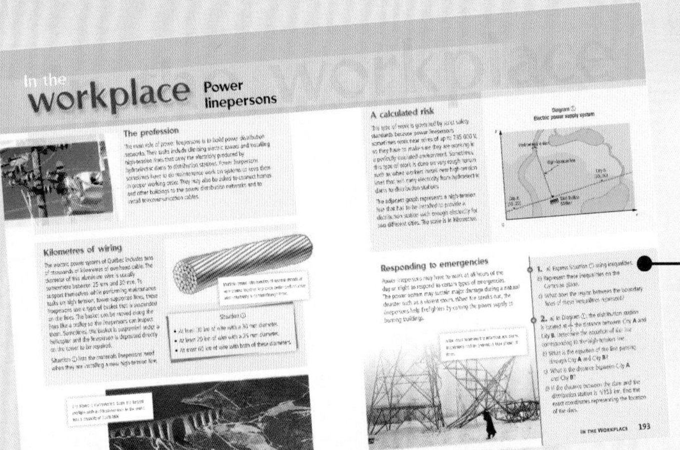

In the workplace
The "In the workplace" feature presents a profession or a trade that makes use of the mathematical notions studied in the related "Vision" chapter. This feature includes a series of questions designed to deepen students' understanding of the subject.

Overview

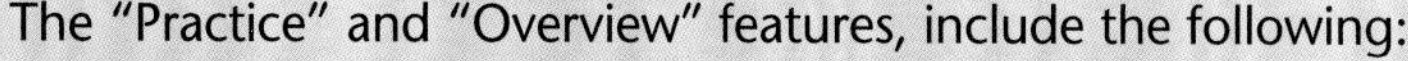

The "Overview" feature concludes each "Vision" chapter and presents a series of contextualized exercises and problems that integrate and consolidate the competencies that have been developed and the mathematical notions studied. This feature ends with a bank of problems, each of which focuses on solving, reasoning or communicating.

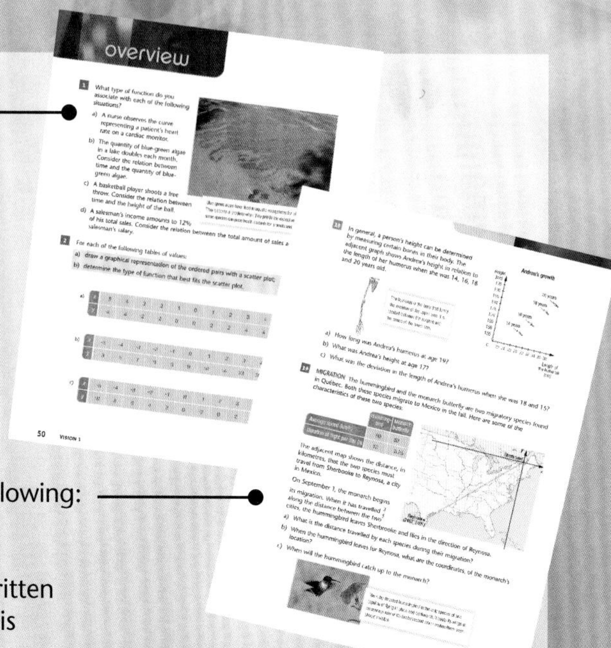

The "Practice" and "Overview" features, include the following:

- A number in a blue square refers to a Priority **1** and a number in an orange square a Priority **2**.
- When a problem refers to actual facts, a keyword written in red uppercase indicates the subject with which it is associated.

Learning and evaluation situations

The "Learning and evaluation situations" (LES) are grouped according to a common thematic thread; each focuses on a general field of instruction, a subject-specific competency and two cross-curricular competencies. The knowledge acquired through the sections helps to complete the tasks required in the LES.

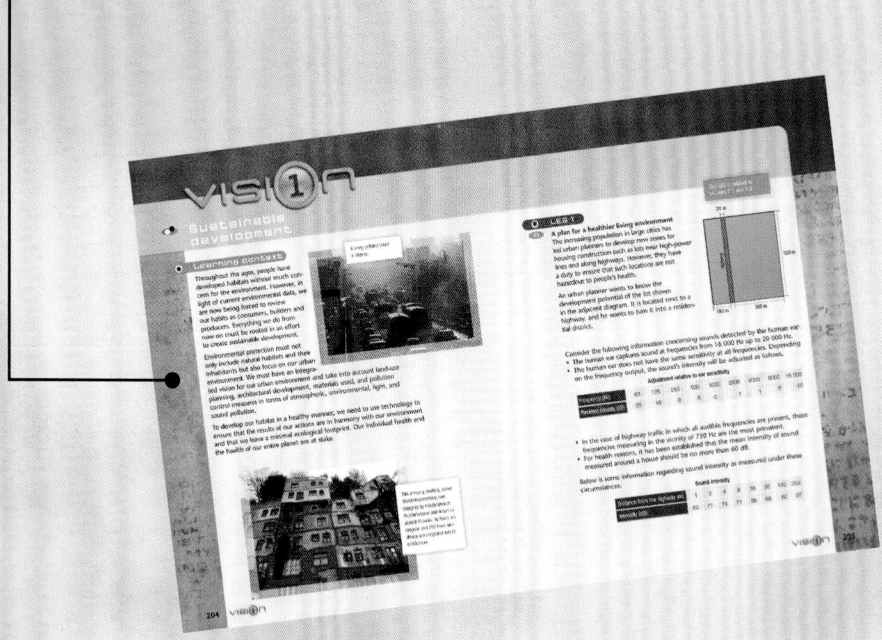

Reference

Located at the end of the *Student Book*, the "Reference" section contains several tools that support the student-learning process. It consists of two distinct parts.

The "Technology" part provides explanations pertaining to the functions of a graphing calculator, the use of a spreadsheet program as well as the use of dynamic geometry software.

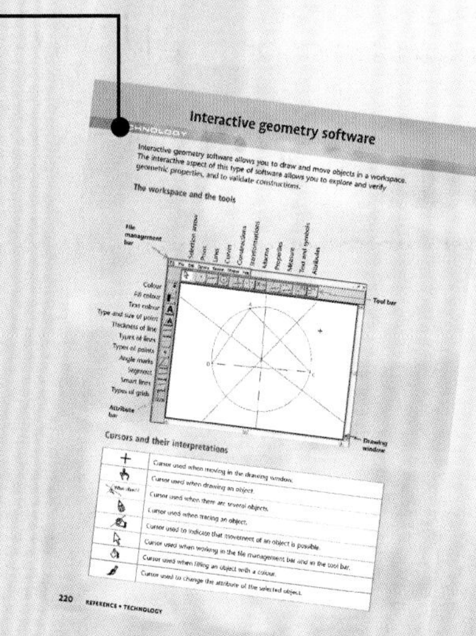

Interactive geometry software

Interactive geometry software allows you to draw and move objects in a workspace. The interactive aspect of this type of software allows you to explore and verify geometric properties, and to validate constructions.

The workspace and the tools

Cursors and their interpretations

+	Cursor used when moving in the drawing window.
	Cursor used when drawing an object.
	Cursor used when there are several objects.
	Cursor used when tracing an object.
	Cursor used to indicate that movement of an object is possible.
	Cursor used when working in the file management bar and in the tool bar.
	Cursor used when filling an object with a colour.
	Cursor used to change the attribute of the selected object.

220 REFERENCE • TECHNOLOGY

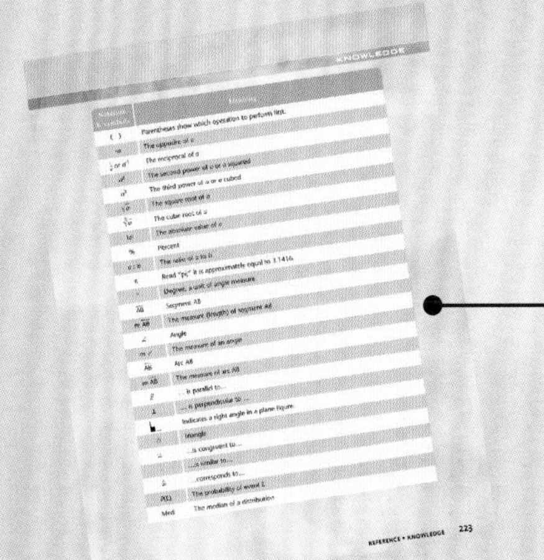

KNOWLEDGE

( )	Parentheses show which operation to perform first.
$-a$	The opposite of a
$\frac{1}{a}$ or a^{-1}	The reciprocal of a
a^2	The second power of a or a squared
a^3	The third power of a or a cubed
$\sqrt{a}$	The square root of a
$\sqrt[3]{a}$	The cube root of a
$\|a\|$	The absolute value of a
%	Percent
$a : b$	The ratio of a to b
π	Read "pi." It is approximately equal to 3.1416.
°	Degree, a unit of angle measure
$\overline{AB}$	Segment AB
$m\overline{AB}$	The measure (length) of segment AB
$\angle$	Angle
$m\angle$	The measure of an angle
$\overset{\frown}{AB}$	Arc AB
$m\overset{\frown}{AB}$	The measure of arc AB
//	... is parallel to...
⊥	... is perpendicular to ...
	Indicates a right angle in a plane figure
△	Triangle
≅	... is congruent to...
	... is similar to...
≙	... corresponds to...
P(E)	The probability of event E
Med	The median of a distribution

REFERENCE • KNOWLEDGE 223

The "Knowledge" part presents notations and symbols used in the *Student Book*. Geometric principles are also listed. This part concludes with a glossary and an index.

Icons

Indicates that a worksheet is available in the *Teaching Guide*.

Indicates that the activity can be performed in teams. Details on this topic are provided in the *Teaching Guide*.

Indicates that some key features of subject-specific competency 1 are mobilized.

Indicates that some key features of subject-specific competency 2 are mobilized.

Indicates that some key features of subject-specific competency 3 are mobilized.

Indicates that subject-specific competency 1 is being targeted in the LES.

Indicates that subject-specific competency 2 is being targeted in the LES.

Indicates that subject-specific competency 2 is being targeted in the LES.

VISION 1

From functions to models

Choosing the telephone plan that best suits your needs, forecasting the market value of shares, managing marine or rail transportation and evaluating the speed of sound through the air in relation to temperature are examples of activities that involve mathematical models. What type of function can be applied to each of these situations? What characteristics are exhibited by each of these functions? In "Vision 1," you will analyze and model various situations linked to economic, social, technical and scientific contexts, as well as everyday activities and determine the characteristics applicable to each of them. You will also study the role of multiplicative parameters associated with the rule of a function and the effects generated by changes in their values.

Arithmetic and algebra

- Relations, functions and inverse of a function
- Various families of functions
- The role of multiplicative parameters
- Step, periodic and piecewise functions
- Modelling and analyzing situations

Geometry

Statistics

Probability

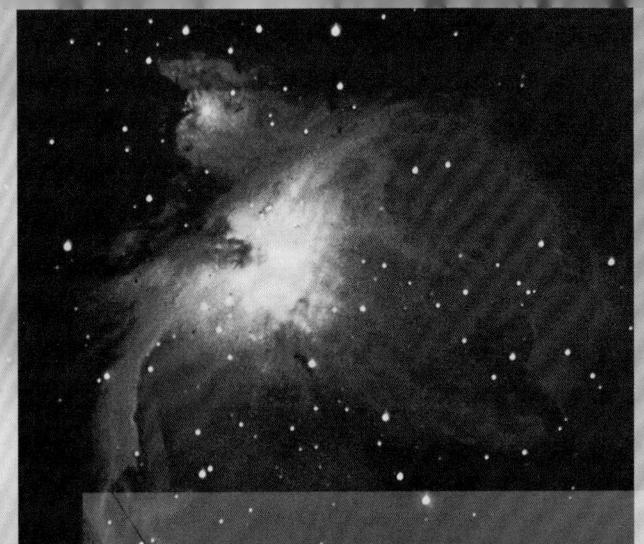

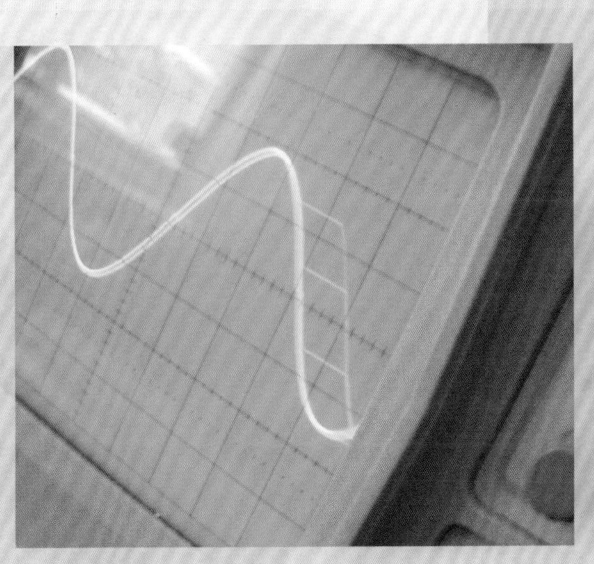

LEARNING AND EVALUATION SITUATIONS — Sustainable development 204

Chronicle of the past — The history of electricity 46

In the workplace — Investment advisors 48

PRIOR LEARNING 1 Stock exchange

A stock exchange is an organized market where shares of companies are bought and sold. One of the first companies to issue shares for exchange was created in Toulouse, France. The Société des Moulins de Bazacle saw its 96 shares fluctuate in price according to economic circumstances and the perceived good or poor performance of its mills.

Before stock exchanges were computerized, shares were bought and sold by open outcry, in other words through shouts and gestures indicating the desired transaction.

Below are the results for one year posted by two companies listed on the stock exchange:

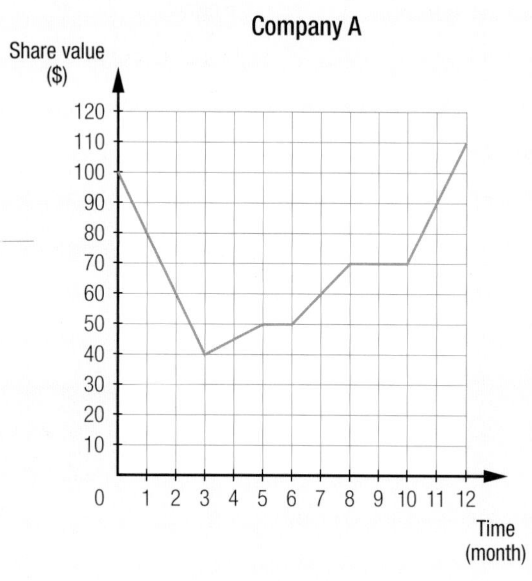

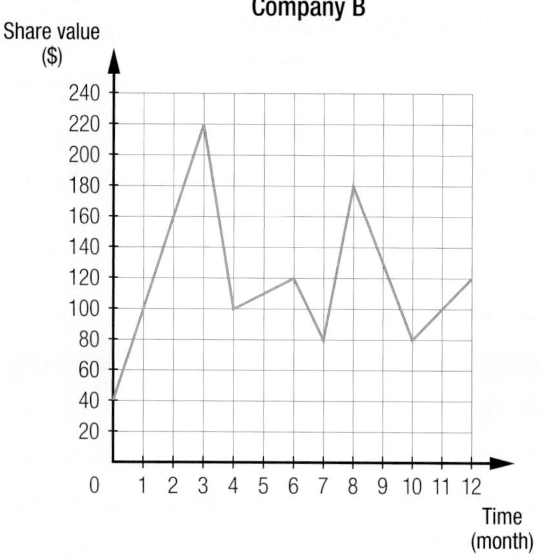

a. What was each company's share value at the start of the year?

b. What was Company **A**'s minimum share value?

c. What was Company **B**'s maximum share value?

d. During what time periods did Company **B**'s share value:

1) increase? 2) decrease?

e. What were each company's share values at month 8?

f. Is the inverse of the function representing Company **A**'s results also a function? Explain your answer.

g. Describe in words the variation in Company **A**'s share value during the year.

h. Which company showed the best stock performance during the year? Explain your answer.

knowledge summary

RELATIONS, INDEPENDENT VARIABLES AND DEPENDENT VARIABLES

The link between two variables is called a **relation.**

In general, in a relation between two variables the following is true:

- The variable whose variation **generates** the other's variation is called an **independent variable**.
- The variable whose variation **reacts** to the other's variation is called a **dependent variable**.

E.g.

Relation	Independent variable	Dependent variable
1) The mass and cost of a frozen turkey.	Mass → Cost The cost of a frozen turkey depends on its mass.	
2) The total area of the walls and ceiling of a room and the time required to paint this room.	Total area → Times The time required to paint a room depends on the total area of the walls and ceiling.	

INVERSE

An inverse relation, or simply an **inverse**, is obtained by exchanging the values of each ordered pair in a relation between two variables.

E.g.

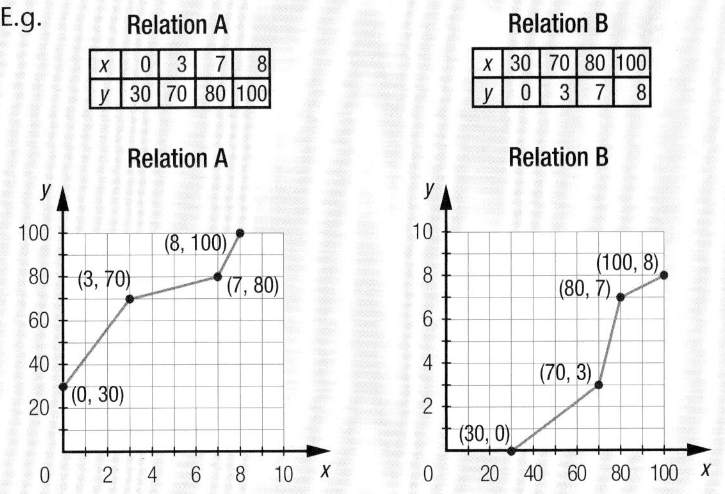

Relation A

x	0	3	7	8
y	30	70	80	100

Relation B

x	30	70	80	100
y	0	3	7	8

Relation **B** is the inverse of Relation **A,** and vice versa.

FUNCTION

A relation between two variables is called a functional relation, or simply a **function** when no more than one value of the dependent variable is associated with each value of the independent variable.

In the graphical representation of a function, no more than one y-coordinate is associated with each x-coordinate.

E.g.

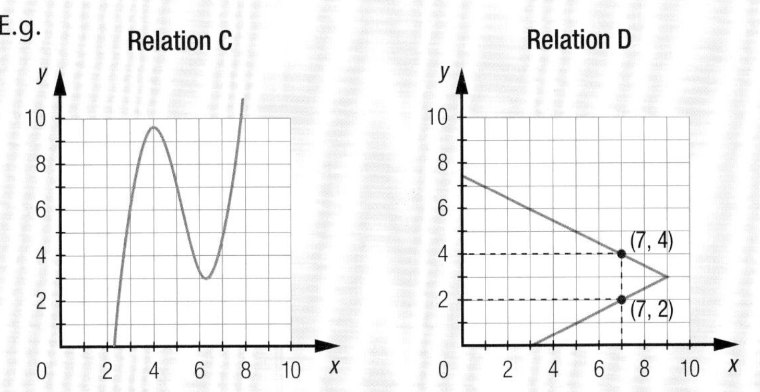

Relation **C** is a function. Relation **D** is not a function since the x-coordinate 7 is associated with more than one y-coordinate, in this case 2 and 4.

PROPERTIES OF FUNCTIONS

Domain and range (image)

The **domain** of a function is the set of all the values of the **independent variable.**

The **range or image** of a function is the set of all the values of the **dependent variable.**

E.g.

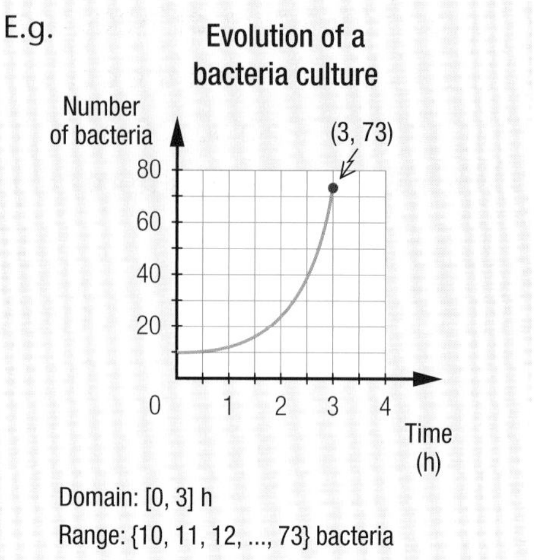

Variation: increase, decrease and constant

Over an interval of the domain, a function is:

- **increasing** when a positive or negative variation of the independent variable generates respectively a positive or negative variation of the dependent variable
- **decreasing** when a positive or negative variation of the independent variable generates respectively a negative or positive variation of the dependent variable
- **constant** when a variation of the independent variable does not generate any variation of the dependent variable

E.g.

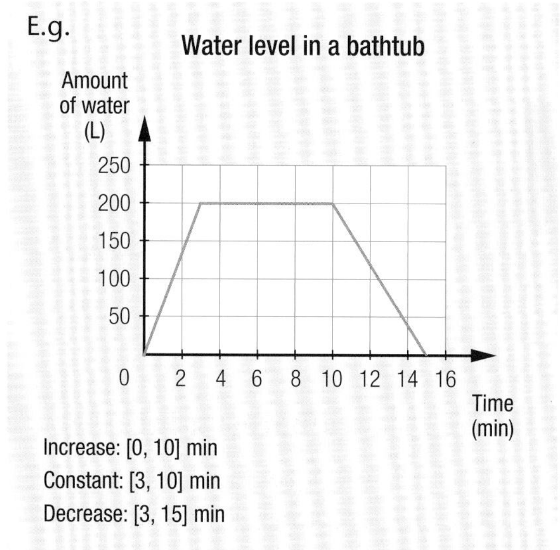

Extrema: minimum and maximum

The **minimum** of a function is the smallest value of the dependent variable.

The **maximum** of a function is the greatest value of the dependent variable.

E.g.

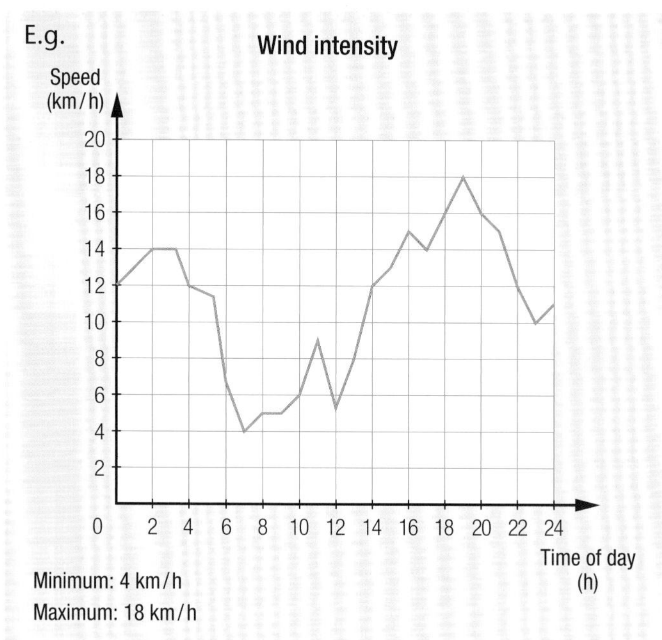

The sign: positive or negative

Over an interval within the domain, a function is:

- **positive** if the values of the dependent variable are positive
- **negative** if the values of the dependent variable are negative

E.g.

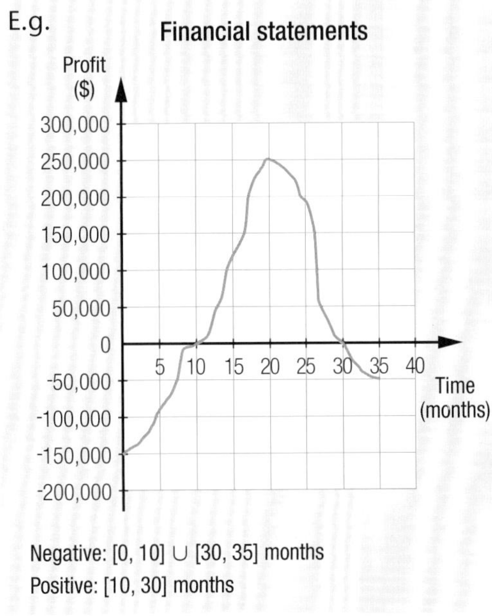

Negative: [0, 10] ∪ [30, 35] months
Positive: [10, 30] months

Intercept coordinates: *x*-intercept (zero) and *y*-intercept (initial value)

The **zero of a function** is a value of the independent variable when that of the dependent variable is zero. Graphically, the zero corresponds to the ***x*-intercept**, meaning the *x*-coordinate of the intersection point(s) of the curve and the *x*-axis.

The **initial value** of a function is the value of the dependent variable when that of the independent variable is zero. Graphically, the initial value corresponds to the ***y*-intercept**, meaning the *y*-coordinate of the intersection point of the curve and the *y*-axis.

E.g.

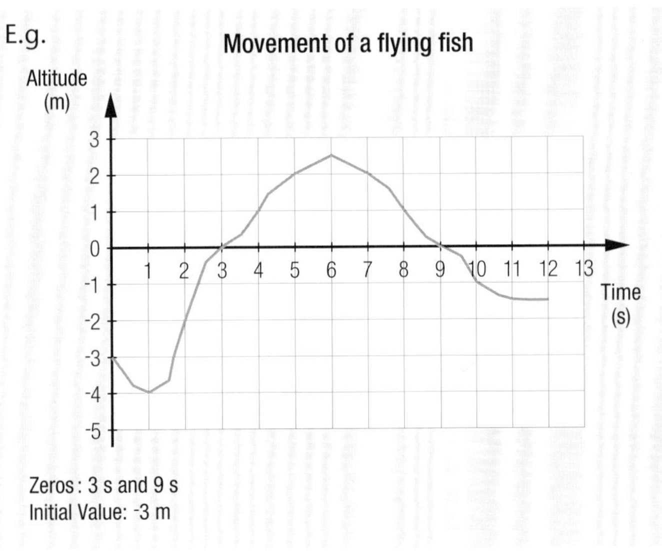

Zeros: 3 s and 9 s
Initial Value: -3 m

knowledge in action

1 For each of the following relations, identify the variable that acts as an independent variable and the one that acts as a dependent variable.

Relations

	Variable ①	Variable ②
a)	the cost of renting a car	the number of days the car is rented
b)	the time required to cook a chicken	the mass of the chicken
c)	the number of members in a family	the weekly grocery bill
d)	the mass of a parcel	the postal mailing cost
e)	the time of day	the Moon's position in the sky
f)	the number of donors at a blood drive	the amount of blood collected

2 Identify which of the following graphs represents a function.

A

B

D

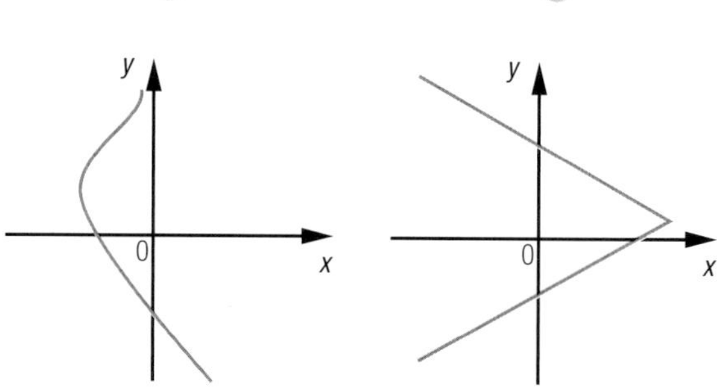

3 Each graph below represents a function.

Graph ①

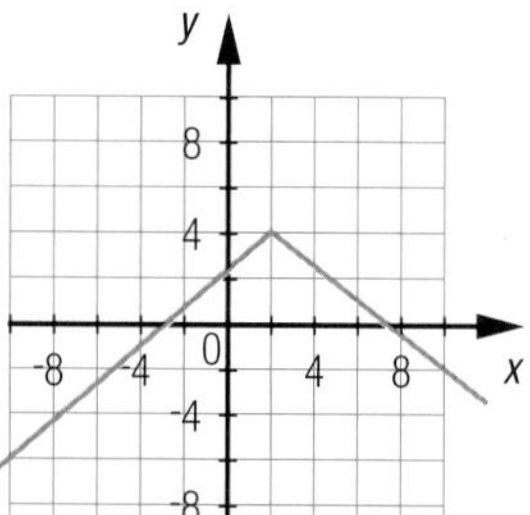

Graph ②

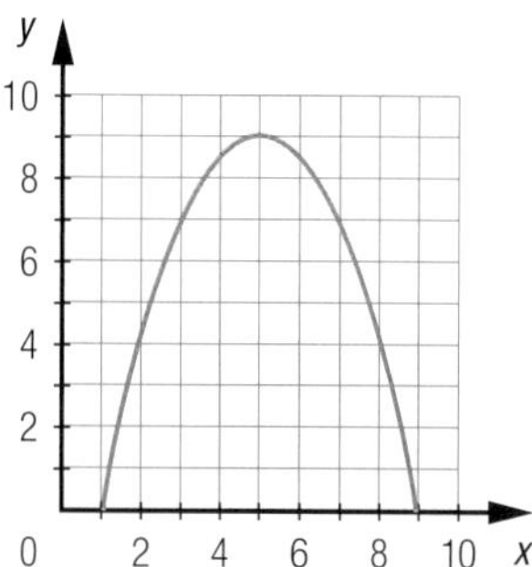

Graph ③

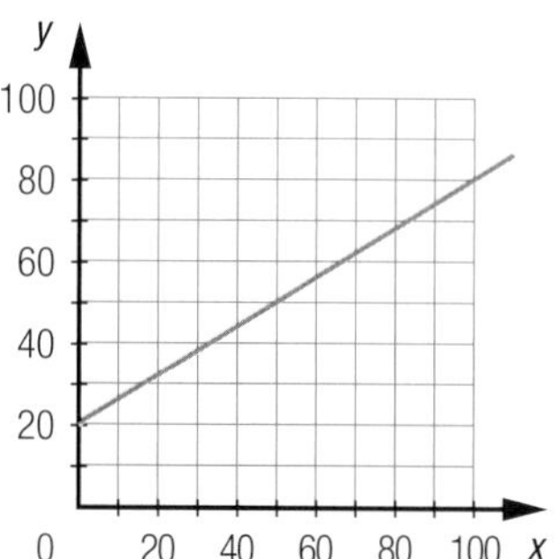

a) Draw a graphical representation of the inverse of each of these functions.

b) Indicate whether the inverse is a function.

4 For each of the functions below, determine:

a) the domain and the range
b) the variation
c) the sign
d) the extrema
e) the initial value
f) the zero(s)

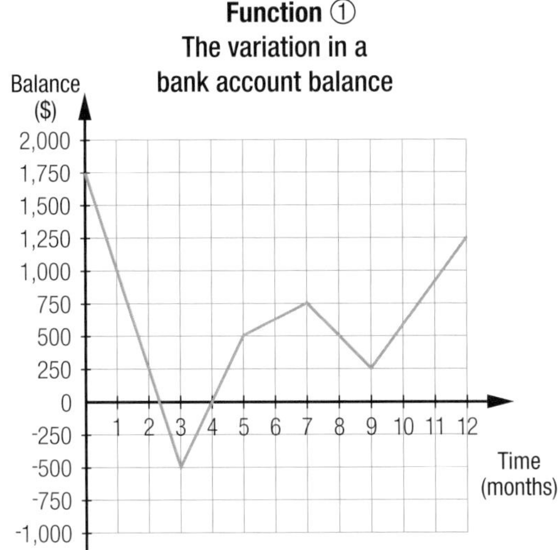

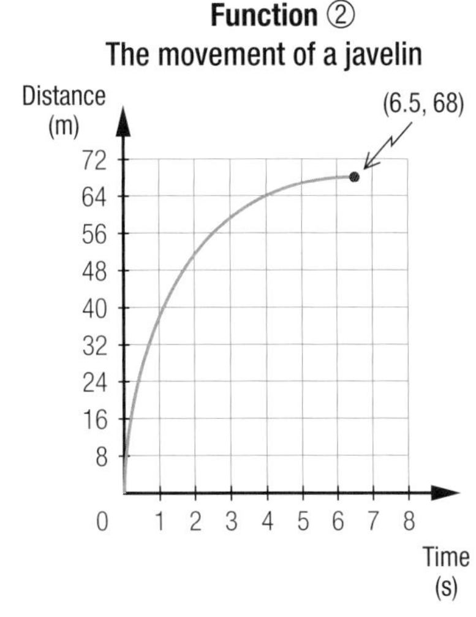

5 The adjacent graph displays the outside temperature recorded during the first 12 hours of an autumn day.

a) What was the lowest temperature?
b) What was the initial temperature?
c) At what time were the following recorded:
   1) the minimum temperature?
   2) the maximum temperature?
d) At what time(s) was the temperature 0°C?
e) Over what time interval(s) was the temperature:
   1) negative?
   2) positive?

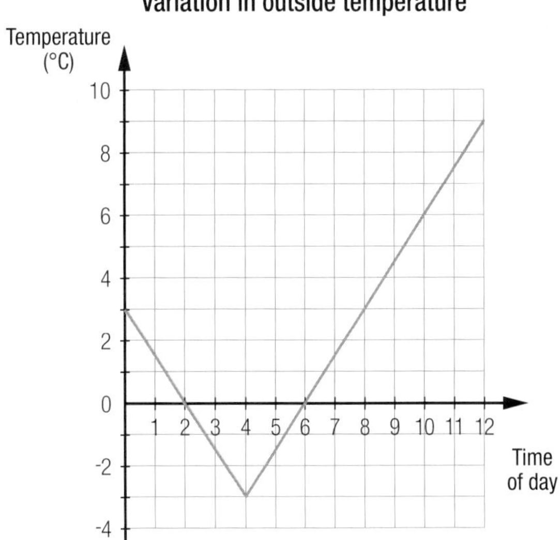

6 Dina borrows $2,785 from her parents to purchase a computer. She pays no interest on this loan. She promises to repay $80 each month during the first year. As her income increases in the years to come, she will increase the amount she repays her parents by 25% per year. How long will it take her to repay the loan?

SECTION 1.1 From real life to a model

This section refers to LES 1.

PROBLEM The Nations Cup

Inaugurated in 2007, the Nations Cup is the first compulsory race for cyclists who want to qualify for international competition. The Nations Cup determines the number of cyclists each country can send to the World Championships.

The table below displays the performance of the winner in one leg of the Nations Cup:

Distance covered by a cyclist during Leg 1

Pit stop	Distance covered (km)	Time elapsed (min)
Start	0	0
A	25	42
B	42	84
C	62	114
D	84	141
E	98	171
Finish		216.8

What is the total distance covered by the cyclists during this leg?

In 2008, one segment of the Nations Cup, U23, took place in Saguenay, Québec. This was the only qualifying event of the Cup to be held outside of Europe, and it involved the best male cyclists aged 18 to 23.

ACTIVITY 1 Different models for different situations

Many situations in daily life can be represented by mathematical models. Models, such as those shown below, make it possible to analyze a situation or to make various predictions.
Below are two possible situations:

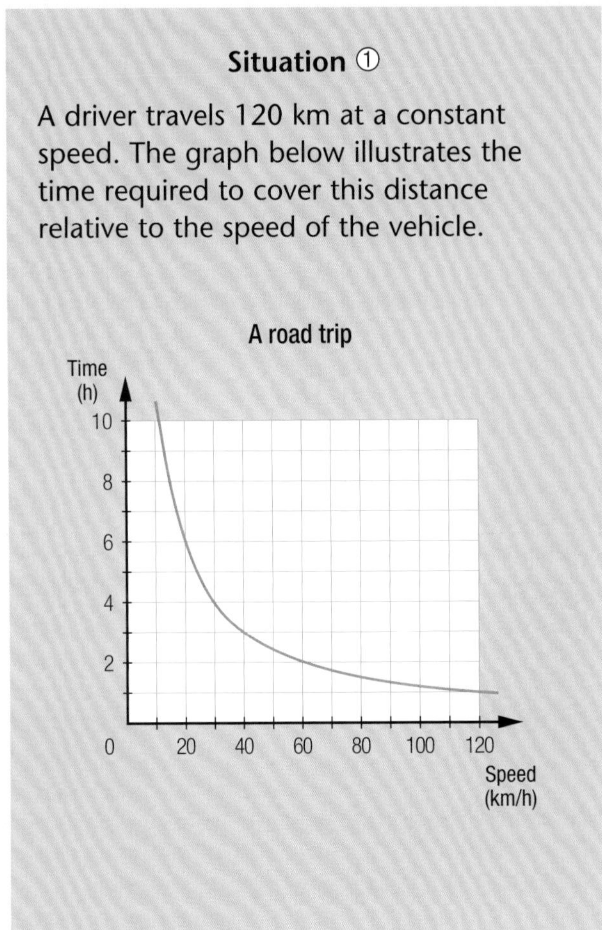

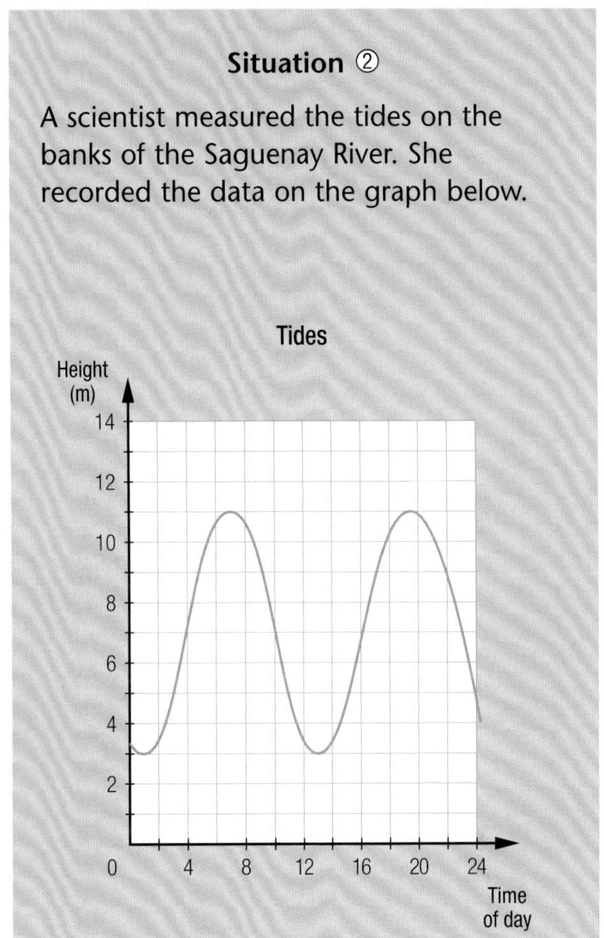

a. Based on Situation ①, answer the following:

1) What is the duration of the trip if the speed of the vehicle is 80 km/h?
2) What is the maximum duration of the trip?
3) What is the minimum duration of the trip?

b. Based on Situation ②, answer the following:

1) What is the height of the tide at the start of the observations?
2) What is the tidal range, meaning the difference in height between consecutive high and low tides?

c. Describe Situations ① and ② in your own words.

Below are two more examples from daily life that can be represented by mathematical models:

Situation ③

Employees are cleaning the reservoir that provides drinking water to a municipality. Using a constant-displacement pump, they empty the reservoir, then clean and refill it. The following information pertains to this operation:

Reservoir water level

Time (h)	Amount of water in reservoir (m^3)
0	6000
2	4500
4	3000
6	1500
8	0
10	1500
12	3000

Situation ④

An amount of $5,000 is invested at an annual compound interest rate of 6%. The table of values below shows the amount *A* accrued over time *t*.

Variation in an investment

Time (years)	Accrued amount ($)
0	5,000
1	5,300
2	5,618
4	≈ 6,312.38
6	≈ 7,092.60
8	≈ 7,969.24
10	≈ 8,954.24
12	≈ 10,060.98
14	≈ 11,304.52
16	≈ 12,701.76

d. Draw graphs to represent Situation ③ and ④.

e. Based on Situation ③, answer the following:

1) How much water was in the reservoir when the clean-up began?
2) What is the hourly displacement of the pump?
3) Considering that the reservoir was filled to capacity at the start of the cleaning job, how long did it take to complete this operation?

f. Based on Situation ④, answer the following:

1) What will be the accrued amount 18 years after the initial investment?
2) When will the amount accrued reach $16,000?

g. Do Situations ③ and ④ represent functions? Explain your answer.

h. Describe Situations ③ and ④ in your own words.

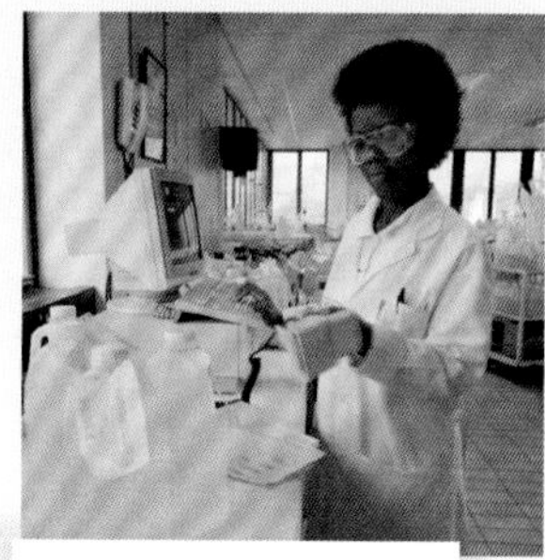

In order for water to be considered safe to drink, the water quality must meet very specific criteria: particular attention must be paid to the presence of bacteria or other substances such as pesticides and heavy metals. Water quality is tested by analyzing samples in laboratories.

ACTIVITY 2 Black gold

As a source of mechanical or thermal energy or as a component in the production of lubricants, oil is a necessity in most of today's economic sectors. In 1970, the cost of a barrel of crude oil was around $5, and by the year 2000 it had stabilized at around $20.

There is an increasing worldwide demand for oil, and this consumption has an impact on the environment. Oil combustion releases tons of carbon dioxide (CO_2) into the atmosphere, thereby contributing greatly to global warming.

The adjacent scatter plot shows the changes in the price for each barrel of crude oil from 2000 to 2008. Each of the curves represents a function.

a. Which of these curves best fits the scatter plot?

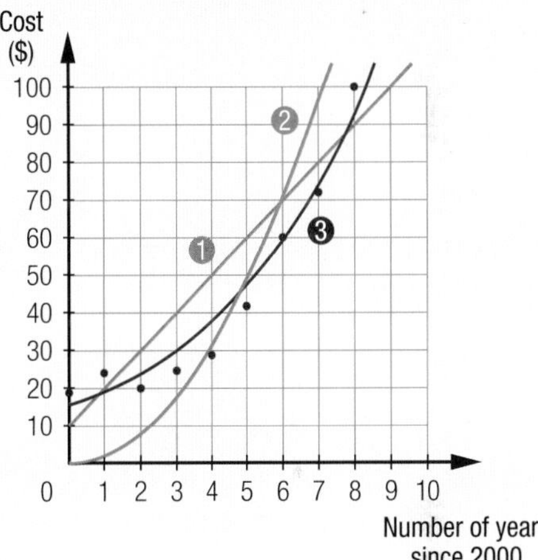

Below are the rules of three functions:

Second-degree polynomial function	First-degree polynomial function	Exponential function
$y = 2x^2$	$y = 10x + 10$	$y = 15.5(1.25)^x$

b. Associate each of the rules above with a curve in the above scatter plot.

c. Which type of function could serve as a mathematical model to describe this situation?

d. If the trend continues, what will be the cost for each barrel of crude oil:

1) in 2025? 2) in 2030?

e. Are the values found in **d.** realistic? Explain your answer.

Techno math

A graphing calculator allows you to draw a scatter plot and to choose the mathematical model that makes it possible to visualize a situation.

This table of values shows the results of an experiment involving two variables.

x	y
40	60
80	105
120	135
160	150
200	145
240	125
280	92

This screen allows you to enter each ordered pair from the table of values.

Screen 1

L1	L2	L3 3
40	60	
80	105	
120	135	
160	150	
200	145	
240	125	
280	92	

L3(1)=

This screen allows you to choose a scatter plot as the display mode.

Screen 2

This screen allows you to view the graphical trend described by the scatter plot.

Screen 3

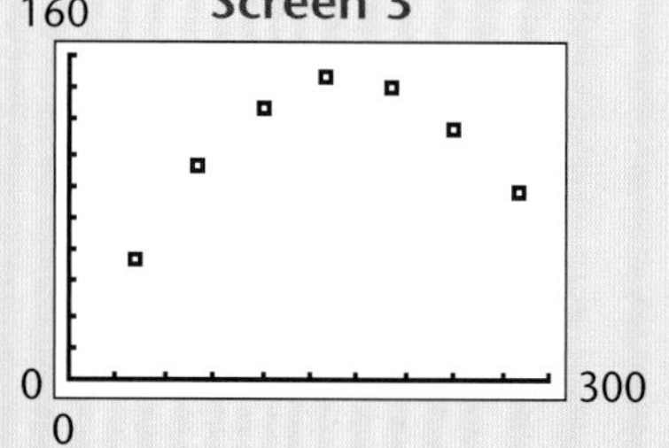

a. What type of function could be associated with the graphical representation in Screen **3**?

b. Based on Screen **1**, what is the maximum value of this function?

c. Using a graphing calculator, do the following:

1) Display the scatter plot for each of the tables of values shown below and describe the graphical trend observed.

2) Determine the type of function that could serve as a mathematical model for the data in each table of values.

Table of values ①

x	1	2	3	4	5	6	9
y	51	24	16	13	10	9	7

Table of values ②

x	0	1	2	3	4	5	6	7	8
y	2.5	3.5	4.1	5	4	3.4	2.4	1.6	0.8

knowledge 1.1

VARIOUS FAMILIES OF FUNCTIONS

Depending on the relationship between two variables, various everyday situations can be represented with mathematical models, in other words, with functions that behave in a manner that is both known and predictable. More specifically, these models make it possible, to analyze a situation or to make certain predictions. Following are a few useful functions that serve as mathematical models for given situations:

Zero-degree polynomial function

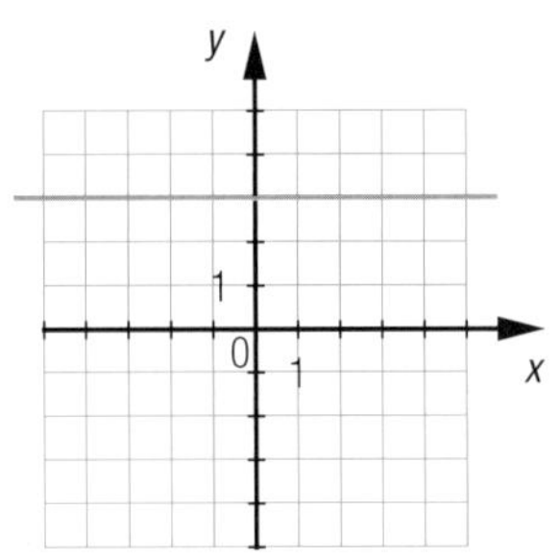

First-degree polynomial function

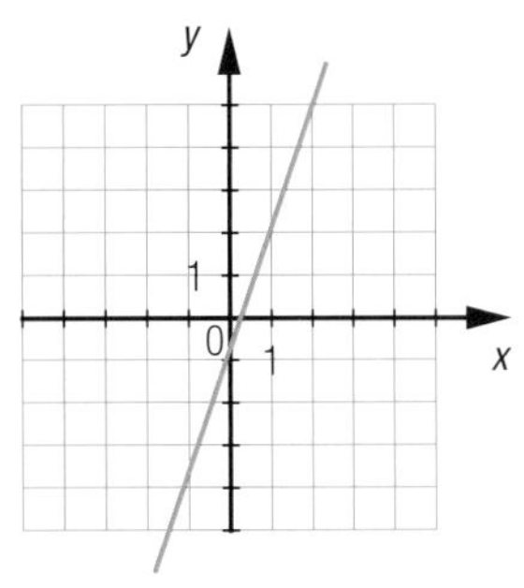

Second-degree polynomial function

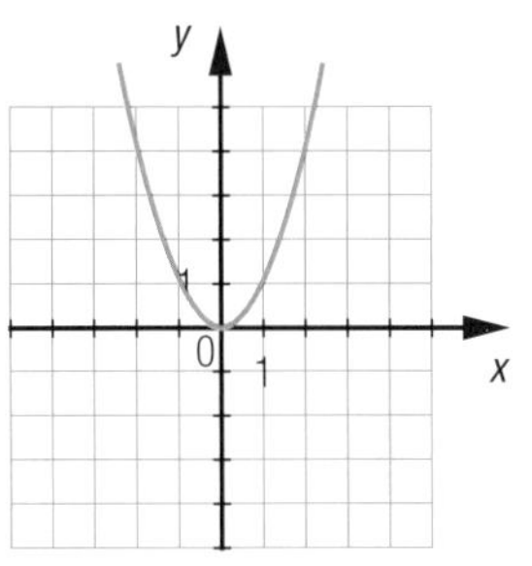

Inverse variation function

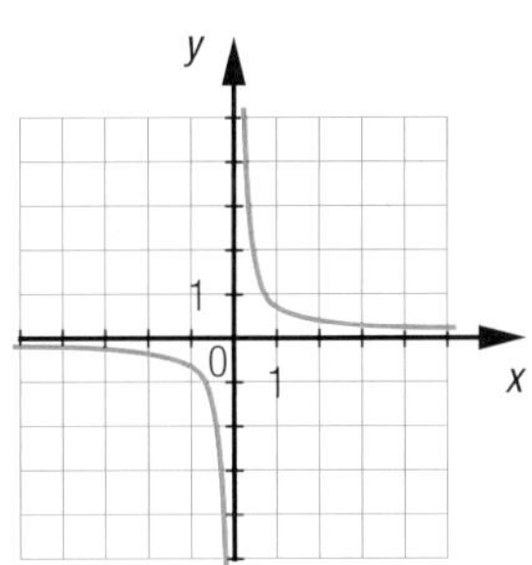

Exponential function

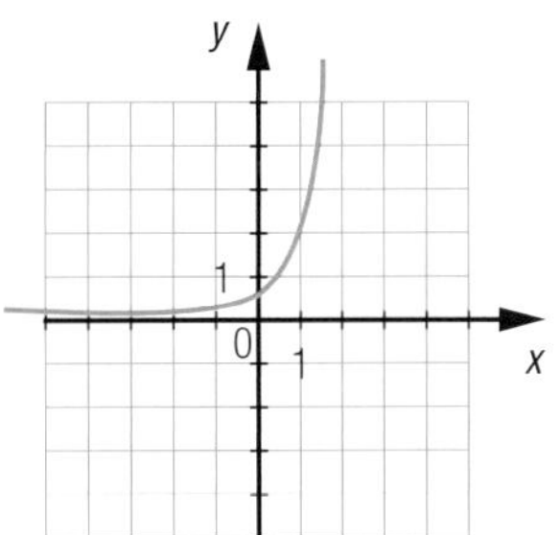

Step function

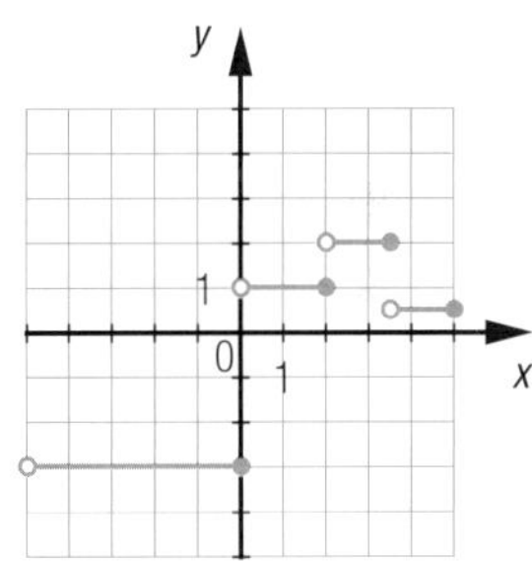

Absolute value function

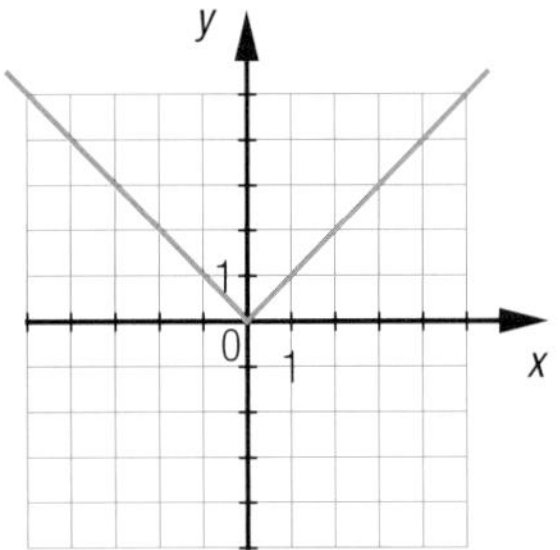

Periodic function

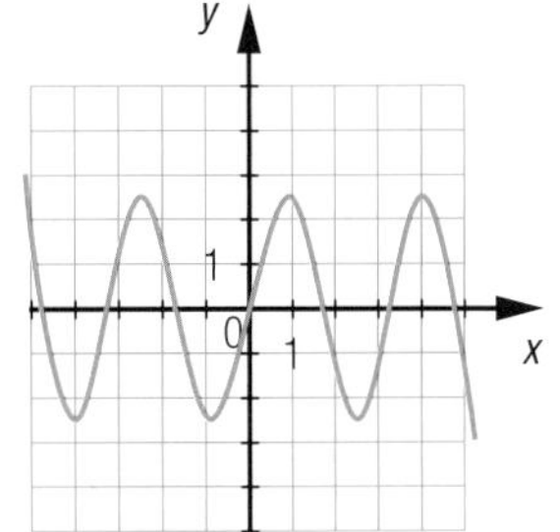

Piecewise function

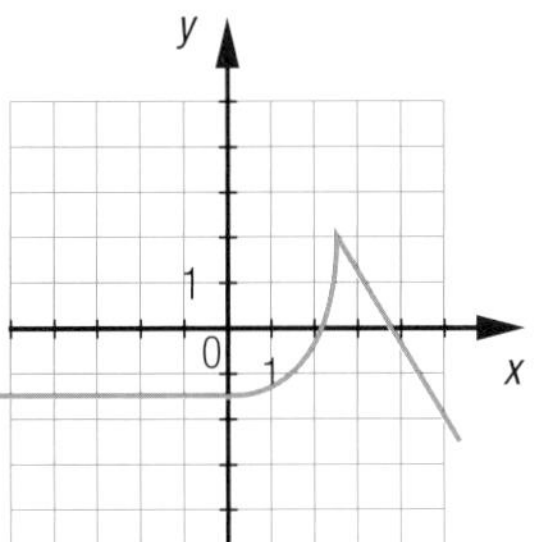

CHOOSING A MODEL

The information in a table of values or a scatter plot associated with a situation involving two variables, does not always exhibit a systematic regularity or result in points arranged according to a perfectly-defined trend. This may be due to errors in manipulation, measurement or to the degree of precision of the instrument used. By analyzing the shape of the scatter plot or by comparing certain properties of functions, you can choose a mathematical model that makes it possible to visualize a situation.

E.g. 1) The information below represents the concentration of greenhouse gases in the atmosphere from 1950 to 2000:

Greenhouse gases

Year	1950	1960	1970	1980	1990	2000
Concentration of CO_2 (ppm)	50	80	100	300	800	1800

The scatter plot representing this situation shows a trend associated with an exponential function, in other words, a function with an increasingly steep curve.

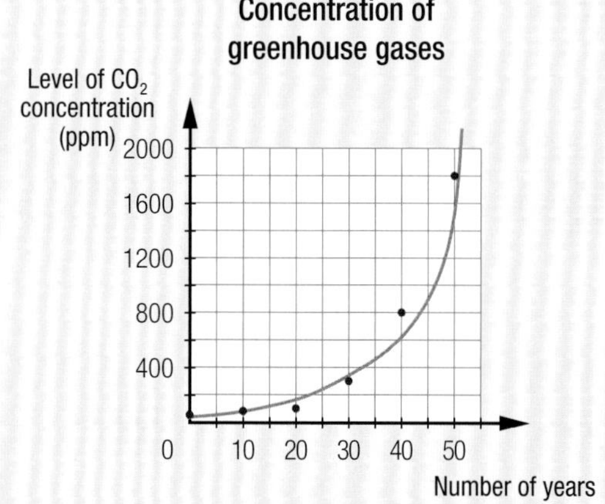

2) Consider the time required to complete a job and the number of workers assigned to the task.

Setting a stage

Number of workers	2	4	6	8	10	12	14	16	18	20
Time (h)	22	12	7	6	4.5	4	3	3	2.5	2.25

The scatter plot representing this situation shows a trend associated with the inverse variation function, in other words, a function where the ends of the curve move more and more slowly towards the axes without ever touching them.

Setting a stage

Time (h)

25
20
15
10
5
0
5
10
15
20
25

Number of workers

practice 1.1

1 Identify the type of function associated with each of the following situations.

a)

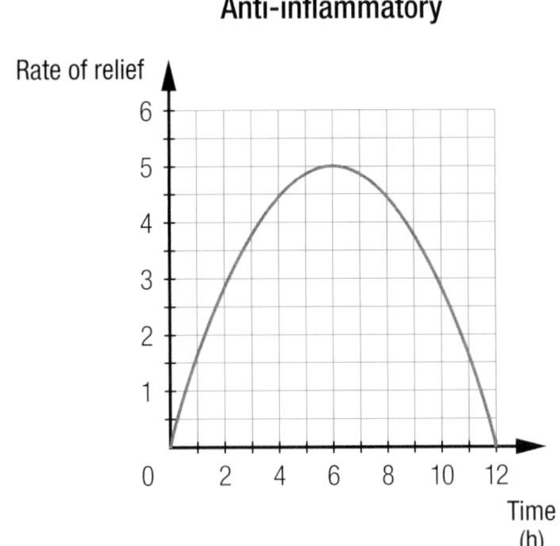

b)

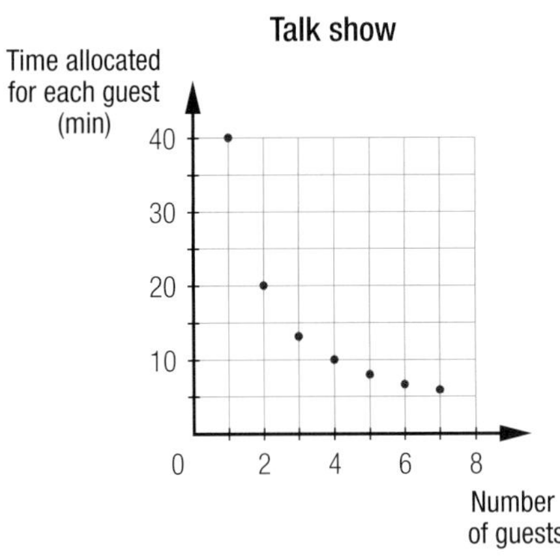

c)

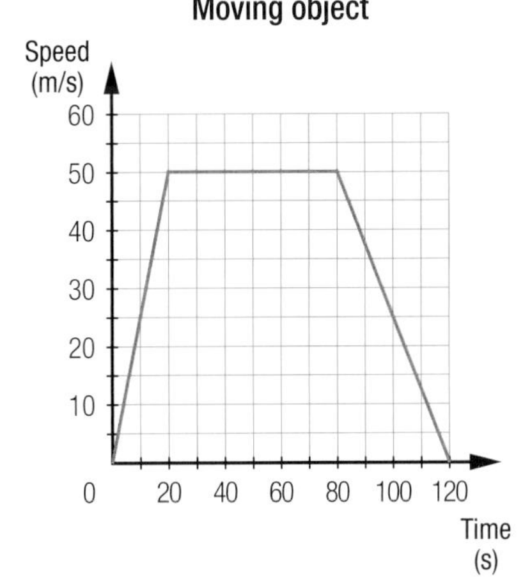

d)

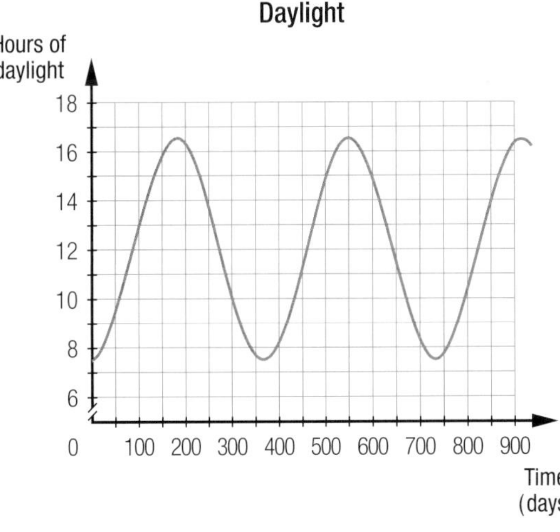

e)

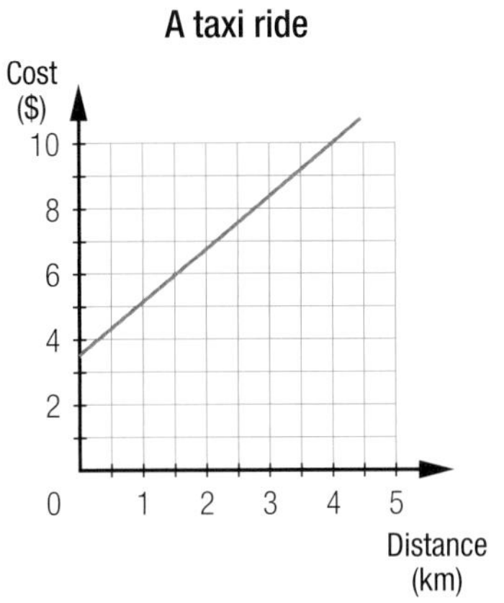

f)

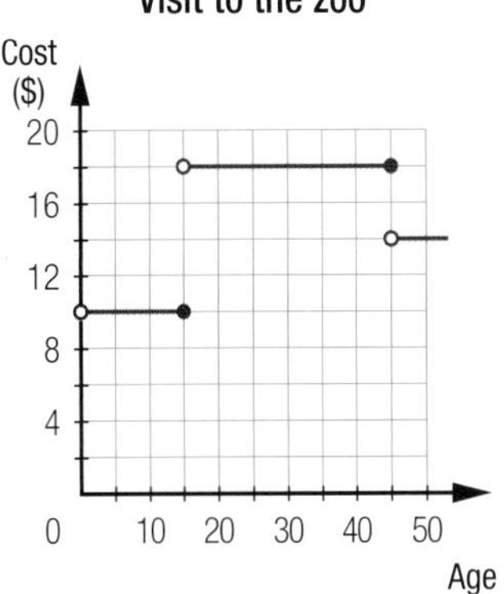

2 Associate each situation in the left-hand column with the corresponding function in the right-hand column.

Situations

- **A** To prepare for the next race, a marathon runner doubles his or her training time every week.
- **B** An individual must spend \$49.99/month for Internet access regardless of the number of minutes in downloads or uploads.
- **C** The area of a circle equals the product of π and the length of its radius squared.
- **D** The hydraulic power output of a Type **A** pump is three times higher than that of a Type **B** pump.
- **E** During a car race, engineers analyze the height of a car cylinder piston according to time.

Functions

1. Zero-degree polynomial function
2. First-degree polynomial function
3. Second-degree polynomial function
4. Exponential function
5. Periodic function

3 For each of the functions below, determine, if possible, the following properties:

a) the domain and the range

b) the variation

c) the sign

d) the extrema

e) the initial value

f) the zero(s)

Function ①

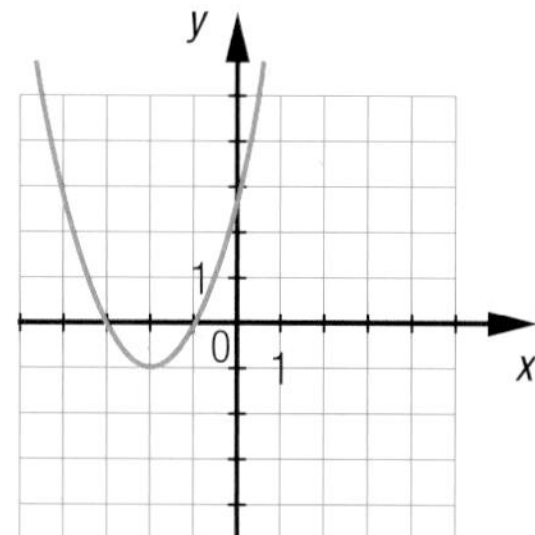

Function ②

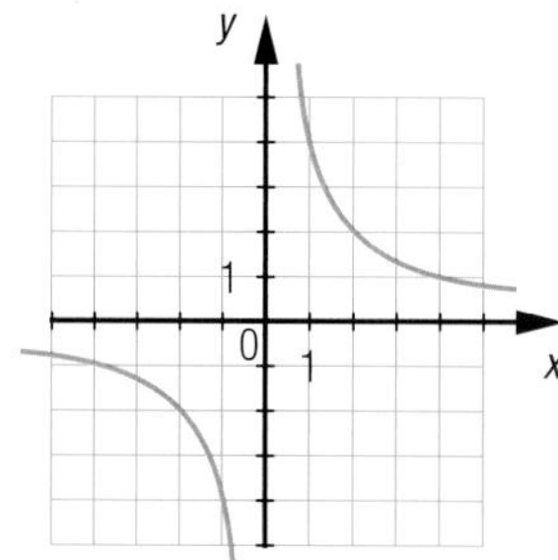

Function ③

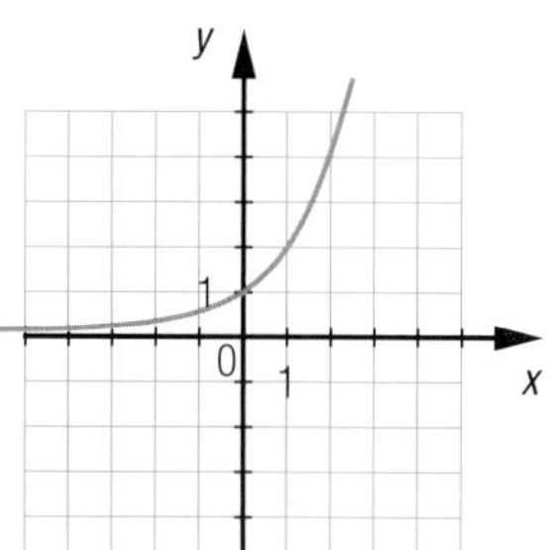

4 For each of the graphs below, do the following:

a) Determine whether the scatter plot can be modelled by a zero-degree polynomial function, by a first-degree polynomial function or by an inverse variation function.

b) Draw a curve of best fit.

c) Determine the rule associated with the curve drawn in **b)**.

Graph ①

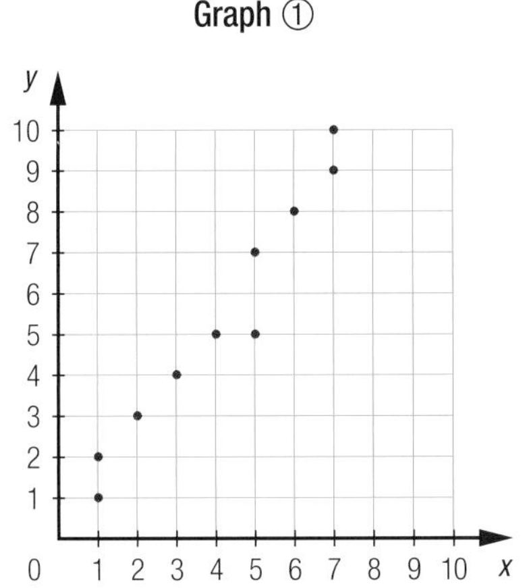

Graph ②

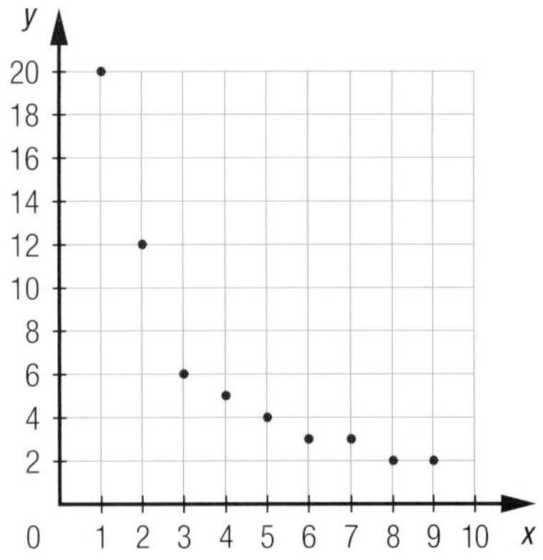

Graph ③

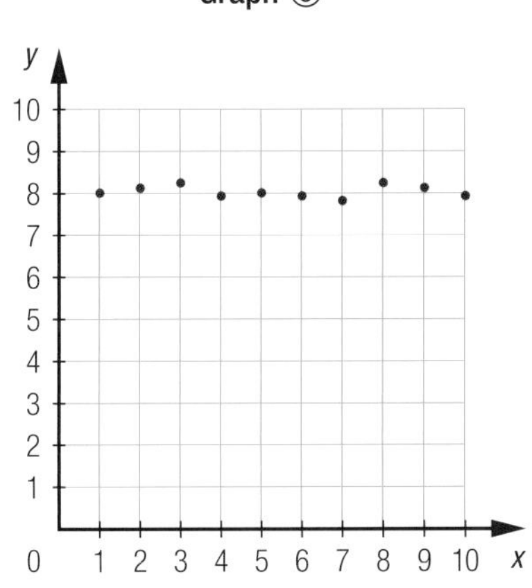

Graph ④

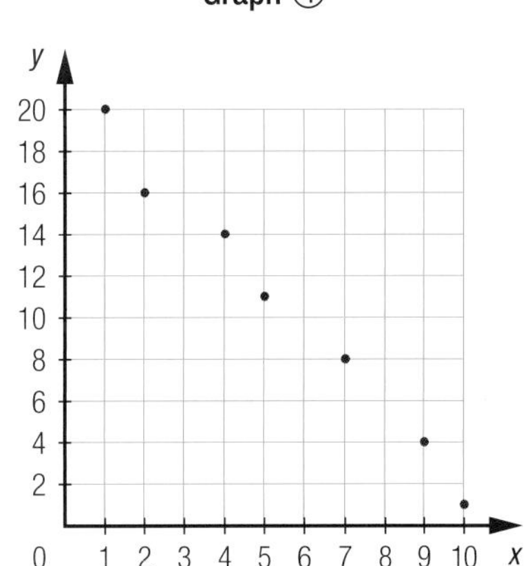

5 For each of the tables of values, do the following:

1) Draw a graphical representation of the ordered pairs with a scatter plot.

2) Determine the type of function that best fits the data displayed on the graph.

a)

x	-3	-1.5	0	1.5	3	4.5	6	7.5	9	11.5	13
y	-2	0	2	0	-2	0	2	0	-2	0	2

b)

x	-2	-1	0	1	2	3	4	5
y	0.5	2	3	5	14	22	50	99

c)

x	-8	-6	-4	-2	0	2	4	6	8	10	12
y	2	0	-1	-5	-6	-8	-6	-3	-3	0	3

6 The graph below shows a company's share value for the first 90 days of a year.

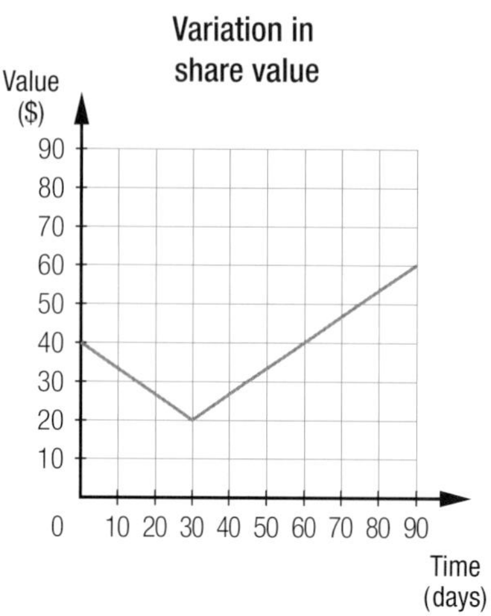

a) What was the minimum share value during this period?

b) What was the share value at the start of the year?

c) Over what time interval did the share value increase?

d) For how many days was the share value $35 or less?

e) What type of function best represents this situation?

f) If the trend of the last days continues, what will the share value be at the end of the year?

7 **TAX BRACKETS** Canadian income tax rates generally vary in terms of tax brackets based on an individual's taxable income. The table below contains information concerning an individual's income tax rates for 2008:

Federal income tax

Taxable income	Tax rate (%)
Over $0 but not exceeding $37,885	15
Over $37,885 but not exceeding $75,769	22
Over $75,769 but not exceeding $123,184	26
Over $123,184	29

For example, an individual whose taxable income is $65,000 will pay 15% income tax on the first $37,885 in earnings and 22% on the additional $27,155.

a) Calculate the federal tax payable for a person whose taxable income is:

1) $35,000 2) $75,769 3) $155,000

8 Finding a mineral deposit requires massive investments and involves science, technology and a lot of luck. Once surface samples have been collected and certain geophysical studies have been conducted, mining companies extract rock cores with a diamond drill. The adjacent graph shows the data collected by a drill's computer during a core extraction.

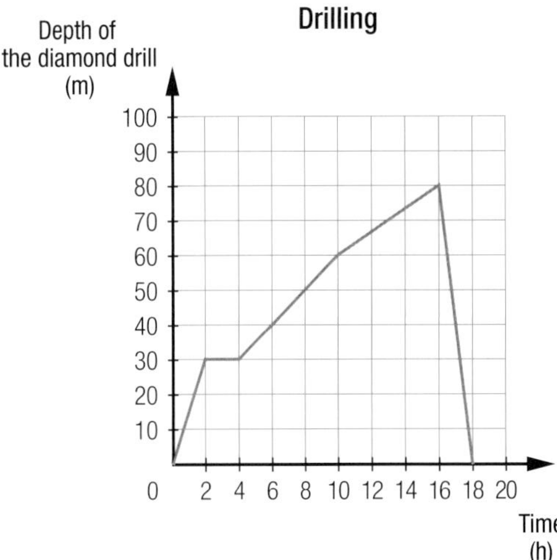

a) What type of function would best represent this situation?

b) Is the inverse of this function still a function? Explain your answer.

c) What was the mean drilling speed during this core extraction?

d) What was the maximum depth reached by the drill?

e) How much time did the drill take to get back to the surface?

f) At one point, a mechanical failure disrupted the drilling operation. What was the duration of this mechanical failure?

Since the 1920s, following the discovery of gold deposits in Abitibi and Temiscamingue, more than 50 mines have been put into operation. Only about ten of these mines are still active today.

9 A restaurant owner has modelled the evening service at his restaurant using the second-degree function shown in the adjacent graph.

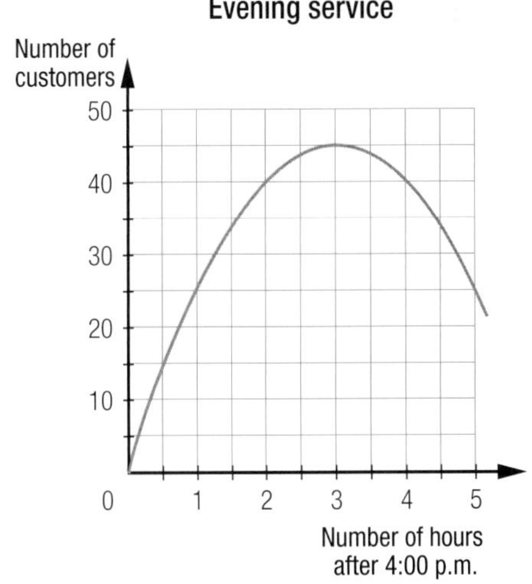

a) What is the maximum number of customers in the restaurant during the evening?

b) What is the length of time during which the number of customers increases?

c) How many customers are in the restaurant at 6:00 p.m.?

d) At what times are there 25 customers in the restaurant?

SECTION 1.2 Multiplicative parameters

This section refers to LES 2.

PROBLEM Zero-G

In the course of their training, astronauts use an aircraft that follows an alternating up and down trajectory in order to simulate the effect of weightlessness. After soaring into the air for several minutes, the pilot cuts the engines creating conditions of microgravity inside the plane for at least twenty seconds. The adjacent table of values describes part of the trajectory followed by the aircraft during a training session.

Altitude of the Zero-G aircraft

Time (s)	Altitude (m)
0	10 000
1	10 380
2	10 720
3	11 020
4	11 280
5	11 500
6	11 680
7	11 820
8	11 920
9	11 980
10	12 000
11	11 980
12	11 920
13	11 820
14	11 680
15	11 500
16	11 280
17	11 020
18	10 720
19	10 380
20	10 000
21	9580

The term "Zero-G" stands for zero gravity. The Airbus A300-0G, a specially refurbished Airbus A300, is the largest Zero-G aircraft in the world.

Based on the model represented in this table of values, at what moment will the aircraft reach an altitude of 7500 m?

ACTIVITY 1 Are you in sync?

Electrical signals are generally characterized by the shape of their wave, their voltage and their frequency. The variations of an electrical signal relative to time can be visualized with a device called an "oscilloscope."

Oscilloscopes are used in various fields such as physics, medicine or engineering.

The adjacent graph shows the variation of two electrical signals.

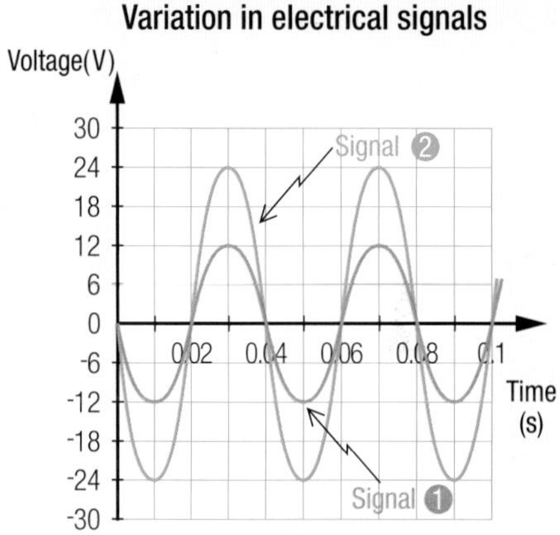

a. What type of function can you associate with the graphical representation of these signals?

b. Considering that voltage is represented by the maximum height of a curve, determine the voltage of each electrical signal.

c. In what way are the two signals:
1) similar?
2) different?

d. In what way are the curves of these two functions:
1) similar?
2) different?

The second graph shows the variation of electrical Signal 1 and a third signal.

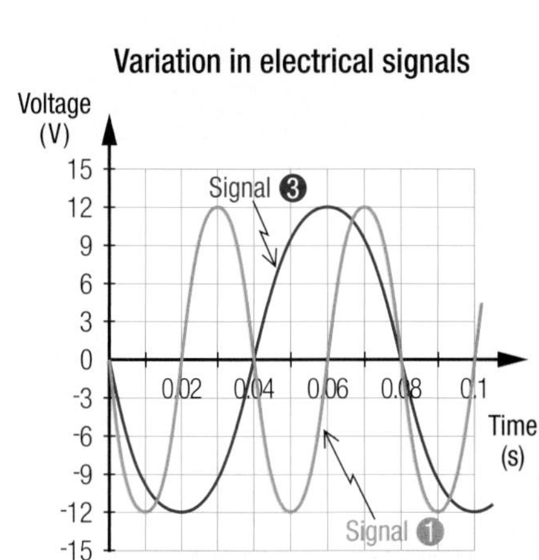

e. Considering that the frequency is represented by the number of alternating movements of the electrical current during one second, determine the frequency, in Hertz, of each electrical signal.

f. In what way are the two signals:
1) similar?
2) different?

g. In what way are the curves of these two functions:
1) similar?
2) different?

ACTIVITY 2 Upside down!

The rules of functions g and h, shown in the adjacent graph, were obtained by modifying the rule of function f.

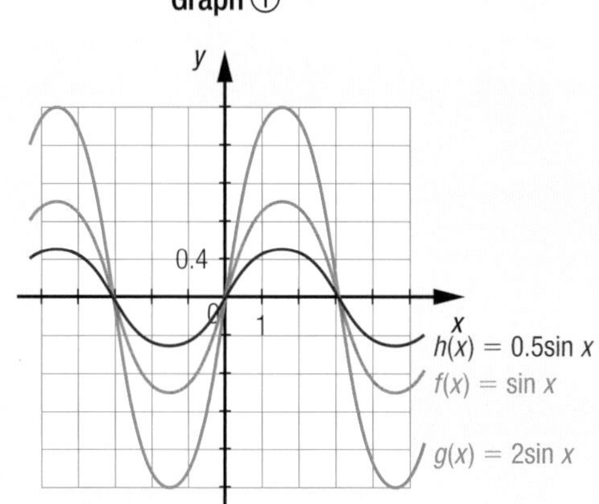

a. In relation to the rule of function f, what change was made to obtain the rule of:

1) function g?
2) function h?

b. In relation to the curve of function f, what change was made to obtain the curve of:

1) function g?
2) function h?

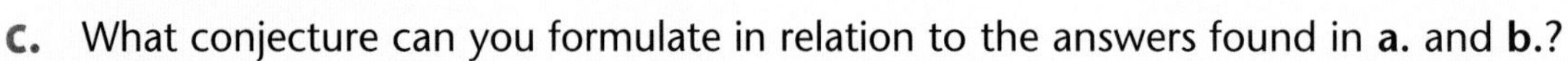

c. What conjecture can you formulate in relation to the answers found in **a.** and **b.**?

The rules of functions j and k, shown in the adjacent graph, were obtained by modifying the rule of function i.

Graph ②

y

j(x) = cos 2x

i(x) = cosx

0.2

0

1

x

k(x) = cos 0.5x

d. In relation to the rule of function i, what change was made to obtain the rule of:

1) function j?
2) function k?

e. In relation to the curve of function i, what change was made to obtain the curve of:

1) function j?
2) function k?

f. What conjecture can you formulate in relation to the answers found in **d.** and **e.**?

The rules of functions r and s, shown in the adjacent graph, were obtained by modifying the rule of function q.

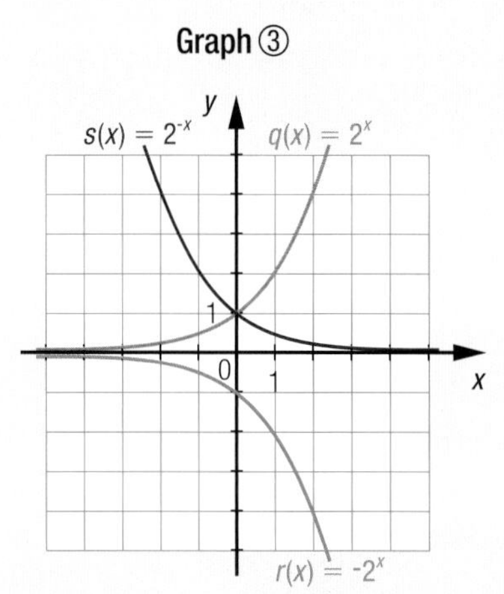

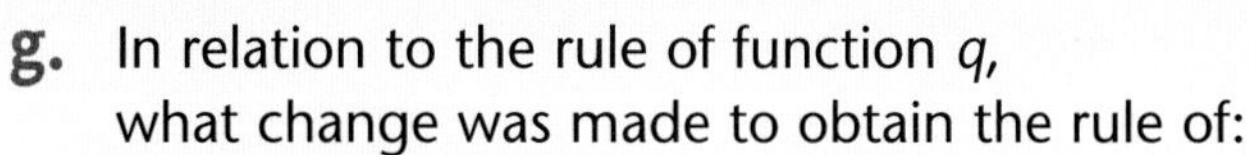

g. In relation to the rule of function q, what change was made to obtain the rule of:

1) function r?
2) function s?

h. In relation to the curve of function q, what change was made to obtain the curve of:

1) function r?
2) function s?

i. What conjecture can you formulate in relation to the answers found in **g.** and **h.**?

Techno math

A graphing calculator allows you to simultaneously display the curves of several functions in the same Cartesian plane. The exploration below allows you to observe the effects of multiplicative parameters in the graphical representation of a function for which the rule is written in the form of $f(x) = a\sin(bx)$.

Screen 1

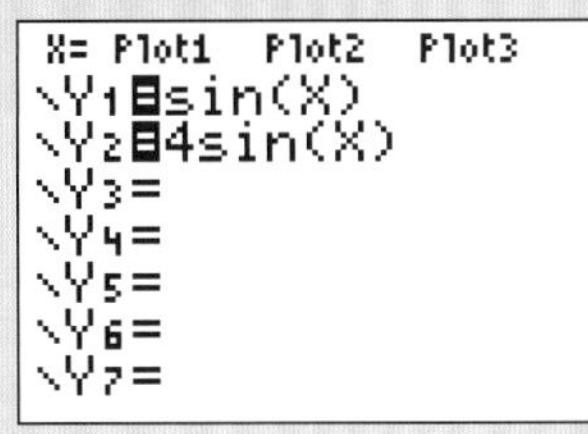

Screen 2

4

-6.28

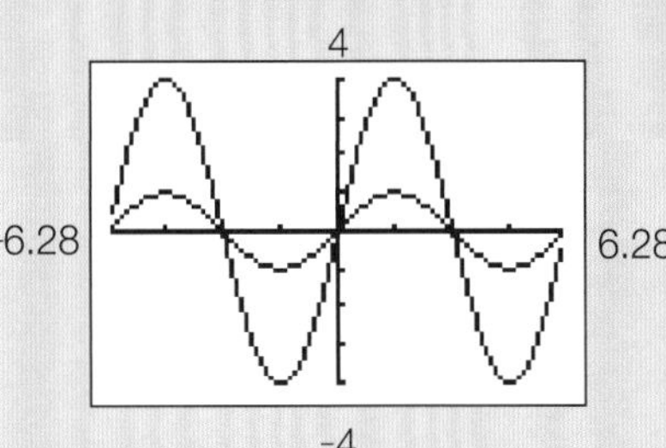

6.28

-4

Screen 3

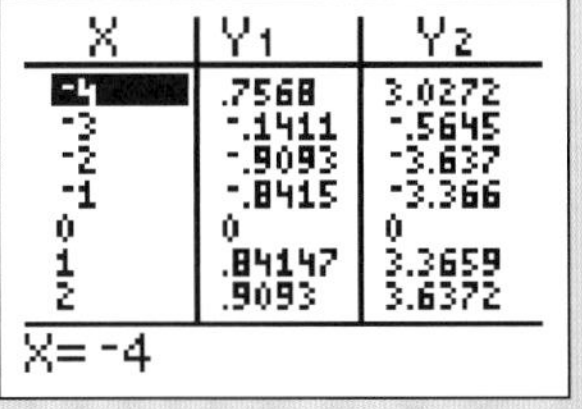

X	Y1	Y2
-4	.7568	3.0272
-3	-.1411	-.5645
-2	-.9093	-3.637
-1	-.8415	-3.366
0	0	0
1	.84147	3.3659
2	.9093	3.6372

X=-4

Screen 4

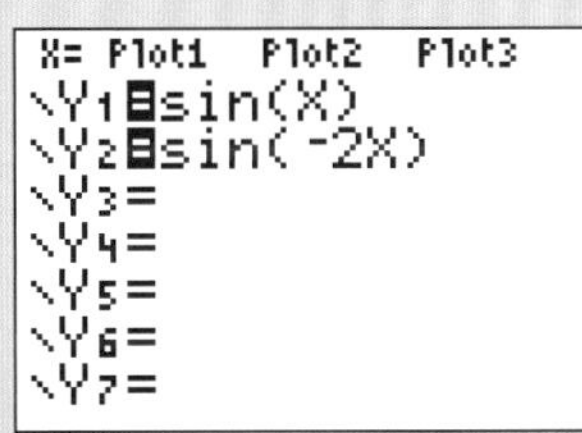

Screen 5

1

-6.28

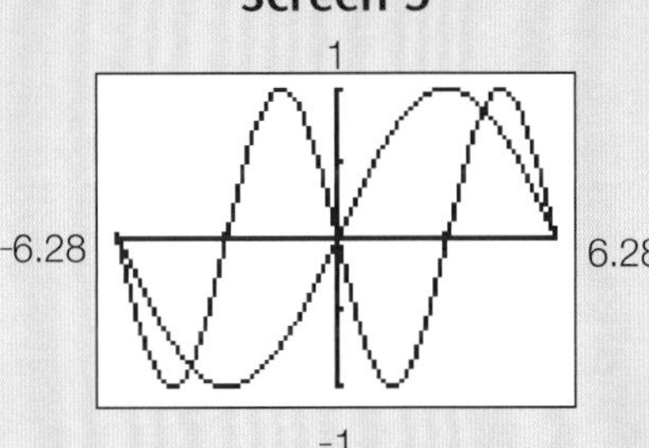

6.28

-1

Screen 6

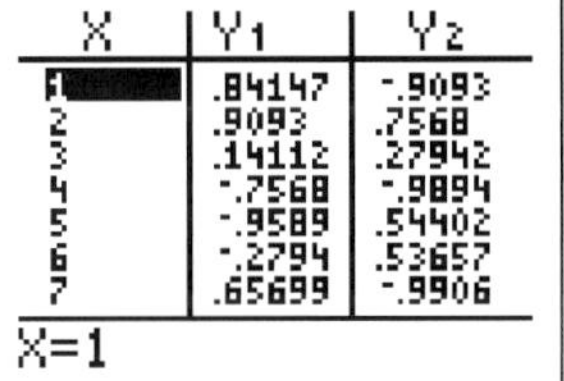

X	Y1	Y2
1	.84147	-.9093
2	.9093	.7568
3	.14112	.27942
4	-.7568	-.9894
5	-.9589	.54402
6	-.2794	.53657
7	.65699	-.9906

X=1

a. What parameter is changed:
1) in Screen **1**?
2) in Screen **4**?

b. What distinguishes the two curves:
1) in Screen **2**?
2) in Screen **5**?

c. For each *x*-coordinate in Screen **3**, calculate the quotient of the *y*-coordinate of Y_2 and the *y*-coordinate of Y_1. What do you notice?

d. What relationship can you identify between the quotients found in **c.** and the curves displayed in Screen **2**?

e. The curve of the basic function $Y_1 = \sqrt{x}$ has been modified by a vertical change in scale in order to get the curve of function Y_2. Among the tables of values below, determine which one represents these two functions.

A

X	Y1	Y2
1	1	1.7321
2	1.4142	2.4495
3	1.7321	3
4	2	3.4641
5	2.2361	3.873
6	2.4495	4.2426
7	2.6458	4.5826

X=1

B

X	Y1	Y2
1	1	10
2	1.4142	14.142
3	1.7321	17.321
4	2	20
5	2.2361	22.361
6	2.4495	24.495
7	2.6458	26.458

X=1

C

X	Y1	Y2
1	1	.70711
2	1.4142	1
3	1.7321	1.2247
4	2	1.4142
5	2.2361	1.5811
6	2.4495	1.7321
7	2.6458	1.8708

X=1

knowledge 1.2

THE ROLE OF MULTIPLICATIVE PARAMETERS

A "basic function" is the simplest function within a family of functions. By introducing certain values, called "parameters," the rule of a basic function can be transformed into a rule of a transformed function. The basic function and all the transformed functions of the same type form a family of functions that have common properties.

E.g.

Name	Basic function	Example of transformed function
First-degree polynomial function	$f(x) = x$	$g(x) = 2x$
Second-degree polynomial function	$f(x) = x^2$	$g(x) = 0.75x^2$
Exponential function	$f(x) = (\text{base})^x$	$g(x) = 7(\text{base})^{-3x}$
Greatest integer function	$f(x) = [x]$	$g(x) = -4[-x]$
Absolute value function	$f(x) = \lvert x \rvert$	$g(x) = -\lvert 6x \rvert$
Sinusoidal function	$f(x) = \sin x$	$g(x) = -2\sin 5x$
Inverse variation function	$f(x) = \frac{1}{x}, x \neq 0$	$g(x) = \frac{5}{x}, x \neq 0$

Multiplicative parameter a

In the rule of a transformed function, the parameter that multiplies the expression for the dependent variable of the basic function of the same type generates a vertical change in scale, meaning a vertical stretch or a vertical compression. In addition to the vertical change in scale, it also may cause a reflection over the x-axis if the numerical value of this parameter is negative. This parameter is generally associated with the letter **a**.

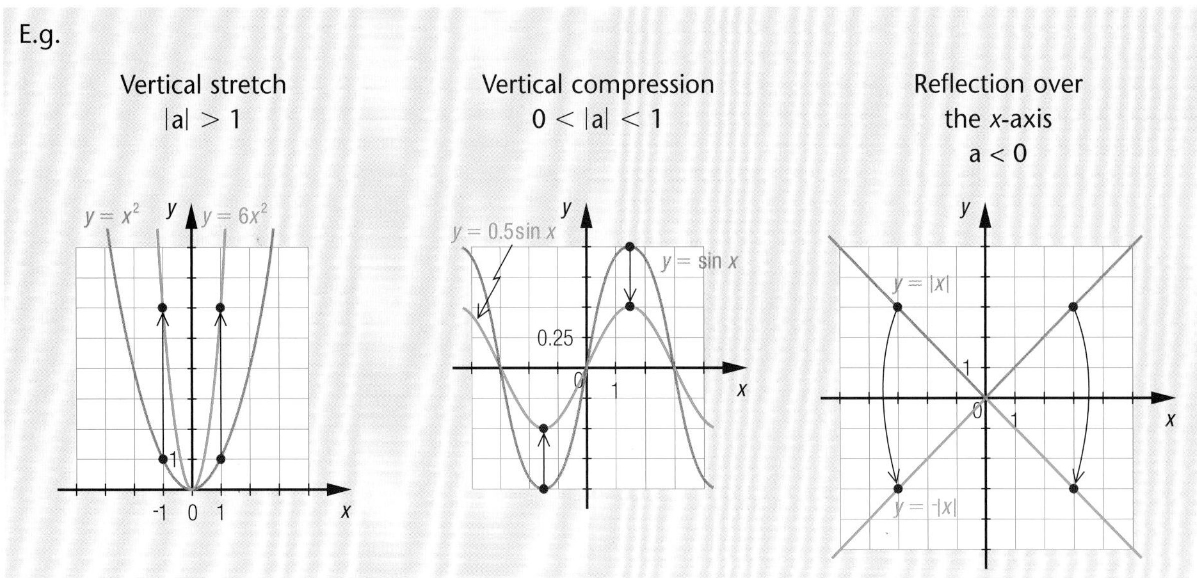

Multiplicative parameter b

In the rule of a transformed function, the parameter that multiplies the expression for the independent variable of the basic function of the same type generates a horizontal change in scale, meaning a horizontal stretch or a horizontal compression. In addition to a horizontal change in scale, it may also cause a reflection over the *y*-axis if the numerical value of this parameter is negative. This parameter is generally associated with the letter **b**.

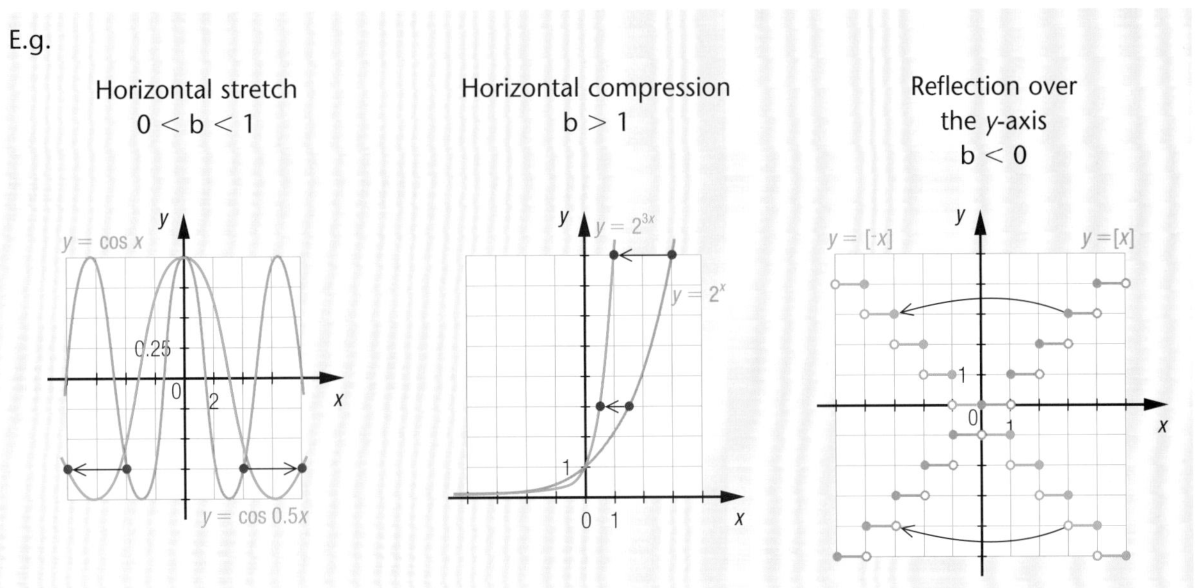

Each ordered pair (x, y) of a basic function is associated with the coordinate $\left(\frac{x}{b}, ay\right)$ of a transformed function of the same type.

E.g.

1) In the rule of function *g*, the value of parameter **a** is 3 and that of parameter **b** is 2.

2) In the rule of function *g*, the value of parameter **a** is 0.5 and that of parameter **b** is 4.

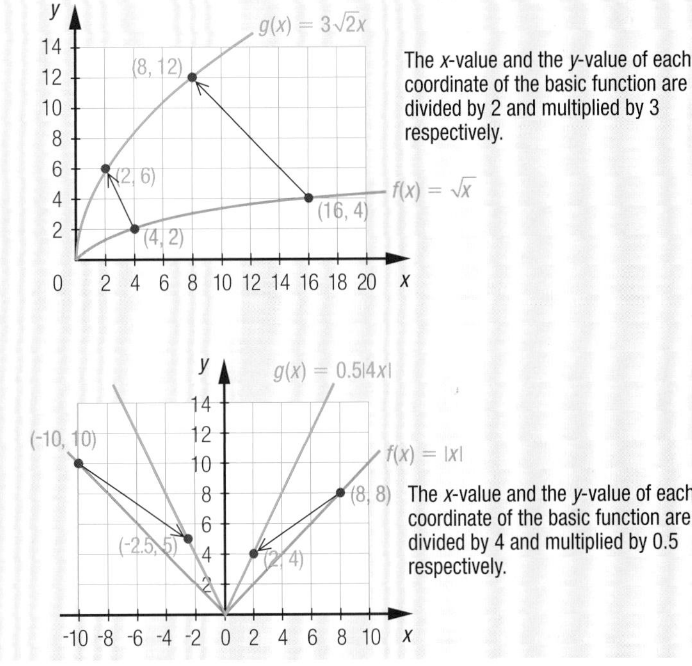

practice 1.2

1 The adjacent graph shows the basic function f whose rule is $f(x) = \sqrt{x}$ in addition to four transformed square root functions. In relation to the rule of the basic function, answer the following:

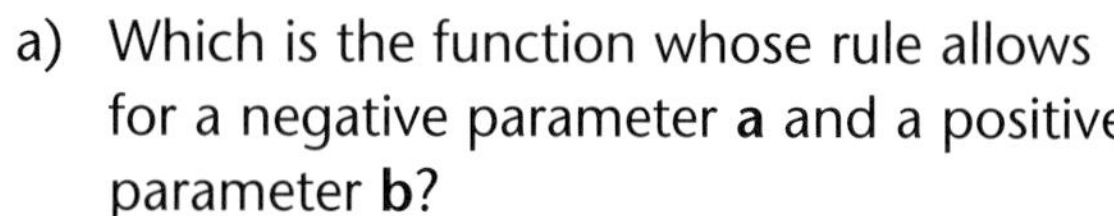

a) Which is the function whose rule allows for a negative parameter **a** and a positive parameter **b**?

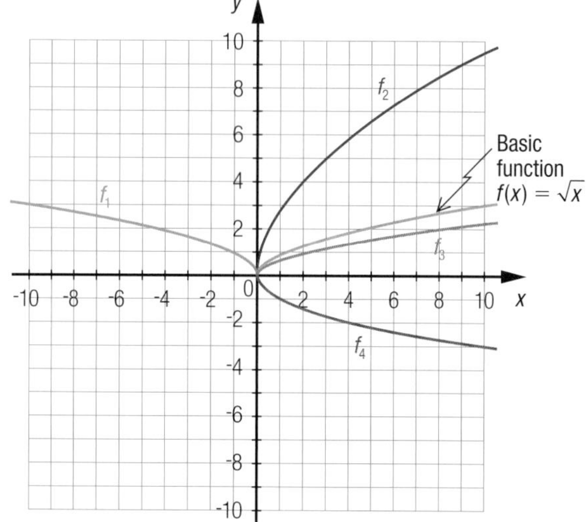

b) Which is the function whose rule allows for a positive parameter **a** and a negative parameter **b**?

c) Which is the function whose rule allows for a parameter **a** that is greater than 1 and a parameter **b** that equals 1?

d) Which is the function whose rule allows for a parameter **a** that equals 1 and a parameter **b** between 0 and 1?

2 Comparing the rule of each of the following functions with the rule of the basic function of the same type, determine the values of parameters **a** and **b**.

a) $h(x) = 3(5x)^2$ b) $g(x) = -2\sin x$ c) $i(x) = -\cos 2x$

d) $s(x) = 7(\text{base})^{3x}$ e) $l(x) = 6[-x]$ f) $g(x) = -x^2$

g) $p(x) = 0.5\sqrt{x}$ h) $f(x) = -7|0.25x|$ i) $v(x) = \frac{9}{x}$

3 Match each of the functions below with its corresponding table of values.

A $f(x) = 3\sqrt{-x}$

1

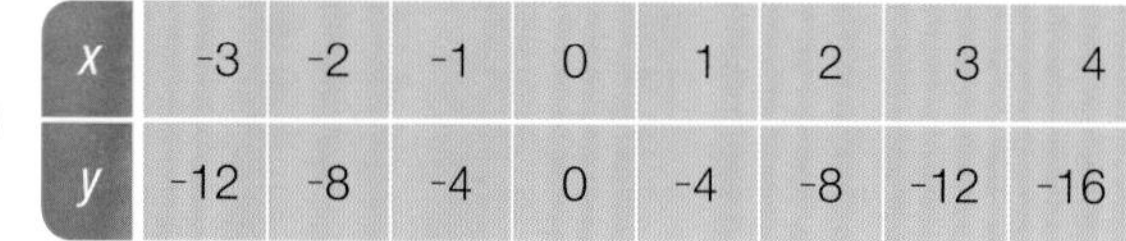

x	-3	-2	-1	0	1	2	3	4
y	-12	-8	-4	0	-4	-8	-12	-16

B $g(x) = 5(3)^x$

2

x	-64	-49	-36	-25	-16	-9	-4	0
y	24	21	18	15	12	9	6	0

C $h(x) = -4|x|$

3

x	0	1	2	3	4	5	6	7
y	5	15	45	135	405	1215	3645	10 935

D $i(x) = 7(2x)^2$

4

x	-3	-2	-1	0	1	2	3	4
y	252	112	28	0	28	112	252	448

4 The adjacent graph is that of basic function f whose rule is $f(x) = |x|$. Match the rule from the choices provided below with graphs **A**, **B**, **C**, and **D**.

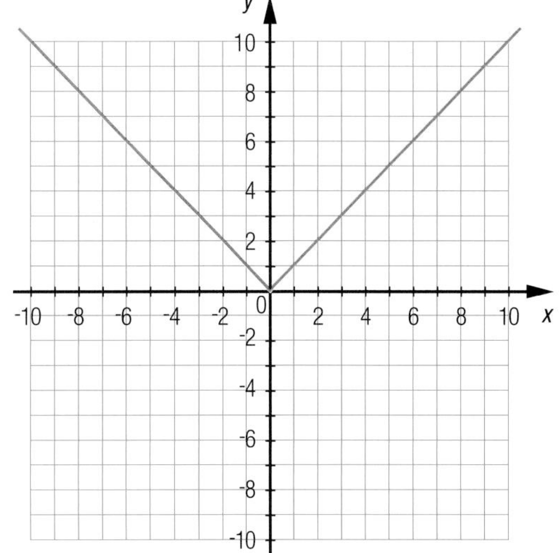

1. 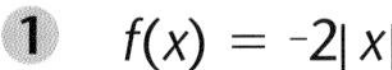 $f(x) = -2|x|$
2. $g(x) = |0.5x|$
3. $h(x) = |-x|$
4. $i(x) = -|-0.75x|$

A

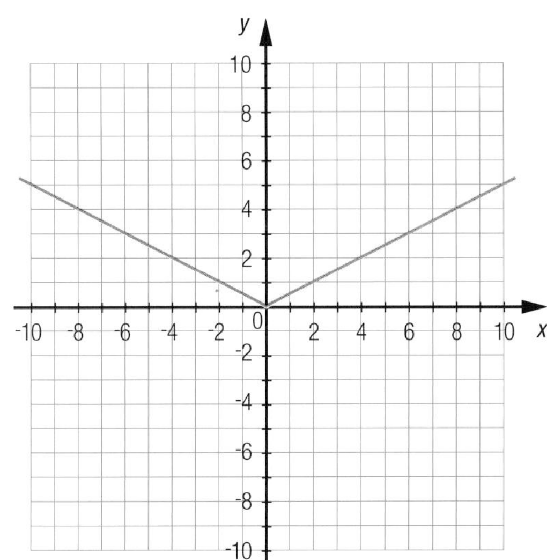

B

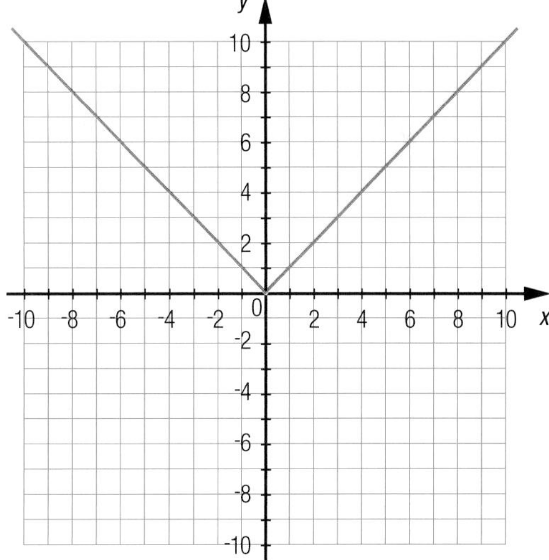

C

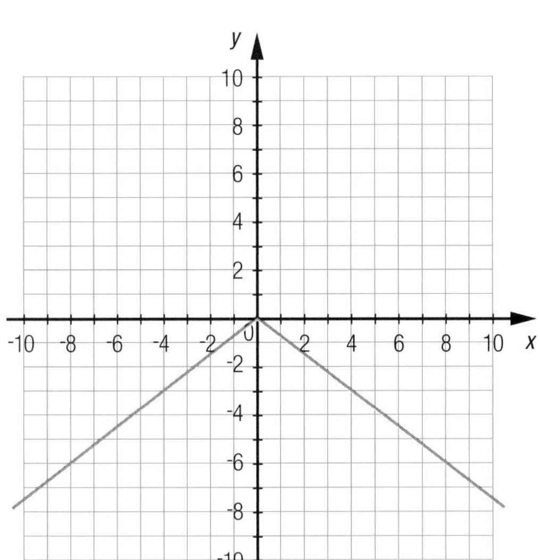

D

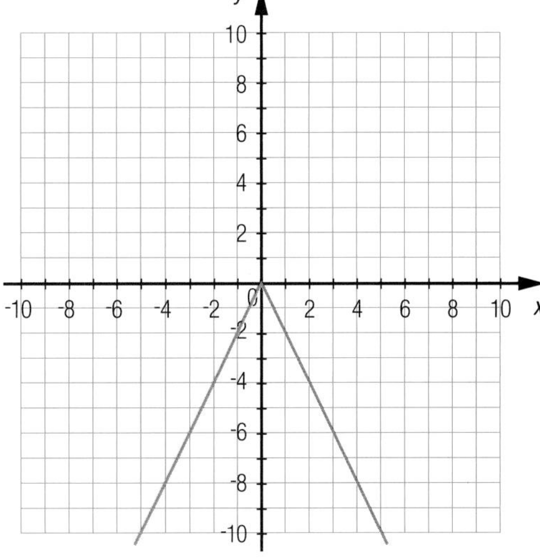

5 Considering the rule of basic function f, determine the rule of the transformed function g in each of the following cases.

a)

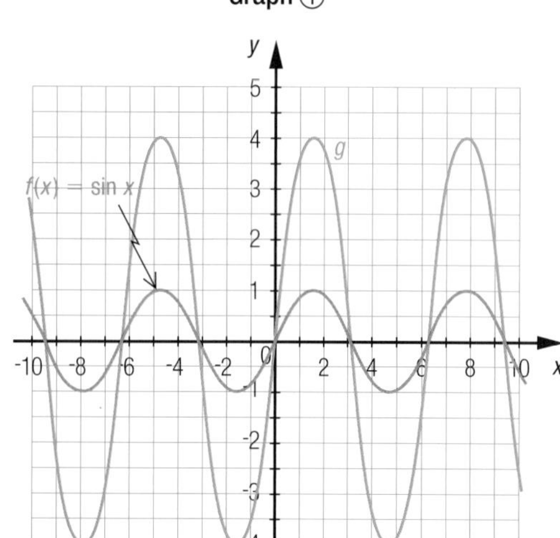

b)

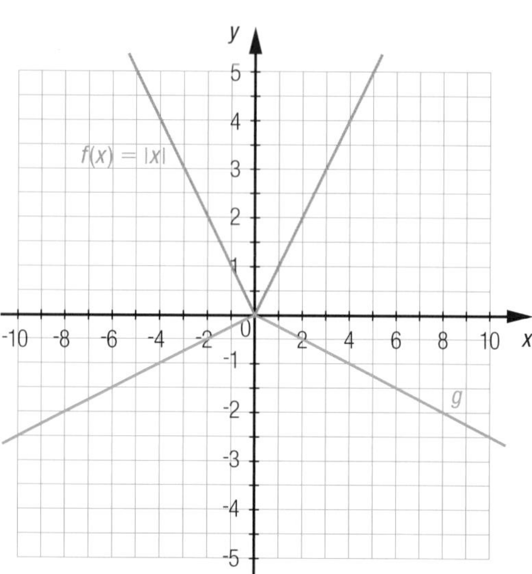

c)

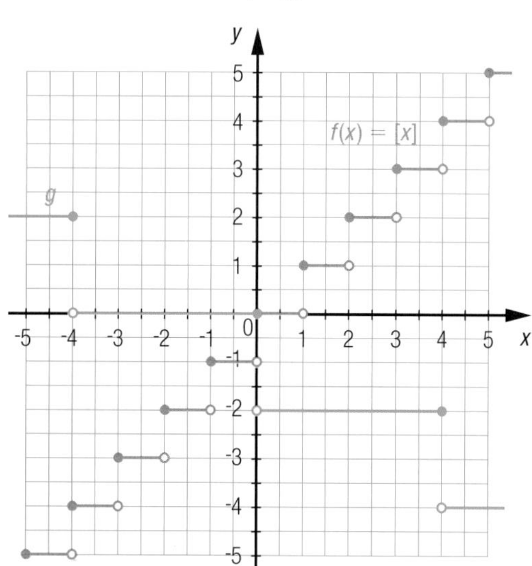

d)

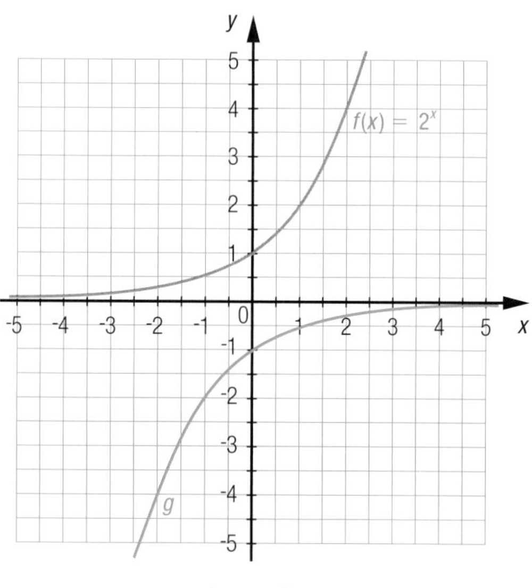

e)

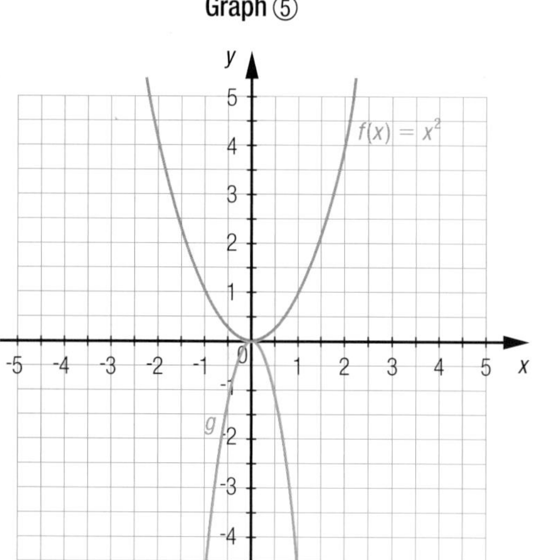

f)

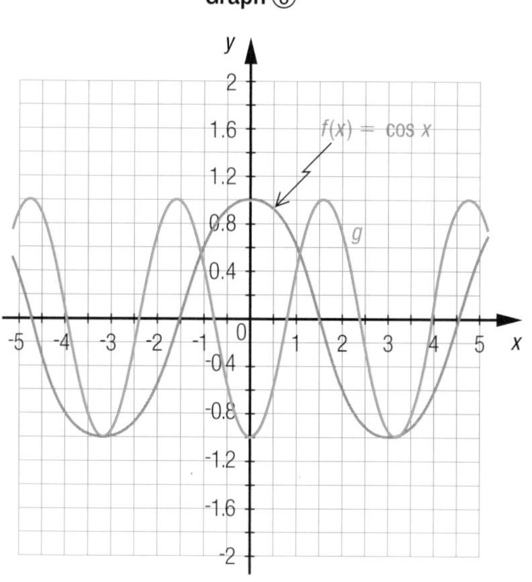

6 In each of the cases below, determine the value of parameter **a** considering that the value of parameter **b** is 1, and that function *f* represents the basic function while function *g* represents a transformed function of the same type.

a)

x	-3	-2	-1	0	1	2	3	4
$f(x)$	9	4	1	0	1	4	9	16
$g(x)$	72	32	8	0	8	32	72	128

b)

x	0	1	4	9	16	25	36	49
$f(x)$	0	1	2	3	4	5	6	7
$g(x)$	0	-0.5	-1	-1.5	-2	-2.5	-3	-3.5

c)

x	-2	-1.5	-1	-0.5	0	0.5	1	1.5
$f(x)$	-2	-2	-1	-1	0	0	1	1
$g(x)$	10	10	5	5	0	0	-5	-5

d)

x	-4	-3	-2	-1	0	1	2	3	4
$f(x)$	16	9	4	1	0	1	4	9	16
$g(x)$	-8	-4.5	-2	-0.5	0	-0.5	-2	-4.5	-8

7 The following prizes were awarded to dance companies at a recent international dance contest.

First place	**Second place**	**Third place**
$30,000	$20,000	$8,000

a) On the same Cartesian plane, represent the relation between the number of individuals in each dance company and each individual's share of the prize.

b) What type of function can you associate with these graphical representations?

c) What change did the curve of the function representing the first place company undergo compared to the one for second place?

d) What change did the curve of the function representing the third place company undergo compared to that of second place?

e) To what multiplicative parameter can you associate each prize amount? Explain your answer.

Les Grands Ballets Canadiens was founded in 1957 by Ludmilla Chiriaeff.

8 The rule of the function representing Y_1 is that of the basic function $f(x) = \cos x$. In each case, determine the rule of the transformed function Y_2, using the following forms:

a) $g(x) = \cos(bx)$ b) $h(x) = a\cos x$ c) $i(x) = a\cos(bx)$

X	Y1	Y2
1	.5403	⁻.4161
2	⁻.4161	⁻.6536
3	⁻.99	.96017
4	⁻.6536	⁻.1455
5	.28366	⁻.8391
6	.96017	.84385
7	.7539	.13674
X=1		

X	Y1	Y2
⁻4	⁻.6536	⁻.3268
⁻3	⁻.99	⁻.495
⁻2	⁻.4161	⁻.2081
⁻1	.5403	.27015
0	1	.5
1	.5403	.27015
2	⁻.4161	⁻.2081
X= ⁻4		

X	Y1	Y2
0	1	⁻2
1	.5403	⁻1.081
2	⁻.4161	.83229
3	⁻.99	1.98
4	⁻.6536	1.3073
5	.28366	⁻.5673
6	.96017	⁻1.92
X=0		

9 **THE ART OF PYROTECHNICS** The Montréal International Fireworks Competition has been held at La Ronde every year since 1985. Each summer, pyrotechnicians from ten competing countries each present a display of fireworks that is carefully synchronized to a musical soundtrack.

The graph below shows the expected height of the fireworks display of three shows relative to the elapsed time during a show.

In addition to fireworks, pyrotechnics have other applications. Matches, for example, work in much the same way as fireworks.

a) What type of function is represented by these curves?

b) How does the curve representing Production ❸ change in relation to that of Production ❶? Explain your answer.

c) How does the curve representing Production ❷ change in relation to that of Production ❶? Explain your answer.

d) What is the effect of a horizontal change of scale in this context?

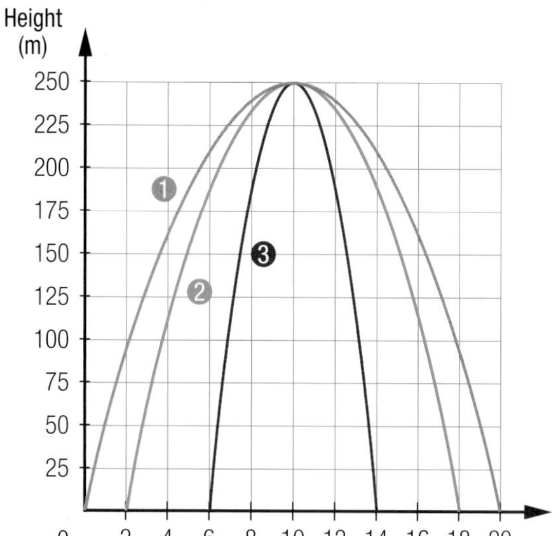

10 **EINSTEIN** Born March 14, 1879 in Ulm, Albert Einstein is best known for his work on relativity.

Around 1905, Einstein discovered that there is a relationship between mass m and energy E. He expressed this relationship with the equation $E = mc^2$ where c represents the speed of light in a vacuum. This speed is somewhere around 299 792 458 m/s.

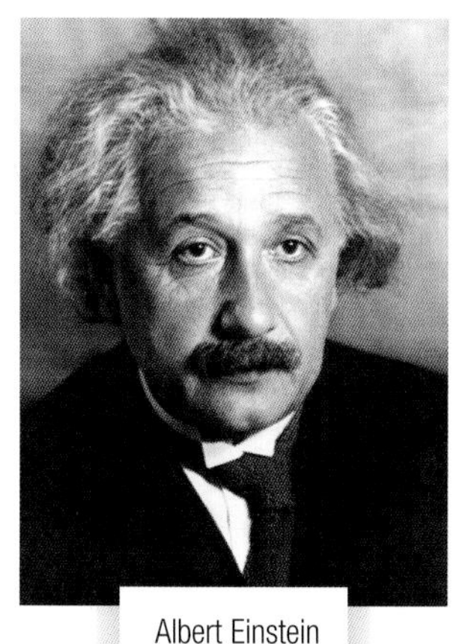

Albert Einstein (1879-1955)

a) What type of function would best represent this situation?

b) What multiplicative parameter can you associate with the rule of this function? Explain your answer.

c) What is the value of the parameter found in **b)**?

11 Because it weighs very little, aluminium is frequently used to make bicycles, automobile parts and various aeronautical components. Before making parts out of aluminium, manufacturers must first fuse the metal. A metallurgical engineer has compared the efficiency of two machines designed for fusing aluminium. The adjacent graph provides information on these two machines.

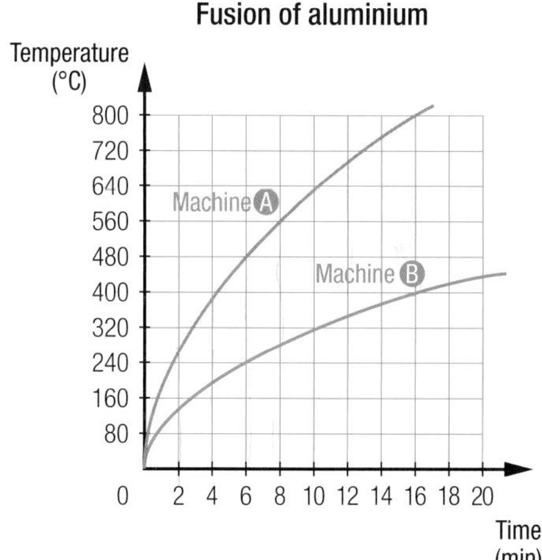

a) What changes in scale could be applied to the curve representing Machine Ⓐ in order to obtain the curve representing Machine Ⓑ?

b) In the course of his work, the engineer concluded that the rule associated with each curve takes the form of $f(x) = a\sqrt{x}$ and that the value of parameter **a** in one rule is double the value of parameter **a** in the other rule. Is he right? Explain your answer.

c) After further verification, the engineer comes to the conclusion that the rule of the function associated with each of the curves takes the form of $f(x) = \sqrt{bx}$ and that the value of parameter **b** in one rule is four times the value of this same parameter in the other rule. What do you think? Explain your answer.

d) Which machine appears to be the most efficient? Explain your answer.

Although abundant in the Earth's crust, aluminium rarely exists in its pure form. This metal must be extracted from certain types of rock. Bauxite is the main source of aluminium ore.

SECTION 1.3 Graphical modelling

This section refers to LES 1 and 2.

PROBLEM Break time

During a training session, Melissa runs at a constant speed, back and forth on a 100-m track for 10 minutes. Using a radar detector, her coach measures her distance from the starting point every 10 seconds. Below is the data collected:

MELISSA'S TRAINING

Time (s)	Distance From Starting Point (m)
0	0
10	41
20	79
30	96
40	56
50	16
60	8
70	49
80	88
90	87

Running was the original event of ancient Olympic games. The first of these competitions was held in Greece in the 8th century BCE and even now, many years later, events that involve running are some of the most popular events at the summer Olympics. These include the 100-m sprint, the middle-distance race, the long-distance race, the hurdle, the relay, the biathlon and the triathlon.

Considering that Melissa gives herself breaks of equal length each time she reaches the end of the track, calculate the total distance she covers during each training session.

ACTIVITY 1 See, and be seen

Many people are opting for bicycles as a means of transportation, yet, very few make safety a priority. In Québec, 29% of all fatal biking accidents occur at night because cyclists aren't visible. The standard bicycle should have at least eight reflectors.

In addition to reflectors, a bicycle must have a white headlight and a red taillight for use after dusk. Bright, colourful, reflective clothing is also recommended to increase visibility.

For a road-safety test, the position of a reflector relative to the ground was observed and data recorded. The initial set-up for the test is as follows:

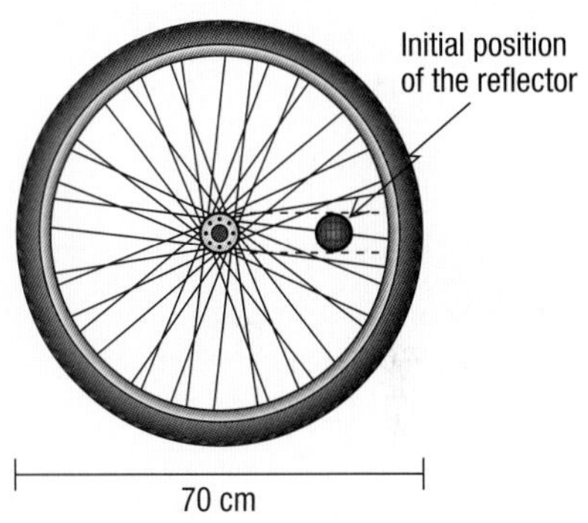

- The reflector placed on one of the spokes of the wheel is 20 cm from the wheel's centre.
- When the test begins, the reflector is the same distance from the ground as the wheel's centre and is positioned on the right side of the centre.
- The wheel rotates at a speed of 10 revolutions/min.

a. Draw a graphical representation of the height of the reflector (in cm) relative to the ground for the first 30 seconds of a bike ride.

b. What type of function best represents this situation?

c. Determine the domain and the range of this function.

d. At what moments is the reflector at its initial position?

e. What properties of the function will change if:

1) the reflector is placed at a distance of 10 cm from the wheel's centre?
2) the wheel turns at a speed of 15 revolutions/min?
3) initially, the reflector is at the same height relative to the ground as the wheel's centre but to the left side of the centre?

ACTIVITY 2 An unpredictable phenomenon

The phenomenon of superconductivity was discovered in 1911 by Kamerlingh Onnes. He discovered that an alloy becomes a superconductor at a certain temperature when its electrical resistance is zero. Today, superconductivity is used to design magnetic levitation commuter trains, to map the brain and to increase computer storage capacity. The graph below provides information on an alloy's superconductivity. The resistance is expressed in ohms, and the temperature is expressed in kelvins.

In 1913, Kamerlingh Onnes was awarded the Nobel Prize in Physics for his work on the liquefaction of helium and on superconductivity.

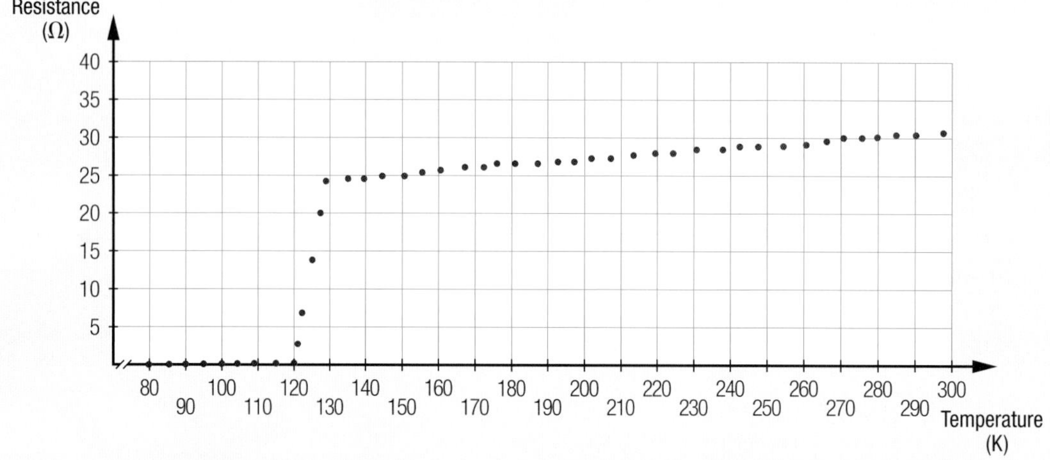

a. What type of function best represents this situation?

b. The scatter plot representing this situation is made up of three distinct parts. What type of function best represents each of these parts?

c. At what temperatures is this alloy superconductive?

d. Does the study of this alloy's electrical resistance at temperatures above 200 K allow you to verify whether it is superconductive? Explain your answer.

The magnet in this photograph is levitating as it is being pushed away by the superconductor placed just below it. Magnetic levitation trains work on this principle.

ACTIVITY 3 Analysis and the competitive edge

Before opening their doors, companies generally submit a business plan to financial institutions to verify whether their proposed endeavour is viable. This plan usually includes a market study and a business plan.

An accounting and management consultant has prepared a business plan for a tool rental company. She used a competitor's rates in order to establish rental rates for her company's market study. Below are the competitor's rental rates:

Tool rental

Duration of rental (h)	Rental rates ($)
]0, 0.5[	4
[0.5, 1[	6
[1, 2[	10
[2, 6[	18
[6, 24[	20
Extra days	20/day

a. Draw a graphical representation of the function corresponding to the competitor's rental rates for a maximum of 10 hours.

b. What type of function best represents this situation?

c. Determine the domain and the range of this function.

d. How much does it cost to rent a tool for:

1) 1.5 h? 2) 2 h? 3) 3 days?

In a business plan, the consultant proposes the use of the equation $C = \frac{10}{3}d$ where C represents the rental rates (in $) and d is the duration of the rental (in h) for a rental of 6 h or less.

e. Will the equation she has chosen attract customers to this new company? Explain your answer.

Techno math

A graphing calculator allows you to display the graphical representation of a piecewise function.

These screens allow you to select inequality symbols and logical connectors.

Screen 1

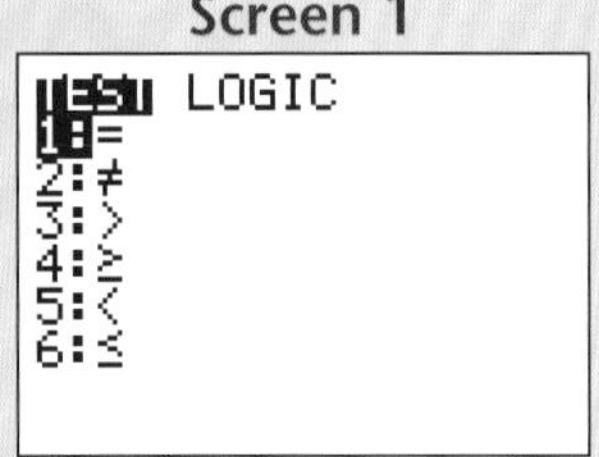

Screen 2

Using inequality symbols and logical connectors, it is possible to define the domain of a function.

Screen 3 **Screen 4**

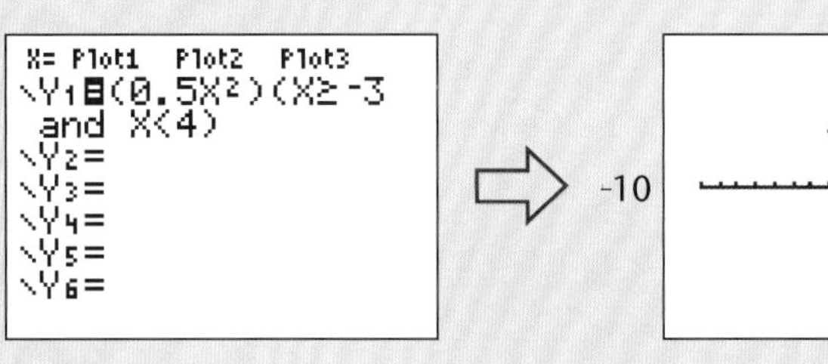

This screen shows an example of how to input the rule that defines a piecewise function.

Screen 5 **Screen 6**

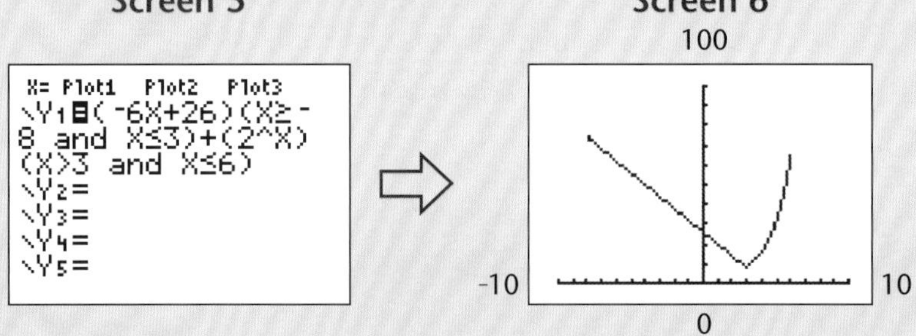

a. According to Screens **3** and **4**, answer the following:

1) What is the domain of the function?
2) What is the range of the function?

b. According to Screens **5** and **6**, answer the following:

1) What is the domain of the first part of the function?
2) What is the domain of the second part of the function?
3) What is the domain of the piecewise function?
4) What is the range of the piecewise function?

c. Using a graphing calculator, display the graphs of:

1) a function whose rule is $y = {}^-2(3)^x$ and the domain is]1, 4]
2) a piecewise function where the rule of the first part is $y = {}^-0.6x^2$ and the domain is [-8, 4]; and the rule of the second part is $y = x - 30$ and the domain is]4, 9[

knowledge 1.3

STEP FUNCTION

A step function is a function that is constant over certain intervals and then varies suddenly at certain values of the independent variable, called "critical values." The graphical representation of this function is made up of horizontal segments. At both extremities of each segment, an opened or closed dot is used to designate an ordered pair that may or may not belong to the function.

E.g.

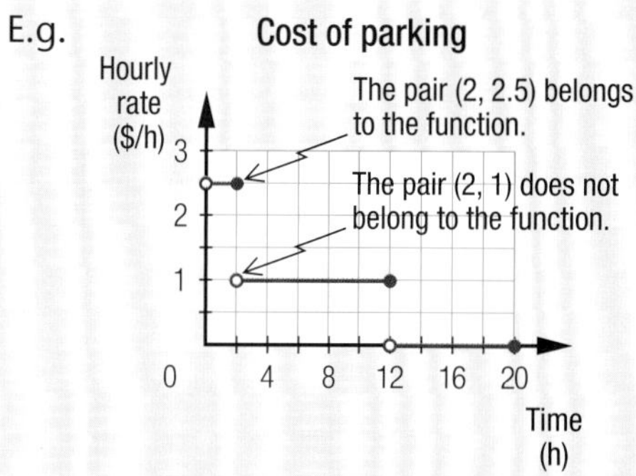

The cost to park a car in a parking lot varies in relation to parking duration. The hourly rate is $2.50 for the two first hours. Then the rate decreases to $1/h for the four subsequent hours, after which there are no extra charges. The parking lot is open 20 h/day.

PERIODIC FUNCTION

A function is periodic when its graphical representation consists of a regularly recurring pattern. The interval between the x-values located at either end of this recurring "pattern" is the period of this function.

E.g.

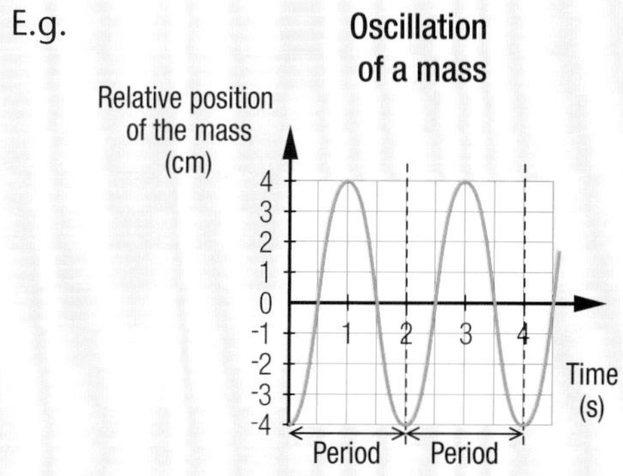

The behaviour of a mass suspended from a spring that oscillates vertically without friction can be modelled by a periodic function.

It is possible to conclude that the mass returns to its original position every 2 s. The period of the function therefore is 2 s.

PIECEWISE FUNCTION

A piecewise function is a function made up of several functions defined over different intervals within the domain. The parts that make up such a function may come from one or several families of functions.

E.g.

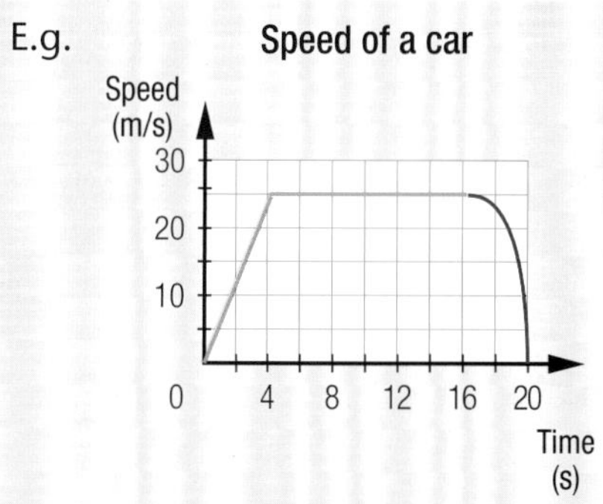

The speed of a car that is accelerating at a constant speed, which it then maintains and then slows down, can be modelled by a piecewise function.

This function is made up of a first-degree polynomial function, a zero-degree polynomial function and a second-degree polynomial function.

practice 1.3

1 Indicate which of the following graphs represents a periodic function, a piecewise function or a step function.

a) 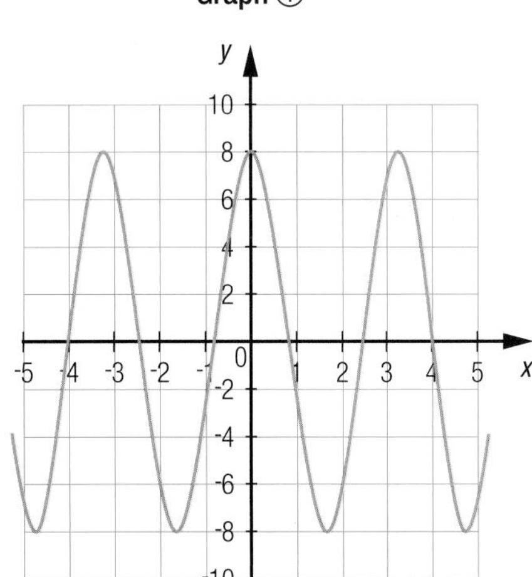

b) 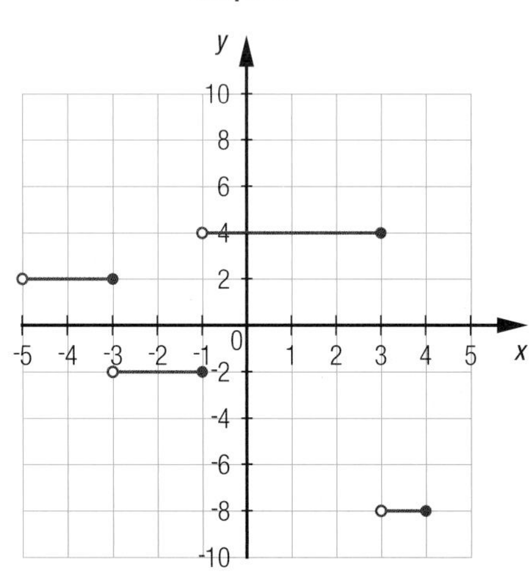

c) 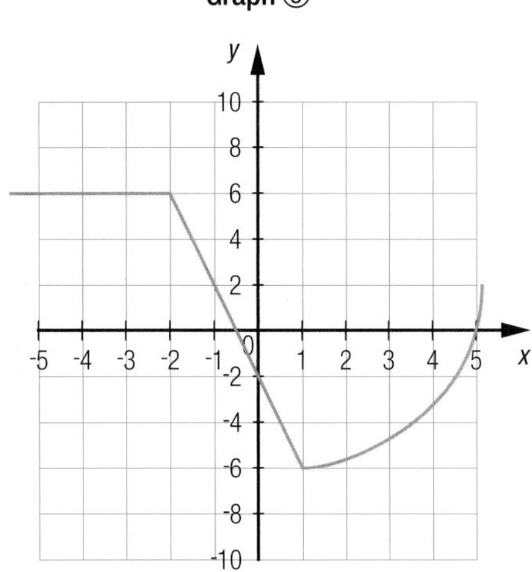

d) 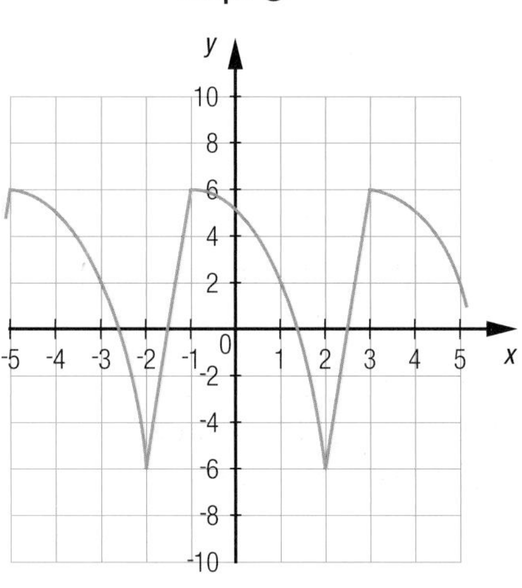

2 Indicate if each the following situations corresponds to a periodic function, a piecewise function or a step function.

a) A business offers its customers a theatre ticket for every \$150 worth of purchases. Consider the relation between the total amount of a customer's purchases and the number of theatre tickets received.

b) An all-terrain vehicle (ATV) covers a few kilometres at a constant speed, slows down, then stops for a few minutes to fill up with gas. Consider the relation between time and the distance covered by this ATV.

c) A metal ball is suspended at the end of a spring that oscillates above the ground with a regular motion. Consider the relation between the time elapsed from the start of the motion and the height of the ball relative to the ground.

3 Below is the rule of a piecewise function:

$$f(x) = \begin{cases} 3 & \text{if } x \in [0, 4[\\ 2x - 5 & \text{if } x \in [4, 8[\\ -1.5x + 23 & \text{if } x \in [8, 12[\end{cases}$$

a) Draw a graphical representation of this function.

b) Determine:

1) the domain and the range
2) the extrema
3) the variation
4) the initial value
5) the zeros
6) the sign

c) Calculate:

1) $f(4)$
2) $f(8)$
3) $f(11)$
4) $f(20)$

4 Is there a periodic function whose inverse is also a function? Explain your answer.

5 A computer technician analyzes the performance of a computer search engine. The adjacent graph shows the results of this research.

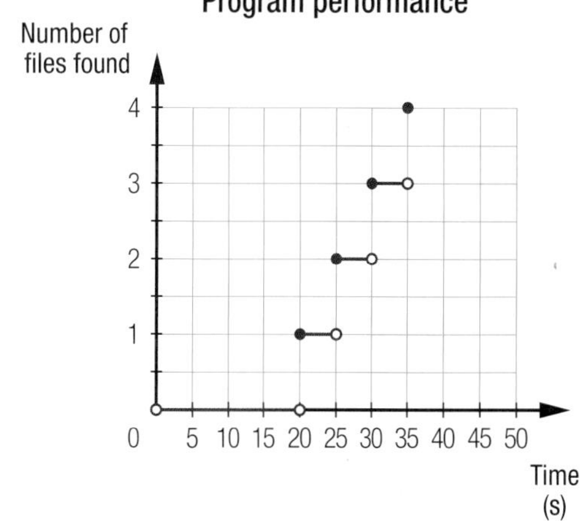

a) In this situation, what do the critical values represent?

b) How many files did the program find during this search?

c) Determine the domain and the range of this function.

d) Determine the zeros of this function and interpret them in relation to the context.

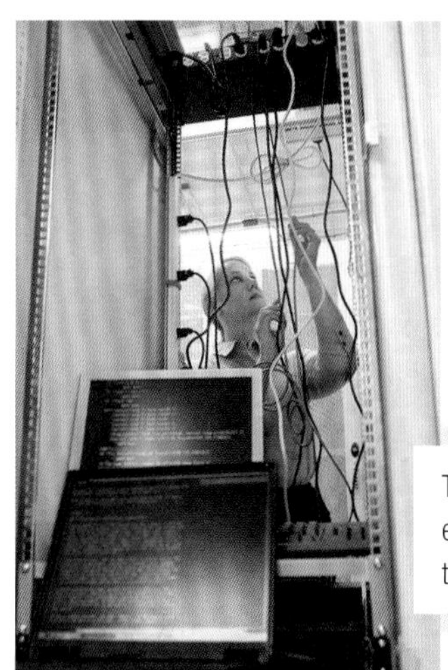

The main responsibilities of a computer technician are to install and to maintain equipment including computers and software, to set up networks and to provide technical assistance to users.

6 For each of the following, indicate whether the graph represents a periodic function. If so, determine its period.

a)

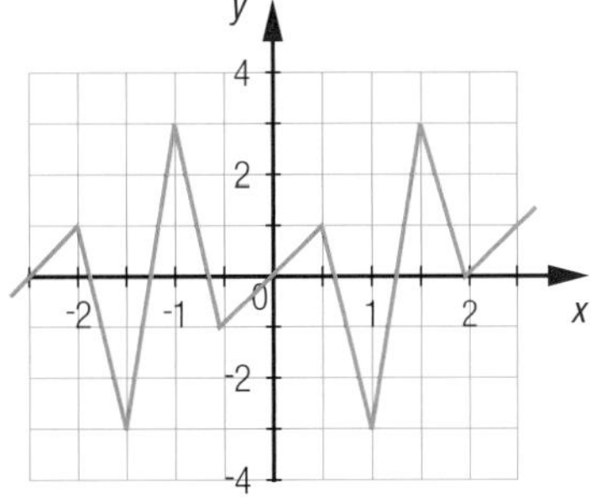

b)

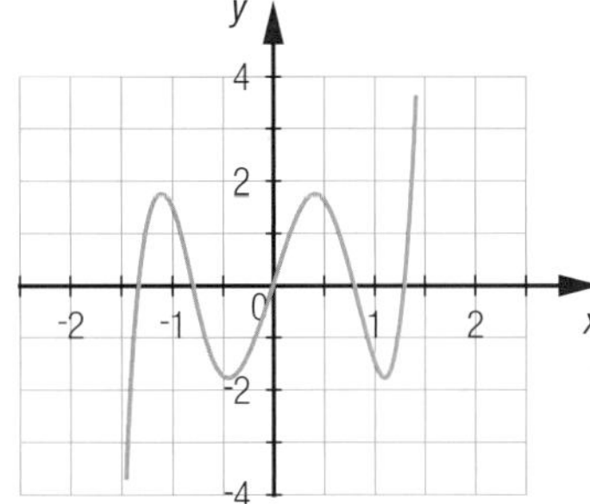

c)

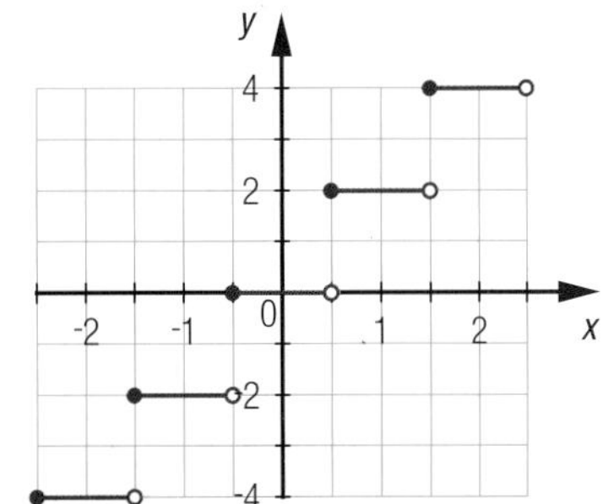

d)

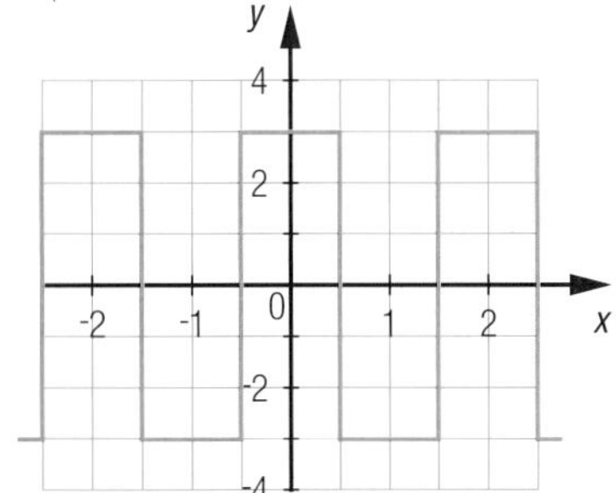

7 **CONSTELLATION** The Delta star in the Cepheus Constellation is the second variable star for which we have been able to measure magnitude, in other words, brightness. The brighter a star, the lower its magnitude. The graph below provides information on the magnitude of the Delta star.

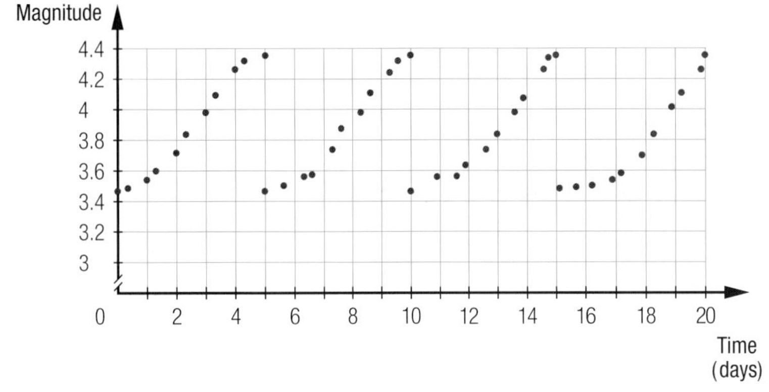

Astronomers use magnitude to describe a star's brightness.

a) Which function best represents this situation?

b) At what times is Delta's brightness at its highest?

c) Considering that observations began during the night of December 1, determine the date on which the Cephei Delta will reach its maximum brightness for:

1) the first time in the coming year

2) the 25th time in the coming year

8 The adjacent graph represents a periodic function defined in $\mathbb{R}$.

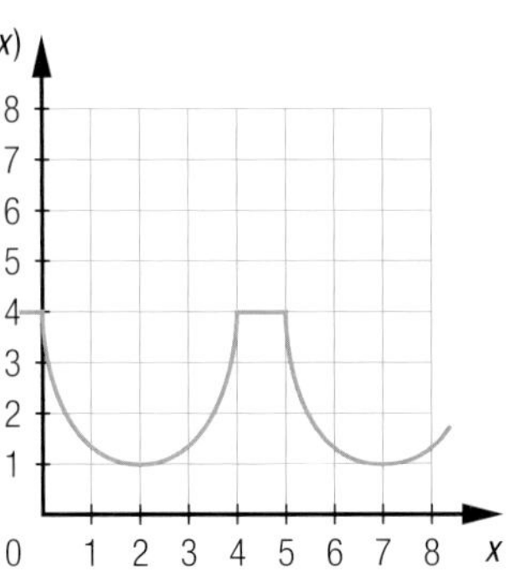

a) What is the period of this function?

b) Determine the domain and the range of this function.

c) Draw a graphical representation of this function over interval [0, 20].

d) In each case, indicate whether the image corresponds to the function's minimum.

1) $f(13)$ 2) $f(37)$ 3) $f(300)$ 4) $f(^-62)$

9 Below are descriptions of three experiments conducted using a marble:

Experiment ①

A marble rolls in a bowl shaped as a hemisphere 15 cm in diameter. The marble is released from the edge of the bowl and always passes through its centre.

Experiment ②

A marble rolls on an inclined board that is 30 cm long; it then continues its course on the floor. At the start, the marble is located at a height of 15 cm.

Experiment ③

A marble rolls down a 3-step staircase. The depth and height of each step measure 30 cm and 25 cm respectively. The marble is released from the top.

Consider the height of the marble (in cm) in relation to time (in s), and do the following:

a) Determine the function that best represents each situation.

b) Draw a sketch of the graph representing each of these situations.

10 It is now possible for anyone to use a computer to collect data on the performance of a vehicle. Below is the information collected during a closed-course test of two mid-sized vehicles:

Performance during a closed-course test

Mean speed of the vehicle (km/h)	5	10	15	20	25	30	35	40	45	50	55	60	65	70
Acceleration vehicule A (m/s^2)	0.1	2	5.8	7.5	6.2	5.3	4.8	4	3.7	3.6	3	2.8	2.4	2.3
Acceleration vehicule B (m/s^2)	0.9	3	6	6.7	5.5	4.6	4.1	3.6	3.1	2.8	2.5	2.4	2.1	2

a) Using different colours on a single Cartesian plane, represent the data for each vehicle with a scatter plot.

b) For each vehicle, determine the function that best models the situation.

c) For each vehicle, calculate:

1) the maximum acceleration

2) the speed where acceleration is the highest

3) the vehicle's acceleration when it reaches a speed of 100 km/h

11 **KREBS CYCLE** During an experiment in the course of his research on the role of citric acid in the Krebs cycle, Boris Pavlovich Belousov discovered that the solution he had prepared would regularly change colour. The graph below shows the results of this experiment.

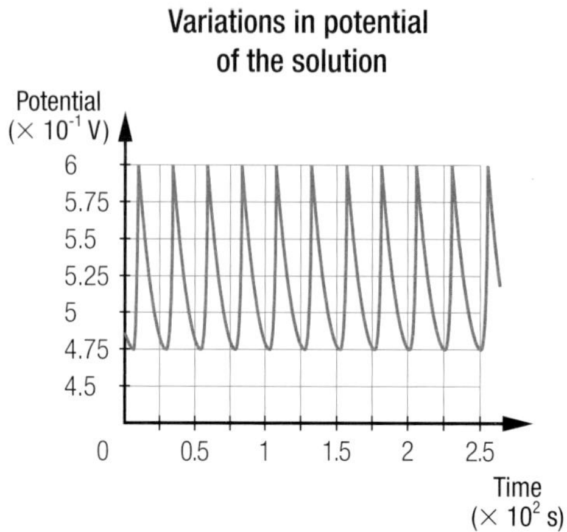

a) Determine the amplitude of this function, in other words half the difference between the function's maximum and minimum.

b) Does this function have at least one axis of symmetry? Justify your answer.

c) What is the period of this function?

d) During another experiment using the same solution, Belousov noticed that the solution changes colour 30 times/min, but the minimum and the maximum potentials remained the same. What multiplicative parameter will change in the rule of the function representing this new experiment?

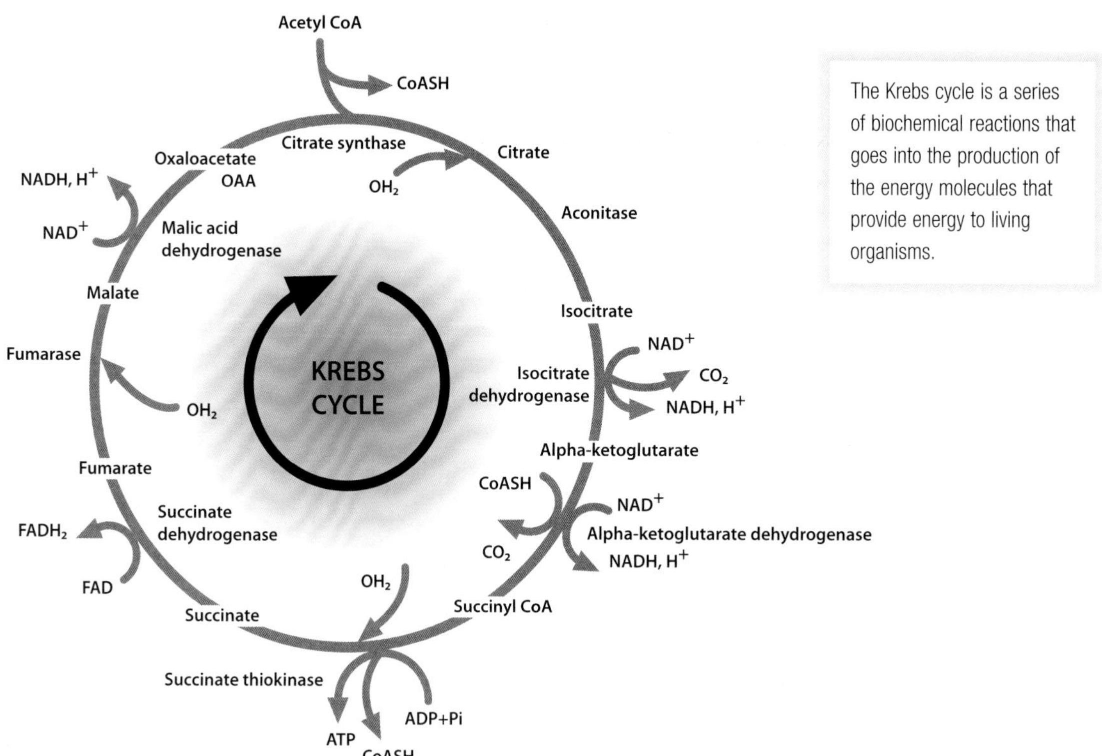

The Krebs cycle is a series of biochemical reactions that goes into the production of the energy molecules that provide energy to living organisms.

12 Jerome programs his treadmill for one-hour interval training sessions. The adjacent table of values provides information on this training program.

a) Draw a graphical representation of this situation.

b) What distance will Jerome cover within each interval?

c) 1) Considering that each segment represents one of the bases of the rectangle and that the other base is located on the x-axis, draw a rectangle under each horizontal segment of the curve.

2) Determine the area of each rectangle.

3) Compare the answer to **b)** with that of **c) 2)**. What do you notice?

d) What distance will Jerome have covered:

1) half-way through his training? 2) by the end of his training?

Speed during training

Start of each interval (min)	Speed (km/h)
0	7
6	12
24	10
54	7

13 During a climb up Mont Blanc, a scientist melts snow and brings the water to a boil. On the right is the graphical representation of the experiment:

a) Determine the type of function that best represents this situation.

b) 1) What is the maximum of this function?

2) What does the maximum represent in this context?

c) How might you interpret the zeros of this function within this context?

d) Using the information provided below, determine the vertical distance between the summit and the camp during this expedition. Keep in mind that Mont Blanc has an altitude of 4810.9 m.

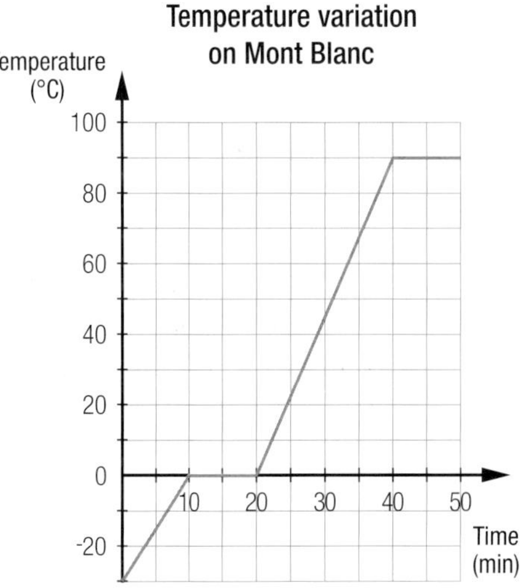

Table of values ①
Variation in boiling temperature

Pressure (kPa)	1.045	2.07	3.88	6.95	11.9	19.2	30.9	47.4	70.4	101.325
Boiling point (°C)	10	20	30	40	50	60	70	80	90	100

Table of values ②
Variation in pressure

Altitude (km)	0	0.5	1	1.5	2	2.5	3	3.5	4	5
Pressure (kPa)	101.3	95.5	90	84.5	79.4	74.6	70	65.8	61.7	54.1

Chronicle of the past

The history of electricity

André-Marie Ampère

André-Marie Ampère (1775-1836)

André-Marie Ampère was a French mathematician and physicist born in Lyon on January 20, 1775. He never went to school because his father insisted on teaching André-Marie himself. At the early age of 13, André-Marie took an interest in mathematics which quickly became a genuine passion for him. In 1801 he became a professor of mathematics and physics in Bourg and Lyon. After his death in 1836, his name was given to the unit of measure for the intensity of an electrical current: the ampere.

Ampère's work

Ampère worked on many projects during his lifetime. His research mainly concerned the fields of probability, magnetism and electricity. He was the first to have defined a theory of electromagnetism. Based on this theory, he and François Arago later invented the electromagnet. He also proposed the existence of electric currents. He is, in fact, the first to have used the term "electric pressure." At the time, Ampère was known as the "Newton of electricity."

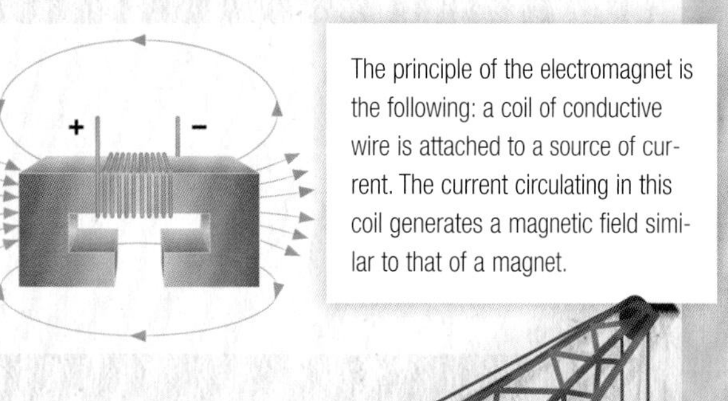

The principle of the electromagnet is the following: a coil of conductive wire is attached to a source of current. The current circulating in this coil generates a magnetic field similar to that of a magnet.

An electromagnet can be used to lift several tons.

Georg Ohm (1789-1854)

Ampère's discoveries would lead another scientist, Georg Ohm, to formulate a law relating the intensity and the voltage of an electric current. Ohm's law verifies that, in an electric circuit, the voltage V in (volts) equals the product of the resistance R (in Ω) by the intensity I of the current (in amperes). The scatter plot below represents the data collected by Ohm in the course of an experiment on electricity.

Resistance of 2.5 Ω

Intensity (A)

Voltage (V)

James Prescott Joule (1818-1889)

In 1841, James Prescott Joule, a British physicist, defined a law verifying that the power *P* dissipated (in watts) in an electric circuit is equal to the product of the resistance *R* (in Ω) by the intensity *I* of the current (in amperes) squared. The power dissipated in an electric circuit translates into heat and is called the Joule effect. Below is the laboratory data collected while studying the Joule effect on an aluminium wire with a resistance of 10 000 ohms:

Joule mainly worked to establish the link between mechanical energy and thermal energy. The unit of work, heat and energy in the international system of units, the joule (J), bears his name.

Joule effect

Intensity (A)	0	0.5	1	1.5	2	2.5	3	3.5	4	4.5	5
Power dissipated (kW)	0	2.5	10	22.5	40	62.5	90	122.5	160	202.5	250

High voltage power lines are used to transport electricity as a means of diminishing energy loss as described by the Joule effect.

1. Based on the scatter plot representing the data collected by Ohm, answer the following:

a) What type of function can you associate with this situation?

b) What is the intensity of the current when the voltage is 20 V?

c) What change would occur to the scatter plot if the resistance was increased to 5 Ω?

2. a) Construct a graph representing laboratory results during the study of the Joule effect on an aluminium wire.

b) What type of function can you associate with this situation?

c) What change will occur to the curve associated with this situation if you decrease the value of the wire's resistance?

In the workplace

Investment advisors

The profession

An investment advisor analyzes a client's financial objectives, suggests products based on these objectives and manages products that allow for maximum return on the client's portfolio. The advisor's work also consists of analyzing and interpreting different graphs whereby the changes in value of certain financial products can be tracked. An investment advisor can work for a bank, credit union, investment company, insurance company, private or brokerage firm.

The stock exchange

A stock exchange is an institution where transactions involving the shares of various companies take place. Based on supply and demand, the price of these shares fluctuates over time, and this allows for companies to be assigned value. Most large companies are listed on a stock exchange. The investment advisor studies the behaviour of shares and investment funds listed on a stock exchange. The graph below provides information on a share's value:

Graph ①
Share value

Value ($)

0.26, 0.24, 0.22, 0.20, 0.18, 0.16, 0.14, 0.12, 0.10, 0.08, 0.06, 0.04, 0.02

0 2 4 6 8 10 12 14 16 18 20 22 24 26 28 30

Number of days elapsed since June 1

A stock exchange also indicates the value of the Canadian dollar in relation to other currencies, the price of oil, the value of an ounce of gold, and many other listings. Below is information concerning the value of the Canadian dollar relative to the US dollar in 2008 over a period of 100 days:

Graph ②
Exchange rate for $1 CAN

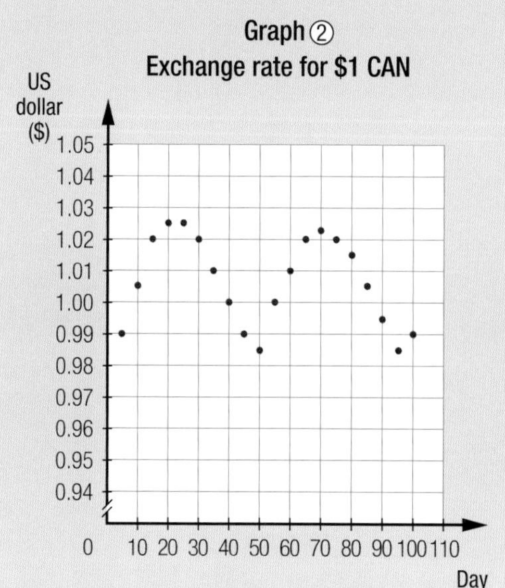

The word "krach" comes from the German word *krak* which means "crash."

Some history

At the end of October 1929, a major financial crisis occurred on the New York Stock Exchange. On that Monday, October 28, also called "Black Monday," the Dow Jones Index fell 13%. In total, the stock index fell by nearly 40% in 2 months. This crisis triggered a general wave of panic that had repercussions on the world economy.

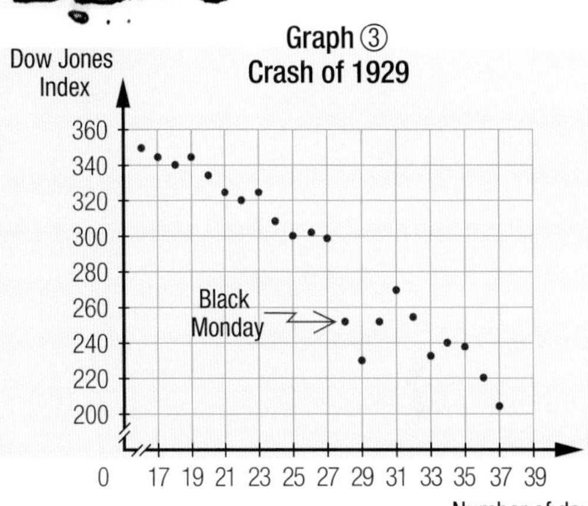

The technological explosion

Towards the end of the 20th century, technological growth skyrocketed. The value of many shares of companies listed on the stock exchange increased considerably, including those of many companies involved in computer technology or the Internet. Consequently, the investment funds of various financial institutions also performed well. Investment advisors kept a close eye on the changes in the value of shares and funds during this period. Below are the results for the "Internology" investment fund over one year:

"Internology" investment fund

Month	Value of the fund ($)
January	4
February	4.2
March	4.5
April	6
May	7.5
June	6.5
July	10
August	11
September	9
October	15
November	18
December	25

1. Determine the type of function that best represents the data from:

a) Graph ①

b) Graph ②

c) Graph ③

2. What was the maximum value of the share associated with Graph ①?

3. a) Draw a graphical representation of the results for the "Internology" investment fund in the form of a scatter plot.

b) What type of function would best represent this graph?

c) If the trend had continued, what would have been the share value in January of the following year?

overview

1 What type of function do you associate with each of the following situations?

a) A nurse observes the curve representing a patient's heart rate on a cardiac monitor.

b) The quantity of blue-green algae in a lake doubles each month. Consider the relation between time and the quantity of blue-green algae.

c) A basketball player shoots a free throw. Consider the relation between time and the height of the ball.

d) A salesman's income amounts to 12% of his total sales. Consider the relation between the total amount of sales and the salesman's salary.

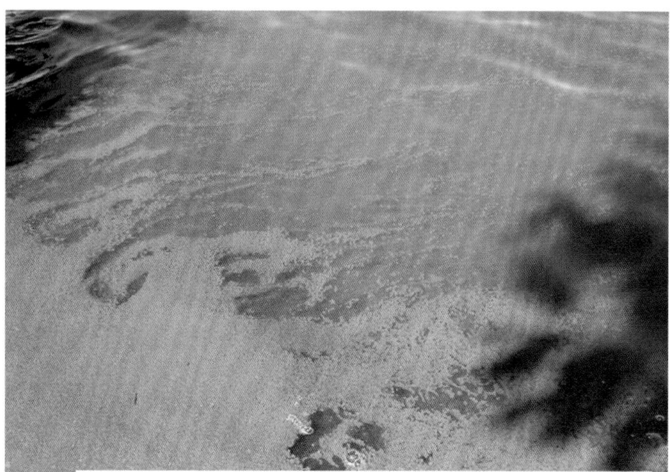

Blue-green algae have lived in aquatic ecosystems for billions of years. They become a problem when they proliferate excessively because some species can pose health hazards for animals and humans.

2 For each of the following tables of values, do the following:

a) Draw a graphical representation of the ordered pairs with a scatter plot.

b) Determine the type of function that best fits the scatter plot.

a)

x	-5	-4	-3	-2	-1	0	1	2	3	4	5
y	-4	-4	-2	-2	0	0	2	2	4	4	6

b)

x	-5	-4	-3	-2	-1	0	1	2	3	4	5
y	3	5	7	9	9	9	10	14	23	39	64

c)

x	-5	-4	-3	-2	-1	0	1	2	3	4	5
y	10	8	6	4	2	0	-2	0	2	4	6

3 Following is a numerical sequence:

1, 2, 4, 8, 16, 32, 64, 128

Consider the relation between each term's rank and its value in the sequence. What type of function could serve as a mathematical model for this situation?

4 For each of the following graphs, do the following:

a) Determine the following properties:

1) the domain and the range
2) the zero(s)
3) the extrema
4) the initial value

b) Determine the type of function that it represents.

c) Specify whether the inverse is a function. Explain your answer.

Graph ①

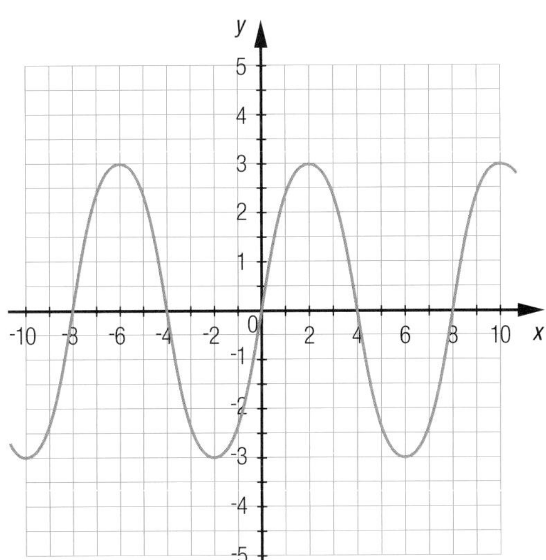

Graph ②

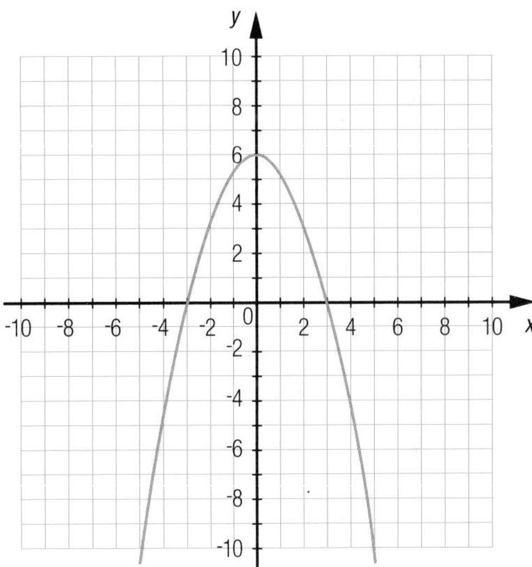

Graph ③

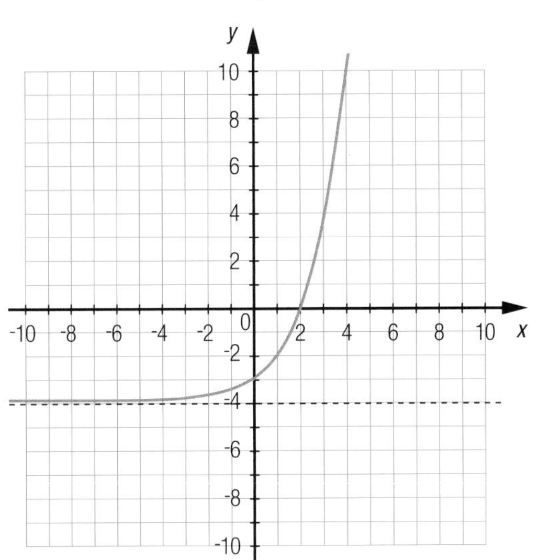

Graph ④

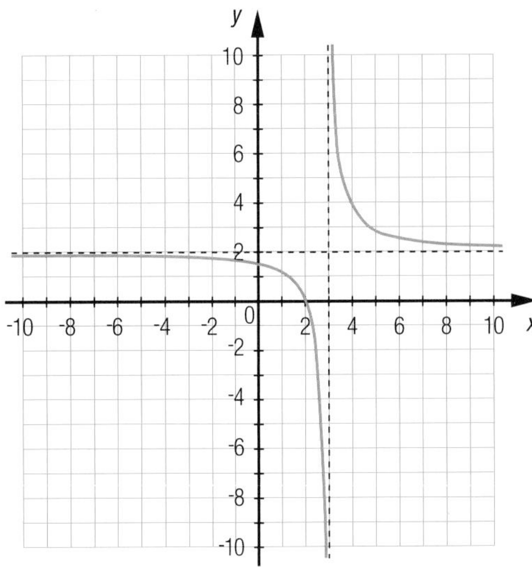

5 Employees at an aquatic park activate an artificial waterfall. A pump brings the water from a basin to the top of the waterfall. The water then falls back into the basin. Below is some information concerning this situation where x represents time (in s) and $f(x)$ represents the height of the water drop (in m).

$$f(x) = \begin{cases} 1.5x & \text{if } x \in [0, a[\\ 9 & \text{if } x \in [a, 13[\\ -(x - 13)^2 + 9 & \text{if } x \in [13, 16] \end{cases}$$

a) What type of function best represents this situation?

b) Draw a graphical representation of this situation.

c) What are the domain and the range of this function?

d) What are the zeros of this function, and what do they represent in this context?

e) How long does the water remain at its maximum height?

In order for the turbine in a hydroelectric power station to revolve under water pressure, the river's flow must be strong enough and its operating head must be high enough. The operating head is the difference in level between the water intake and the water outtake. Certain streams require retaining structures such as dams in order for the operating head to be high enough.

6 **ADRENALINE** When a person experiences a strong sensation of fear, a hormone called adrenaline is secreted in the body by the adrenal glands. When this substance circulates in the human body, the heart rate speeds up, and blood pressure increases. Below is some information on the quantity of adrenaline produced by an individual during testing of a new ride in an amusement park:

Adrenaline production

Time elapsed from the start of the ride (s)	5	10	15	20	25	30	35	40	45	50	55	60	65	70	75
Quantity of adrenaline (mL)	10	13	15	19	22	24	25	24	23	24	20	16	11	8	5

a) Represent this situation with a scatter plot.

b) Draw a curve that best represents all the points on the scatter plot.

c) What is the quantity of adrenaline in this person's body at the start of the ride?

7 In the graphical representations below, function f represents a basic function.

a) In Graph ①, determine the value of the vertical change of scale that allows you to associate the curve of function f to the curve of function:

1) g 2) h 3) i

b) In Graph ②, determine the value of the horizontal change of scale that allows you to associate the curve of function f to the curve of function:

1) g 2) h 3) i

Graph ①

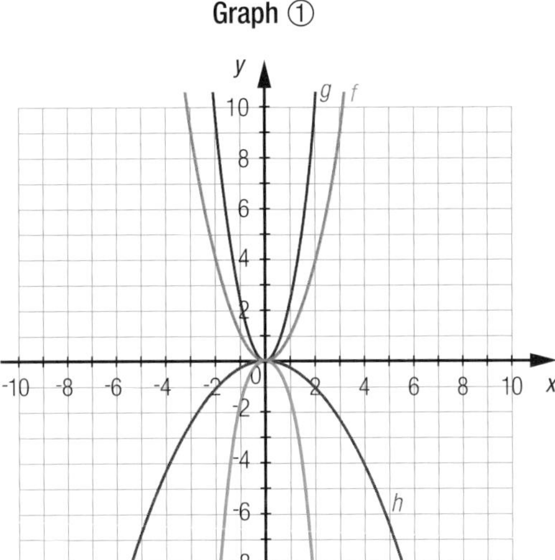

Graph ②

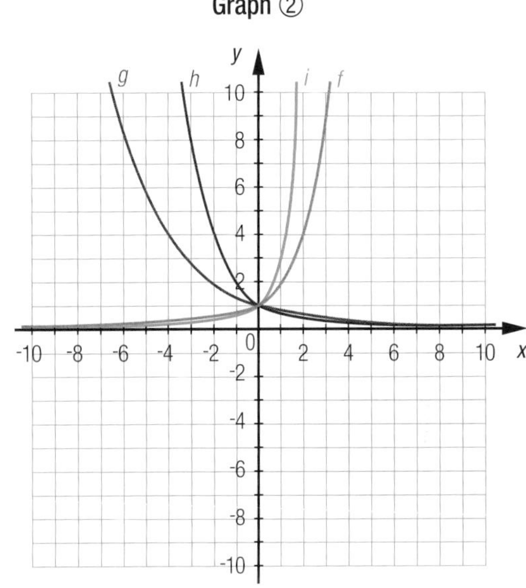

8 In the following graphical representations, the orange curve represents a basic function. Find the rule for each of the three transformed functions of the same type:

a) For Graph ①, assign a possible value to each of the multiplicative parameters.

b) For Graph ②, determine whether **a** > 0 or **a** < 0 considering that **b** > 0.

Graph ①

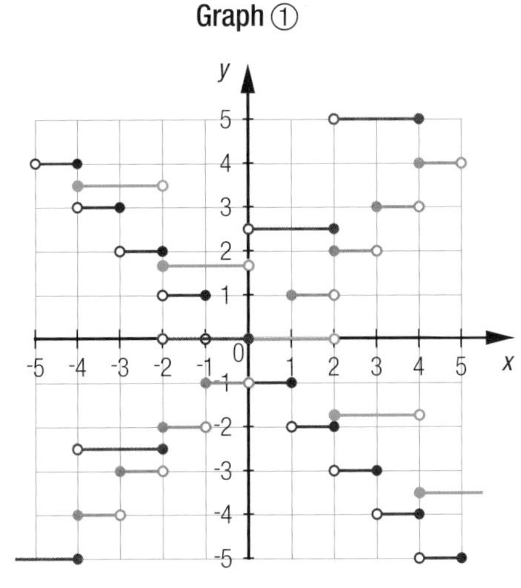

Graph ②

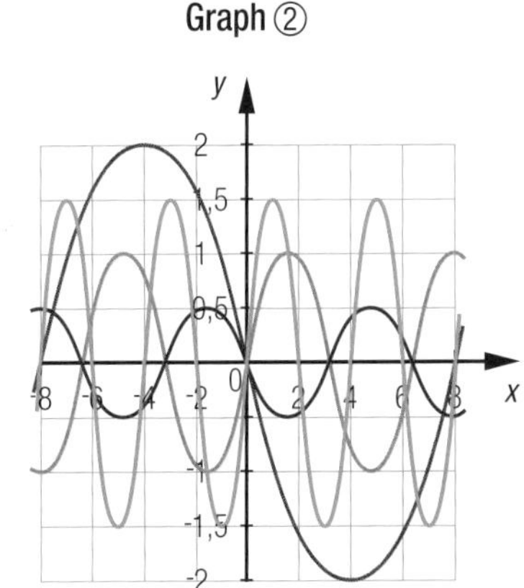

9 The adjacent graph represents the basic function f whose rule is $f(x) = x^2$. From the choices below, associate the rules with each of the graphs.

A $m(x) = -0.5x^2$

B $n(x) = (-0.5x)^2$

C $p(x) = -0.5(0.5x)^2$

D $q(x) = 0.5(-0.5x)^2$

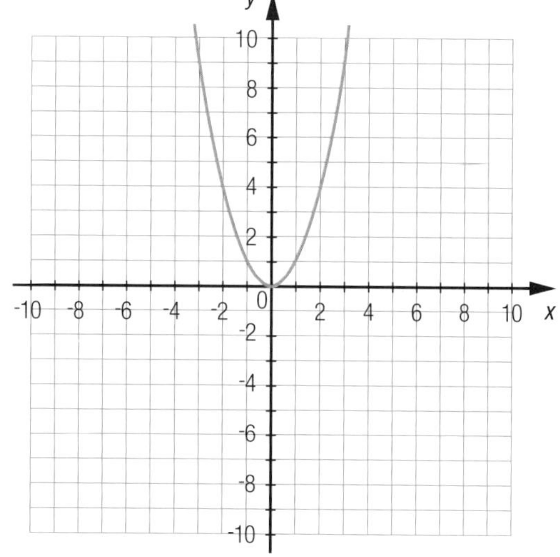

1

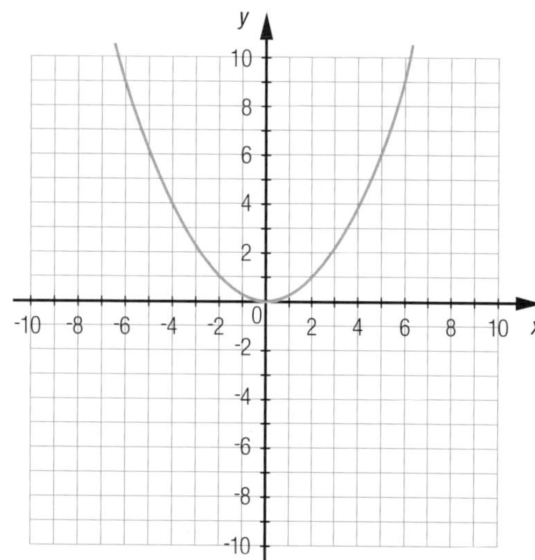

2

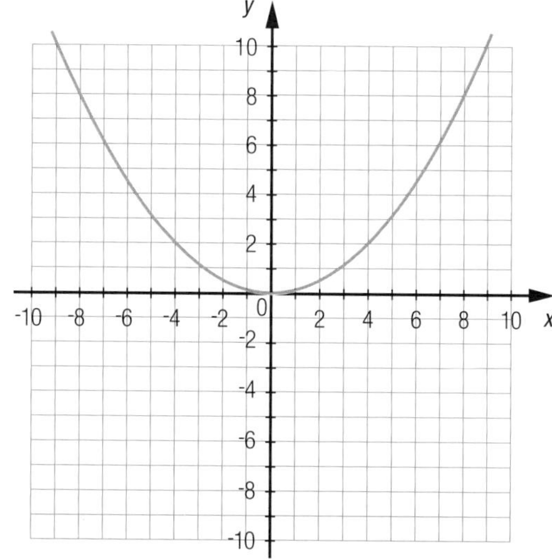

3

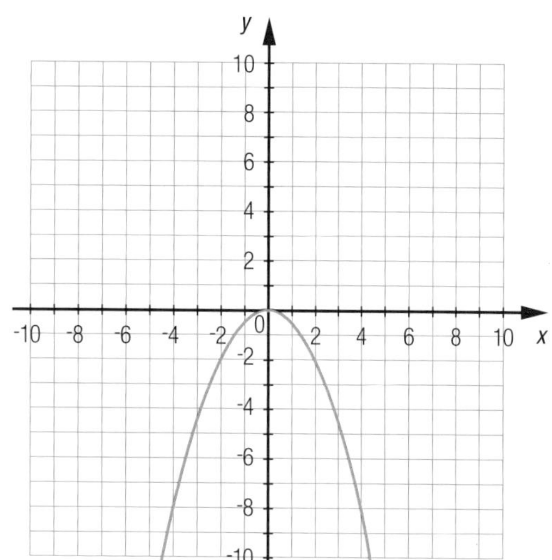

4

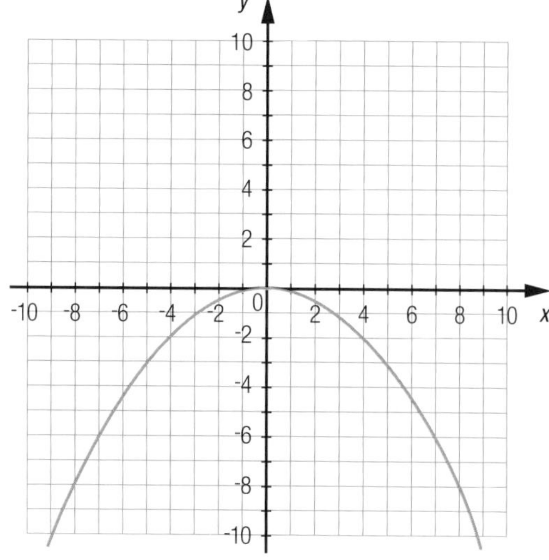

10 In the adjacent Cartesian plane, the "pattern" represents an interval of a periodic function.

a) Determine the period of this function.

b) Calculate:

1) $f(8)$ 2) $f(22)$ 3) $f(36)$

c) What does $f(-18)$ represent?

d) Draw a graphical representation of this function over the interval [15, 35].

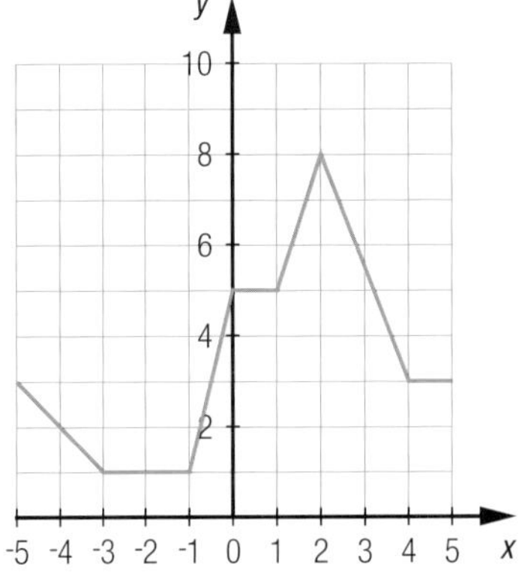

11 The adjacent graph shown represents the height of a golf ball in relation to time.

a) What type of function can you associate with this situation?

b) What was the maximum height reached by the ball?

c) When will the ball hit the ground?

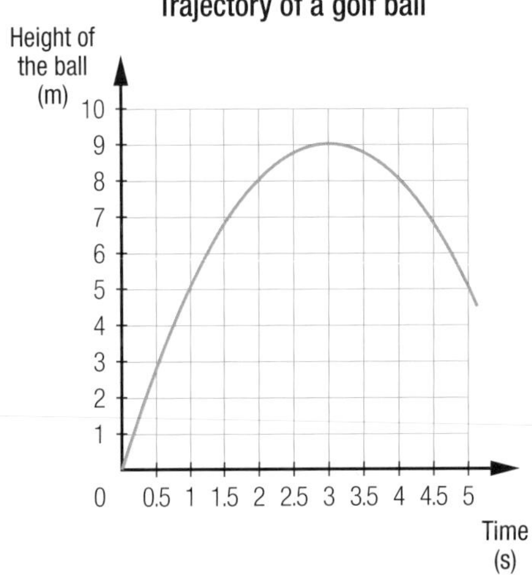

12 An Internet service provider offers subscribers a choice between two monthly plans. Below are the details of these two plans:

Plan ①

$10/month, including 4 Gb in downloads

$3/Gb for the 5 subsequent Gb

$5/additional Gb

Plan ②

$5/Gb in downloads for the first 5 Gb

$2/Gb for the subsequent 5 Gb

No charge for any additional Gb

a) Consider the relation between the number of Gb in downloads and the cost (in $); draw a representation of both these plans on a single Cartesian plane.

b) What type of function best represents these plans?

c) Which of these plans is the most appealing? Explain your answer.

13 The managers of a festival want to give away a certain amount of money in a raffle. This amount will be distributed equally among the winners. The adjacent table of values provides information on the raffle.

Raffle

Number of winners	Each winner's share ($)
1	3,600
2	1,800
3	1,200
5	720
6	600
8	450
12	300

a) What total amount of money will be given away as prize money during this festival?

b) Draw a scatter plot representing this situation.

c) What type of function best represents this situation?

d) If 15 people win during the raffle, what will each winner's share be?

14 A biologist observing the evolution of the hare population within a given forest area notices that the number of these small mammals varies according to the periodic model shown in the adjacent graph.

a) During the first year of observation, over how many days did the hare population decrease?

b) What was the highest hare population in this area?

c) If the trend continues, how many hares will there be in this area 540 days after the beginning of observation?

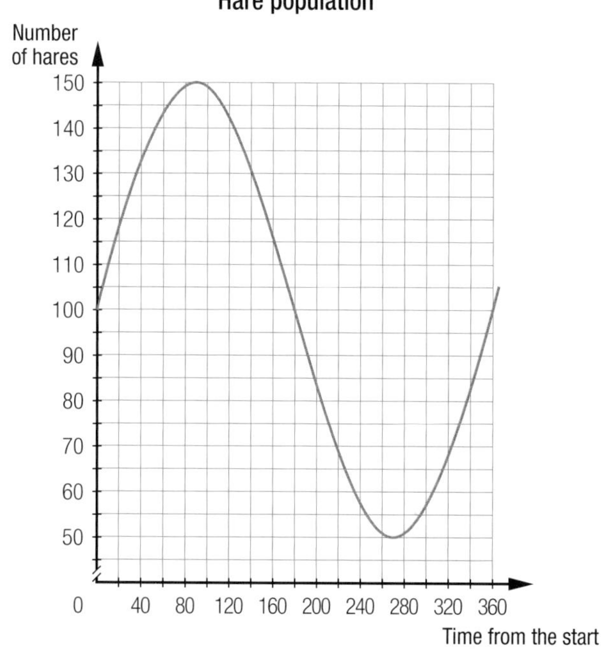

15 The adjacent graph shows the admission price, according to age, to the Science Museum.

a) What type of function best represents this situation?

b) What will the price of admission be for someone who is:

1) 21 years old?

2) 50 years old?

c) In what age group are the people who must pay the maximum admission price?

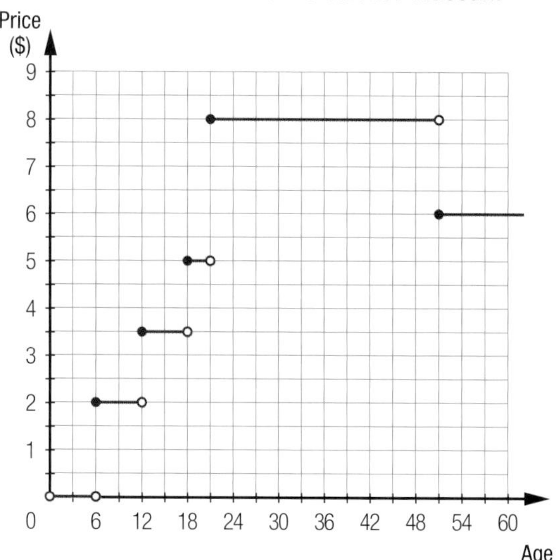

16 **BRAKING DISTANCE** The distance required for a car to come to a full stop is directly proportional to the square of its speed. Below are different functions with which you can determine the stopping distance d (in m) relative to the speed s of an automobile (in km/h):

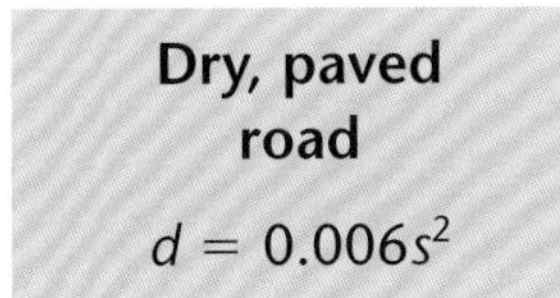

Wet, paved road

$d = 0.009s^2$

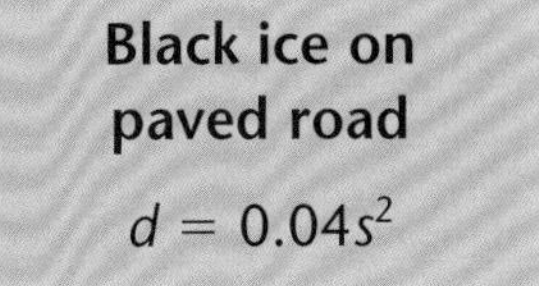

a) Complete the table of values below.

Braking distance

Speed (km/h)	10	20	30	40	50	60	70	80	90	100
Distance on dry pavement (m)										
Distance on wet pavement (m)										
Distance on black ice (m)										

b) Comparing the above functions with the basic function $d = s^2$, determine the value of parameter **a** in each rule.

c) What are the consequences of an increase or decrease in the value of parameter **a** in this context?

17 **METRONOME** A metronome is a tool that indicates the speed or rhythm at which a musical piece is to be played. A small mass is attached to a metal rod that oscillates more or less quickly depending on the height of the mass on this rod. The adjacent graph shows different musical tempos based on the oscillations of a metronome.

a) For each tempo, identify the height of the mass relative to the base of the metronome.

b) Considering that Tempo ❶ is a *larghetto*, name the two other tempos based on the data provided in the table below.

Metronome oscillations

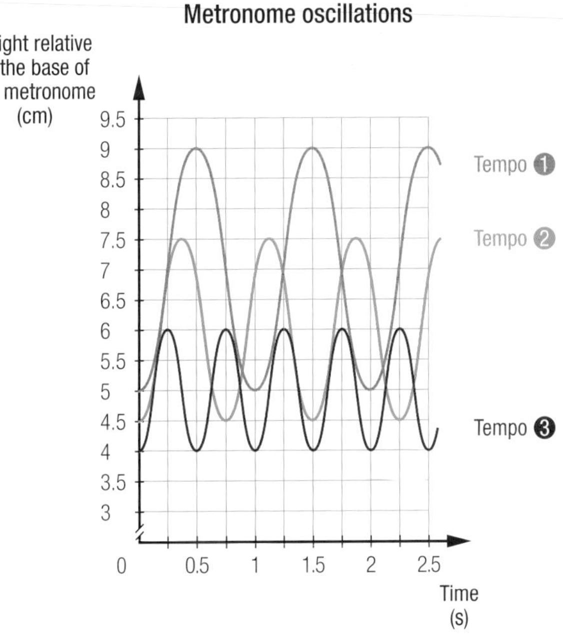

Musical tempos

Tempo	*Largo*	*Larghetto*	*Adagio*	*Andante*	*Moderato*	*Allegro*	*Prestissimo*	*Presto*
Number of beats per minute	[40, 60[	[60, 66[	[66, 76[	[76, 108[	[108, 120[	[120, 168[	[168, 200[	[200, 208]

bank of problems

18 **COMBAT AIRCRAFT** The F-18 is a military aircraft used in aerial combat by the Canadian Air Force. This aircraft can travel almost double the speed of sound, at about 666 m/s or 2400 km/h. The table below shows the data collected by an F-18's onboard computer during a training flight. Determine the duration of each phase of this training flight.

Training flight

	Take-off phase	Flight phase	Descent phase	Landing phase
Duration (min)				
Equation where x represents time (in min) and y represents the altitude of the aircraft (in m)	$y = 100x$	$y = 600$	$y = {}^{-}80x + 1880$	$y = {}^{-}100\sqrt{x - 21} + 200$

19 During an air show, a hot air balloon takes off at the same time as a small airplane begins its descent. The graph shown provides information on these two manoeuvres.

The first hot air balloons were made of silk and paper, and they could easily catch fire. Today, they are made of nylon coated with a light, strong, and most importantly, fireproof coat of polyurethane varnish.

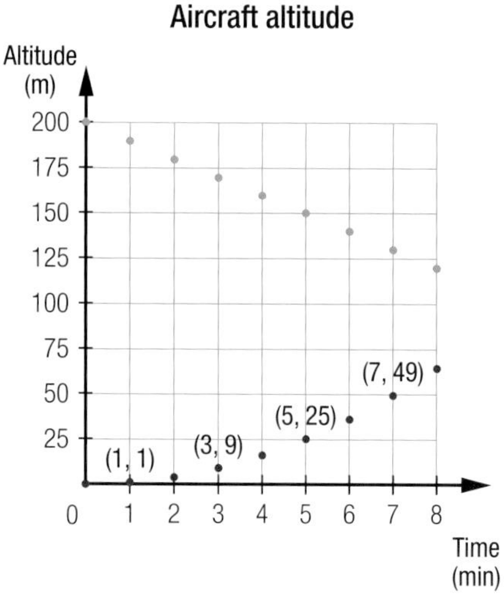

Is it true that these two aircraft will be flying at the same altitude at some point during the air show? Explain your answer.

20 A painter submits a bid to paint an enormous sculpture made of eight spheres. In order to determine the total cost of the project, the painter must estimate the amount of paint needed to complete the work. Based on the following information for paint costs, calculate the total cost of this part of the painting project.

Paint-related costs

Sculpture	Radius (m)	Paint quantity (dL)	Cost ($)
Sphere ❶	1	23	27.60
Sphere ❷	2	90	108
Sphere ❸	3	200	
Sphere ❹	4	360	
Sphere ❺	5	565	
Sphere ❻	6		
Sphere ❼	7		
Sphere ❽	8		

21 Two students have collected data relative to the frequency of certain sound waves. Two of these have been graphically represented. They notice, in particular, that the curve associated with the frequency of 0.32 hertz reflects that of a basic function.

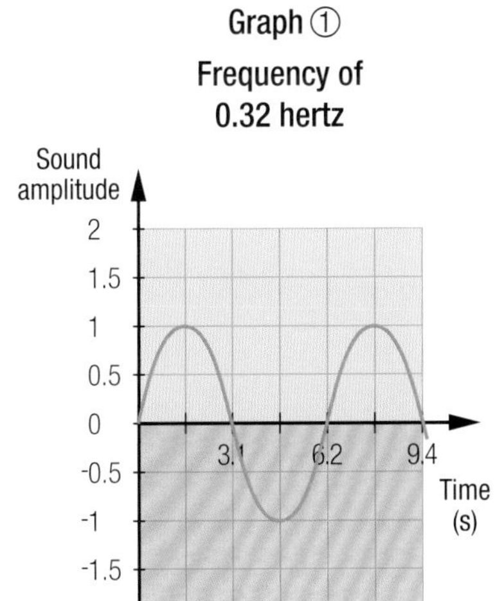

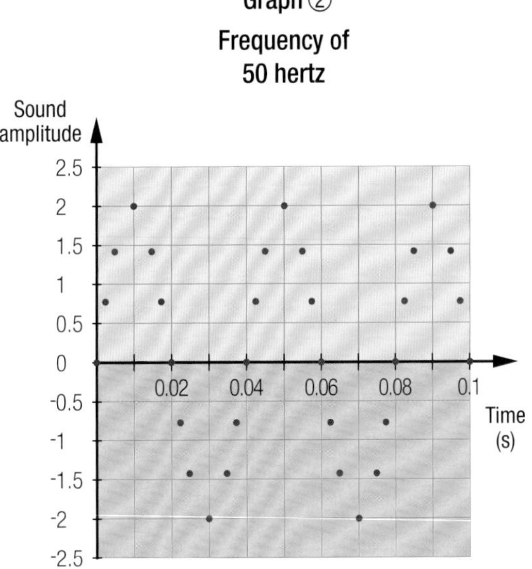

Determine the value of the vertical change of scale and the value of the horizontal change of scale that the students need to apply to the curve in Graph ① in order to obtain that of Graph ②.

VISION 2

Congruent figures and similar figures

Can you use the properties of light to enlarge an image? How do you verify whether the structure of the roof of a house matches the original blueprints? How is it possible to use a mirror to measure the height of a building? Can you use a stroboscope to check the alignment of the shaft in a rotary tool? In "Vision 2," you will discover the geometric statements that allow you to conclude that triangles are congruent or similar. Various situations will lead you to manipulate algebraic expressions written in fraction form. You will learn to factor an algebraic expression and solve problems involving distance. You will also study the metric relations in a right triangle, and explore Thales' theorem.

Arithmetic and algebra	Geometry	Statistics	Probability
• Factoring: grouping, difference of squares, perfect square trinomials • Dividing a polynomial by a binomial • Operations on rational expressions	• Congruent triangles • Similar triangles • Optimizing a distance • Metric relations in a right triangle • Finding missing measurements		

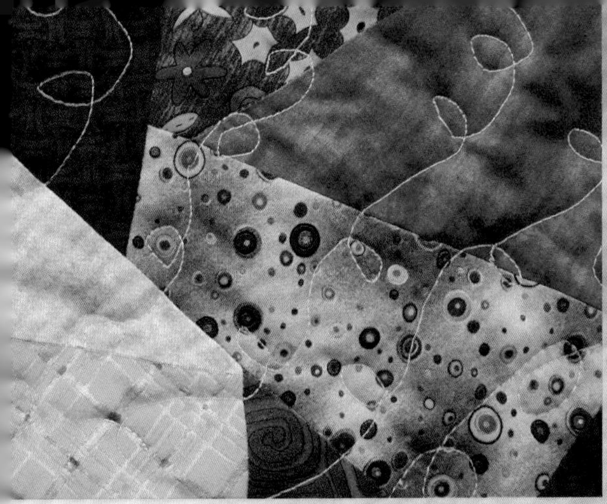

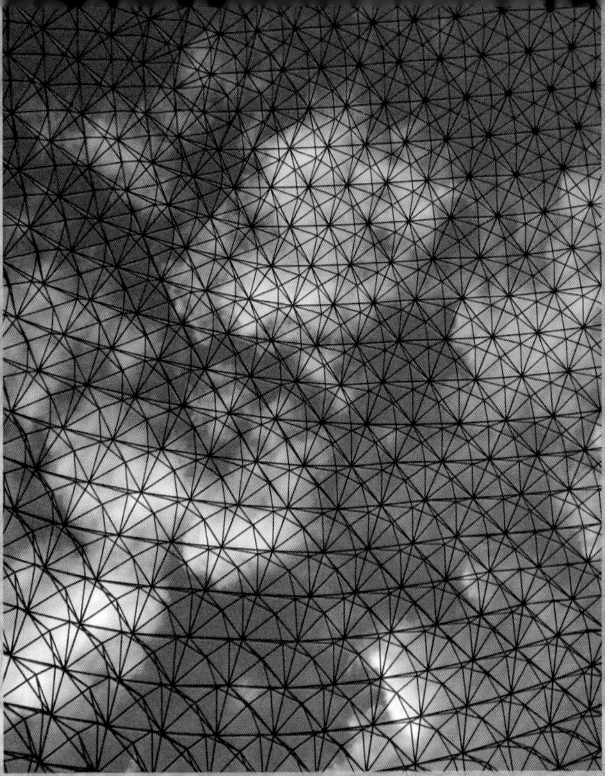

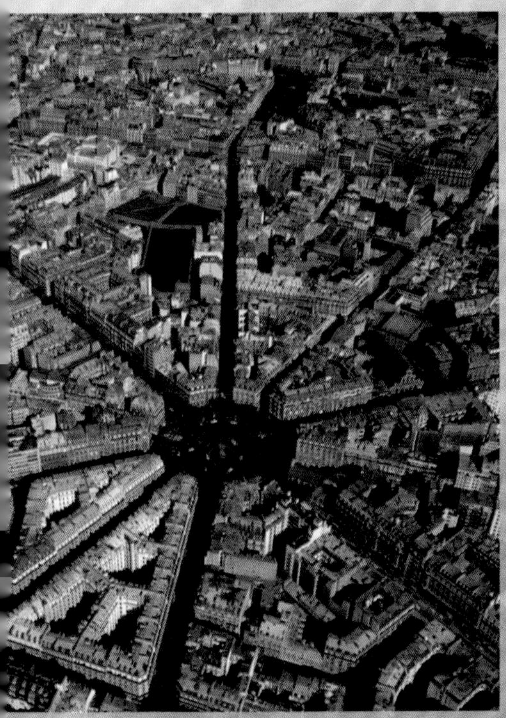

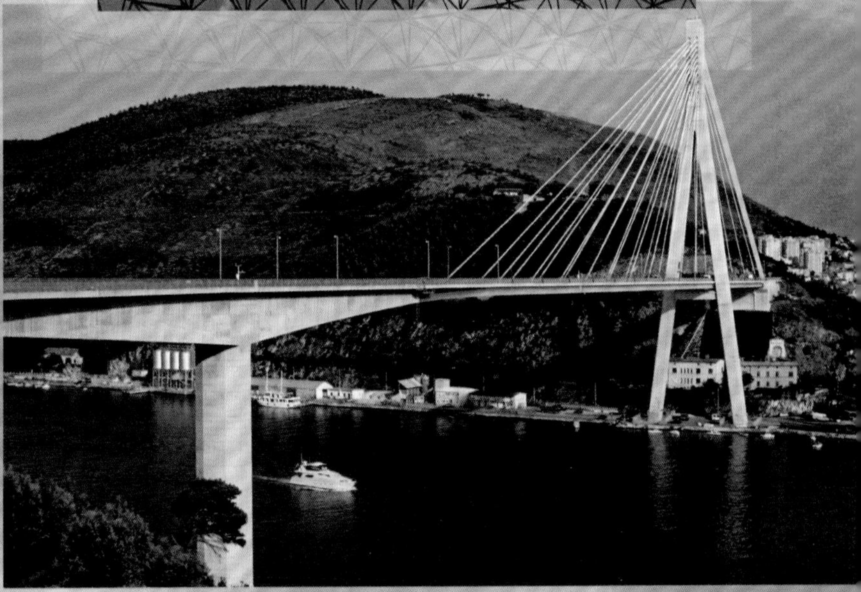

LEARNING AND EVALUATION SITUATIONS — The logging industry 209

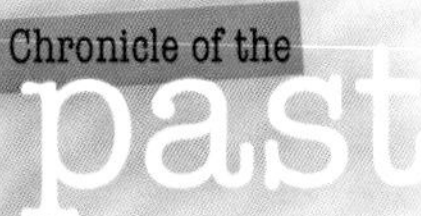

Chronicle of the past — The Ionian School 120

In the workplace — Civil engineering technicians 122

REVISION 2

PRIOR LEARNING 1 Heat deflectors

Many barbecue models have heat deflectors between the burners and the grill in order to enhance heat distribution and prevent fat from dripping onto the burners.

Below is information on a type of deflector made with a triangular sheet of stainless steel:

- The stainless steel plate ACF is an isosceles triangle.
- The quadrilateral BDFG is a parallelogram.

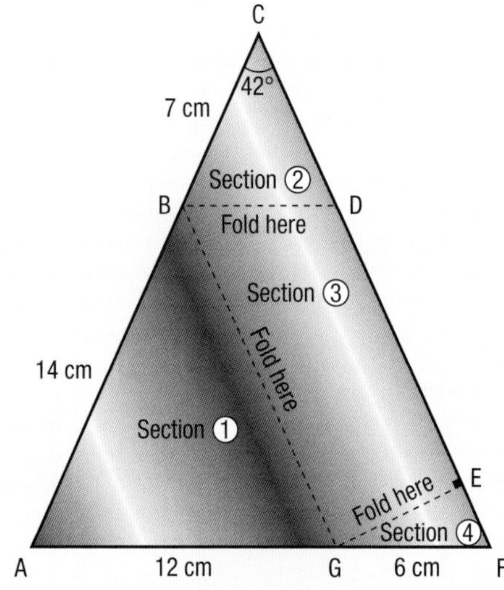

a. What type of triangle corresponds to Section ①?

b. What is the measure of angle EGF?

c. What is the area of:

1) the stainless steel plate?
2) Section ①?
3) Section ②?
4) parallelogram BDFG?
5) Section ④?

The heat generated by a barbecue or any heating or refrigerating unit is measured in BTUs which stands for British Thermal Units.

knowledge summary

CLASSIFICATION OF TRIANGLES

Illustration	Sides: Characteristic	Sides: Name
	No congruent sides	Scalene triangle
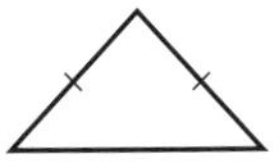	Two congruent sides	Isosceles triangle
	Three congruent sides	Equilateral triangle

Illustration	Angles: Characteristic	Angles: Name
	One obtuse angle	Obtuse triangle
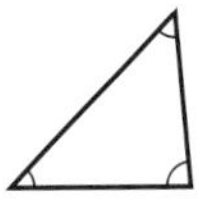	Three acute angles	Acute triangle
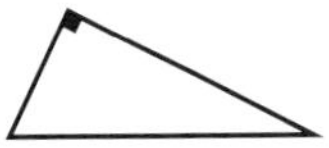	One right angle	Right triangle
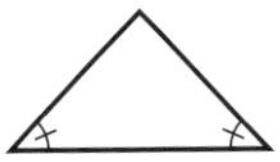	Two congruent angles	Isoangular triangle
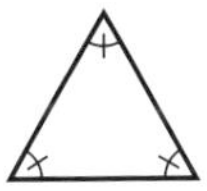	Three congruent angles	Equiangular triangle

PROPERTIES OF QUADRILATERALS

		Properties			
Illustration	Name	Sides	Angles	Diagonals	Axes of symmetry
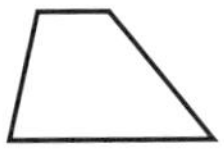	Irregular trapezoid	One pair of parallel sides			
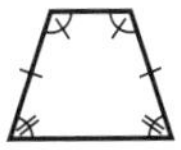	Isosceles trapezoid	One pair of parallel sides One pair congruent sides	Two pairs of congruent angles	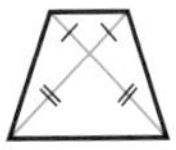	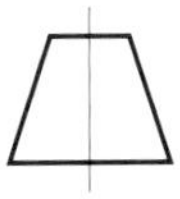
	Right trapezoid	One pair of parallel sides	Two right angles		
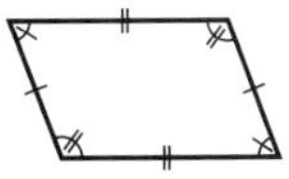	Parallelogram	Two pairs of opposite sides are parallel and congruent	Congruent opposite angles Supplementary adjacent angles	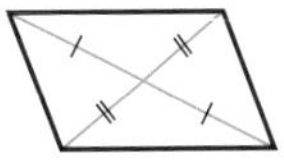	
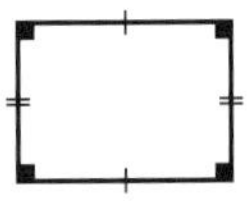	Rectangle	Two pairs of opposite sides are parallel and congruent	Four right angles Supplementary adjacent angles	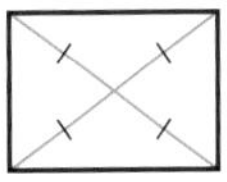	

(*Continued next page*)

Illustration	Name	Properties			
		Sides	Angles	Diagonals	Axes of symmetry
	Rhombus	Two pairs of opposite parallel sides Four congruent sides	Congruent opposite angles Supplementary adjacent angles		
	Square	Two pairs of opposite parallel sides Four congruent sides	Four right angles Adjacent angles are supplementary		

REGULAR POLYGON

A polygon is **regular** if all of its sides have the same length and all interior angles are congruent.

E.g.

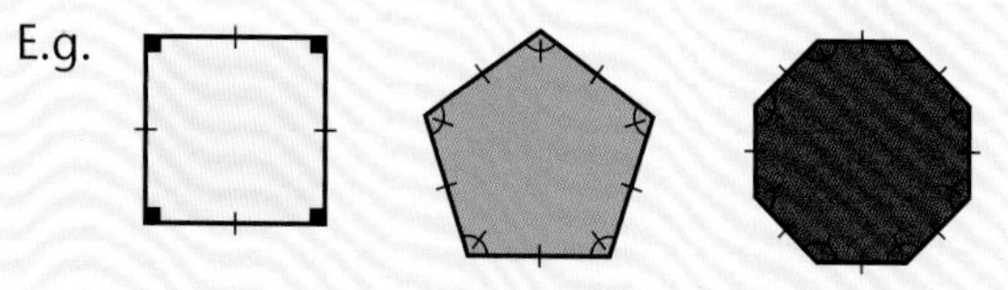

AREA: TRIANGLE, QUADRILATERAL, REGULAR POLYGON AND CIRCLE

Shape	Area
b, h	$A_{triangle} = \frac{b \times h}{2}$
b, h, B	$A_{trapezoid} = \frac{(B + b) \times h}{2}$
b	$A_{parallelogram} = b \times h$
D, d	$A_{rhombus} = \frac{D \times d}{2}$

Shape	Area
h, b	$A_{rectangle} = b \times h$
s	$A_{square} = s^2$
Apothem	$A_{regular\ polygon} = \frac{\text{perimeter} \times \text{apothem}}{2}$
r	$A_{circle} = \pi r^2$

ANGLES CREATED BY A TRANSVERSAL INTERSECTING TWO PARALLEL LINES

When two parallel lines are intersected by a transversal, the following is true:

- The alternate interior angles are congruent: $\angle 4 \cong \angle 6$ and $\angle 3 \cong \angle 5$.
- The alternate exterior angles are congruent: $\angle 1 \cong \angle 7$ and $\angle 2 \cong \angle 8$.
- The corresponding angles are congruent: $\angle 1 \cong \angle 5$ and $\angle 2 \cong \angle 6$ $\angle 4 \cong \angle 8$ and $\angle 3 \cong \angle 7$.

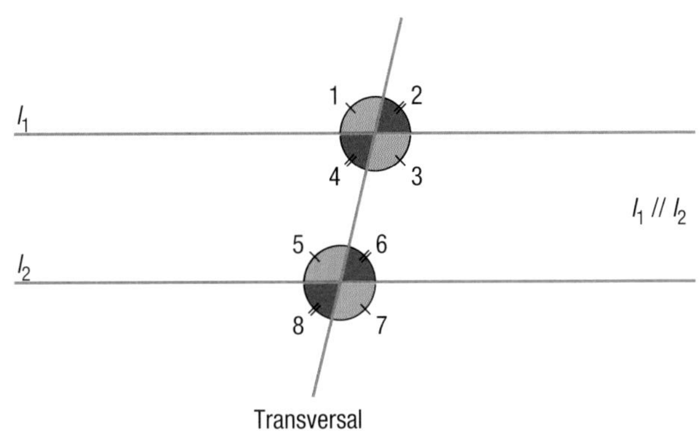

In summary $\angle 1 \cong \angle 3 \cong \angle 5 \cong \angle 7$ and $\angle 2 \cong \angle 4 \cong \angle 6 \cong \angle 8$.

FACTORING: REMOVING A COMMON FACTOR

Factoring an algebraic expression consists of writing it as a product of factors. In algebra, factoring is often used to simplify expressions, to solve equations and to demonstrate equivalence between expressions.

E.g.

Standard form	Factored form	Factors
1) $5a + 35$	$5(a + 7)$	5 and $a + 7$
2) $b^2 - 11b$	$b(b - 11)$	b and $b - 11$
3) $6c^2 + 15c$	$3c(2c + 5)$	$3c$ and $2c + 5$

There are various methods used to factor an algebraic expression, one of which is **removing a common factor.** This method consists of doing the following:

1. Determine the greatest common factor of all the terms of the algebraic expression.	E.g. In the expression $6a^2 + 15a$, the greatest common factor is $3a$.
2. Divide the algebraic expression by the greatest common factor.	$\frac{6a^2 + 15a}{3a} = \frac{6a^2}{3a} + \frac{15a}{3a} = 2a + 5$
3. Write the product of the factor obtained in "1." by the quotient obtained in "2."	The factored form of $6a^2 + 15a$ is: $3a(2a + 5)$.
Verify the result by expanding the factored form using the distributive property.	$3a(2a + 5) = 3a \times 2a + 3a \times 5$ $= 6a^2 + 15a$

knowledge in action

1 The lengths of the three sides of various triangles are shown below. Which of these triplets would allow you to construct a right triangle?

A (6, 8, 10)	**B** (1, 1, 2)	**C** (10, 28, 30)	**D** (14, 28, 50)
E (3, 4, 5)	**F** (50, 50, 75)	**G** (11, 12, 13)	**H** (65, 72, 97)

2 For each of the following, determine the greatest common factor of the monomials.

a) 5 and 35 b) $3n$ and $30n$ c) $6b$ and $9b$

d) $7p^2$ and $21p$ e) $11t^2$, $44t$ and 121 f) $18z^4$, $36z^2$ and $6z$

3 Factor the following polynomials.

a) $8m^2 + 24m$ b) $72s^3 - 18s^2$ c) $-14t^3 - 2t$

d) $a^2b^3 + ab^2 - ab$ e) $5(y - 1) - 3(y - 1)$ f) $5r(2r + 1) - 2r(2r + 1)$

4 For each of the pairs of polygons on the right, state one property associated with Polygon ② but not with Polygon ①.

	Polygon ①	Polygon ②
a)	Rhombus	Square
b)	Trapezoid	Parallelogram
c)	Rectangle	Rhombus
d)	Parallelogram	Isosceles trapezoid

5 List the following regular polygons in ascending order, according to their areas.

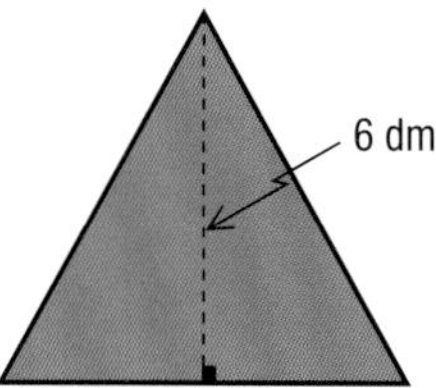

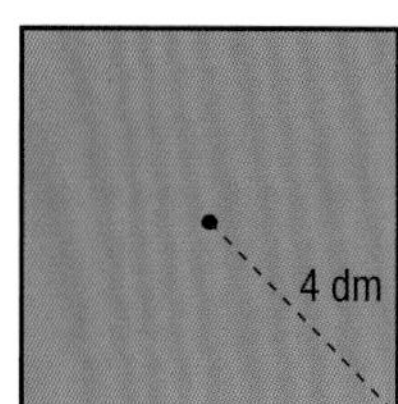

3

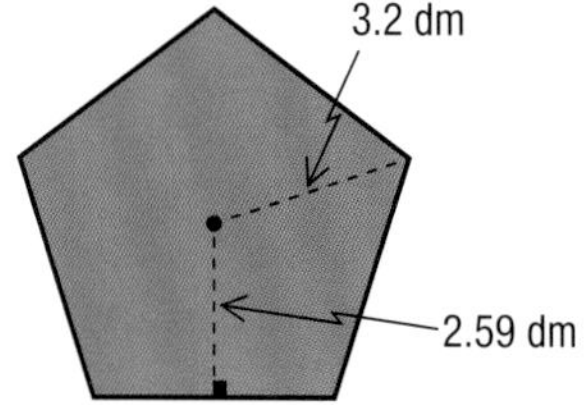

4

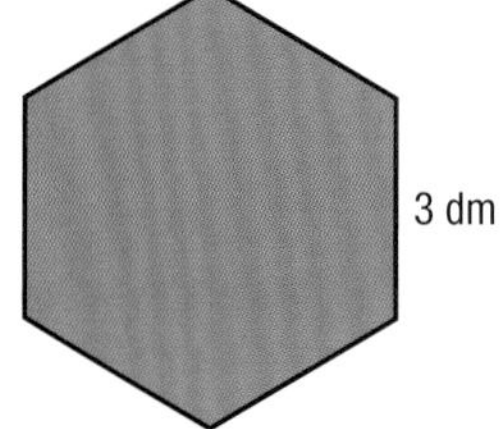

6 Show that the adjacent right triangle cannot exist for any chosen value of y.

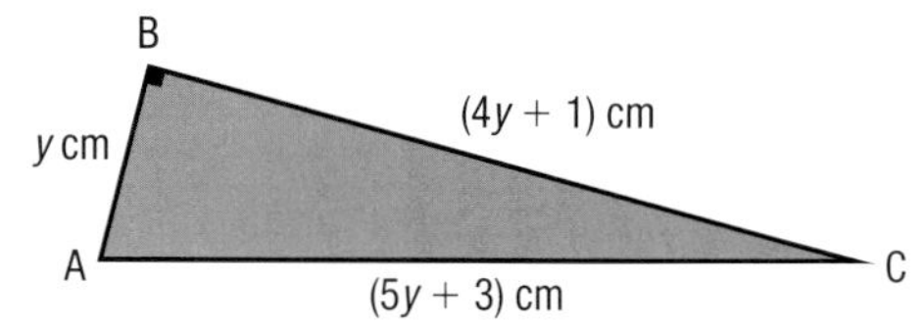

7 In the adjacent figure, lines l_1 and l_2 are parallel. Provide a geometric statement that would make it possible to conclude that:

a) m∠FGH = 35°

b) m∠EFC = 45°

c) m∠CBH = 100°

d) m∠FHB = 80°

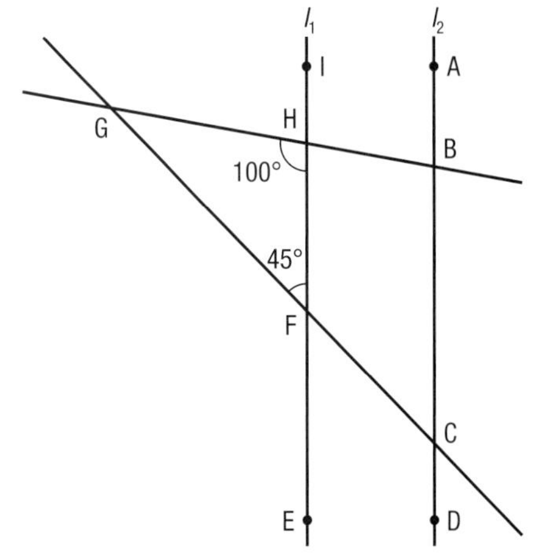

8 In each of the figures below, determine the area of the coloured section.

a)

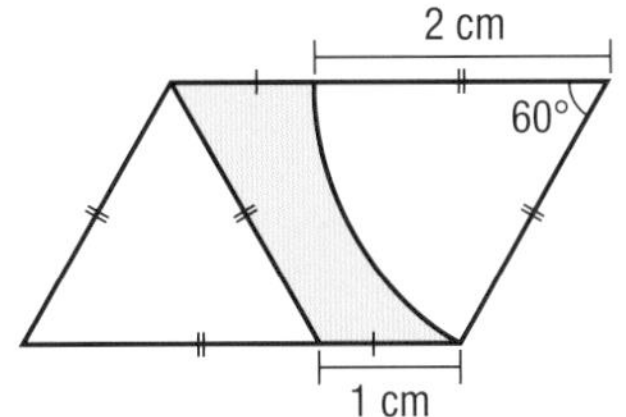

b)

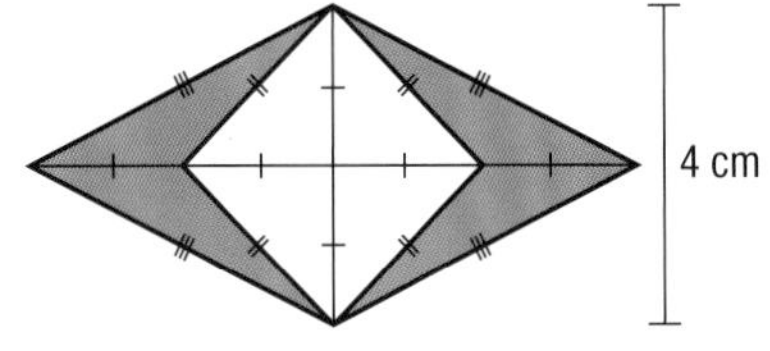

9 In the adjacent diagram, classify triangle ABC:

a) based on the length of its sides

b) based on the measures of its angles

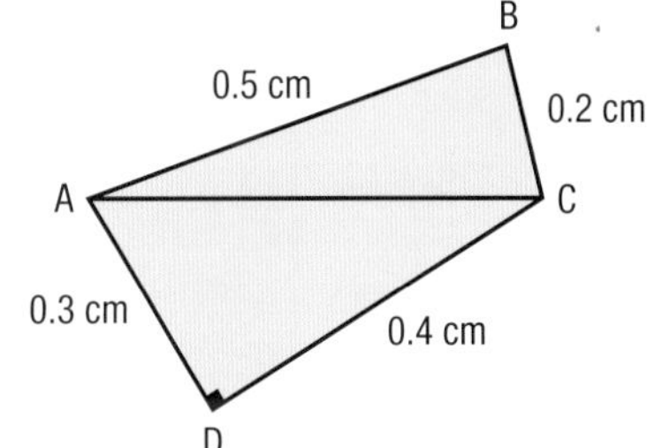

10 Determine the area of the coloured section, considering that the adjacent rectangle ABCD is inscribed in a circle with centre O whose radius measures 8 cm.

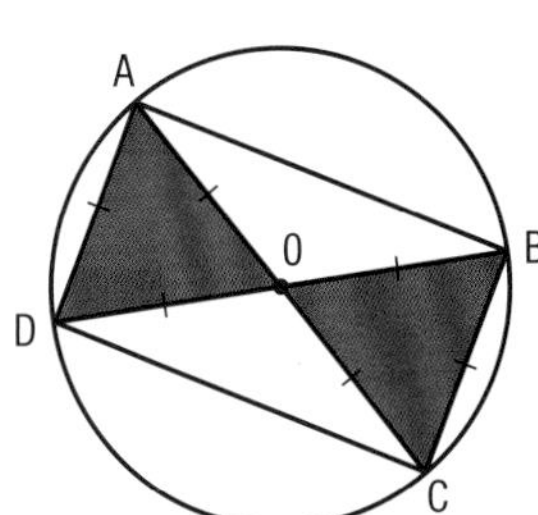

11 Three spotlights illuminate a stage to form pentagon ABCDE as shown. Determine the measures of all the interior angles of this pentagon.

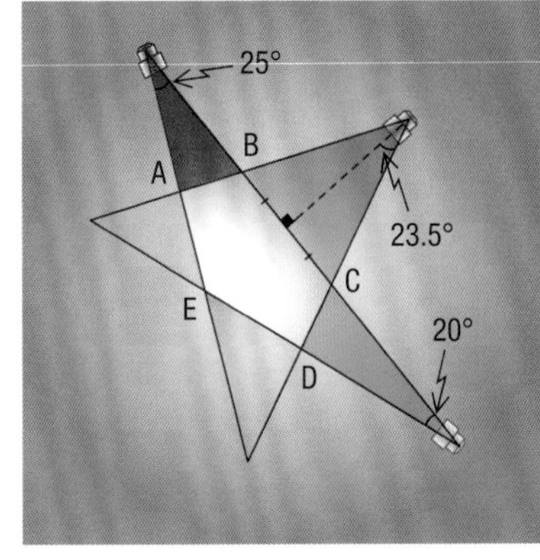

SECTION 2.1 Congruent triangles

This section is related to LES 4.

PROBLEM Playing squash

Squash is a raquet sport dating back to 19th century, England. During a match, the players alternately hit the ball, either directly or indirectly, against the front wall until one of the two misses. Players can use the side and rear walls to execute a shot.

During practice, a player is training to hit the ball through a hoop that has been placed on the court.

View of squash court from above

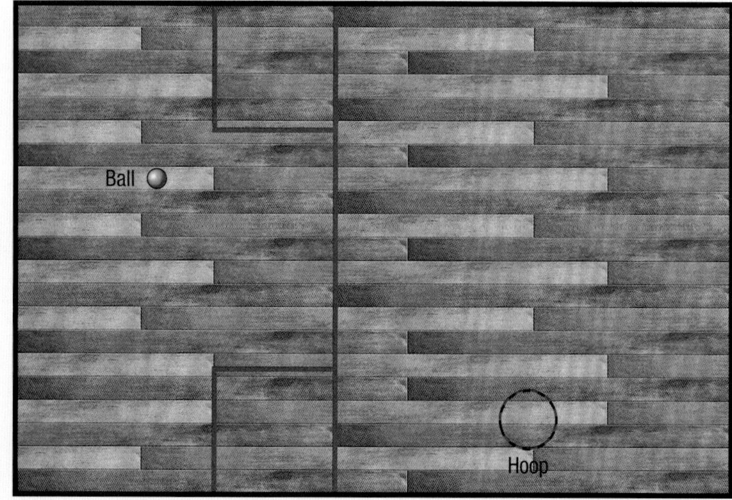

Describe the path that the ball must take to strike two walls, including the front wall, before falling into the hoop.

Squash is a relatively recent sport. The first squash court was built in 1883 in Oxford, England, and the first official international rules were set in 1928.

ACTIVITY 1 Analyzing rapid motion

A stroboscope is a device that produces intermittent light. It is used, in particular, to observe movements "in slow motion," that is movement that is too rapid to be seen with the naked eye.

A stroboscope makes it possible to track the motion of a coin thrown in the air.

To check the alignment of a rotary tool, a technician uses a stroboscope to view the movement of a triangular blade.

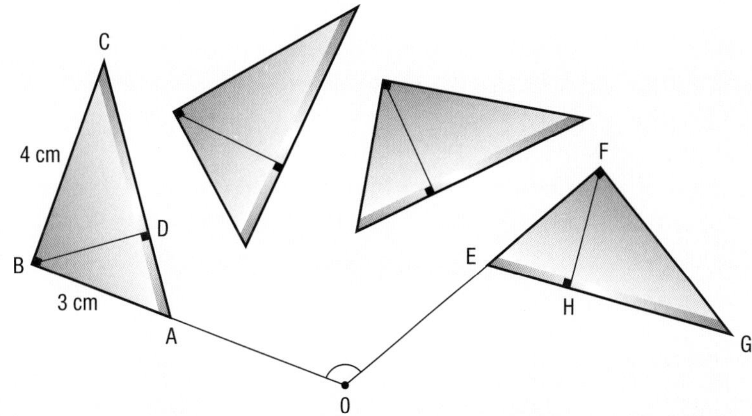

a. What geometric transformation is observed in the series of images shown above?

b. In triangle EFG, what side corresponds to side:

1) AB?
2) BC?
3) AC?

c. Determine the length of side:

1) EF
2) FG
3) EG

d. What could be said about:

1) the heights BD and FH?
2) the perimeter of each of these triangles?
3) the area of each of these triangles?

The shaft in a rotary tool is the axis that transfers the rotational motion to the mobile part of the tool.

ACTIVITY 2 Measuring the inaccessible

To determine distances, land surveyors can use an electronic tachometer as well as a tetrahedral reflecting prism. The electronic tachometer sends a laser beam which is then redirected back to it by the tetrahedral reflecting prism.

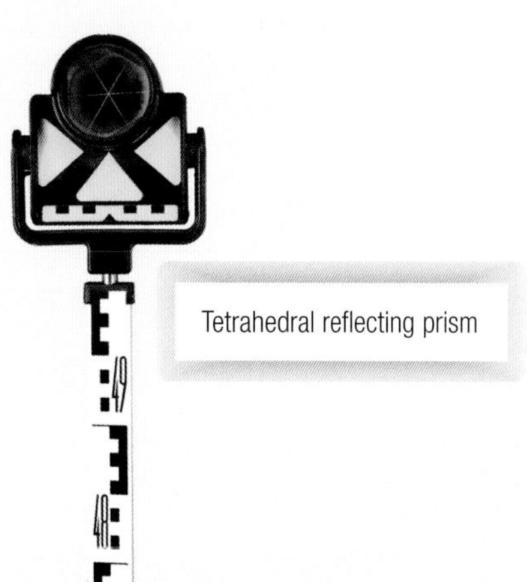
Tetrahedral reflecting prism

Electronic tachometer

Two land surveyors are on the shore of the river as illustrated below. When the first surveyor is stationed at A, she sights the reflecting prism P at an angle of deviation of 30° in relation to the shore. When she is stationed at B, she sights the reflecting prism at an angle of deviation of 60°.

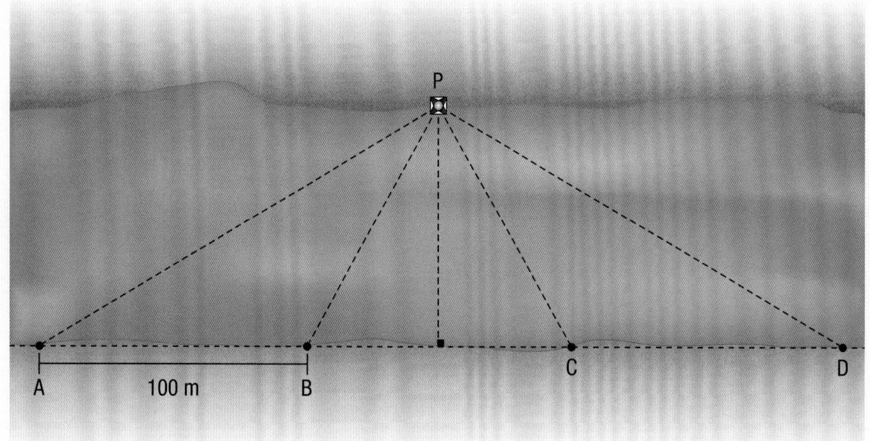

a. What are the measures of the angles of triangle ABP?

b. When she is stationed at C, the second land surveyor sights the reflecting prism at an angle of 60° in relation to the shore. When she is stationed at D, she sights the reflecting prism at an angle of 30°.

1) What is the length of segment CD?
2) What are the measures of the angles of triangle CDP?

c. What statement can you make about triangles ABP and DCP?

d. How many different triangles can be constructed if you know the length of one of the sides and the measure of two angles at each end of this side?

Techno math

Dynamic geometry software allows you to compare geometric figures. By using the tools TRIANGLE, DISTANCE and ANGLE MEASUREMENT, you can construct triangles and verify if they are congruent.

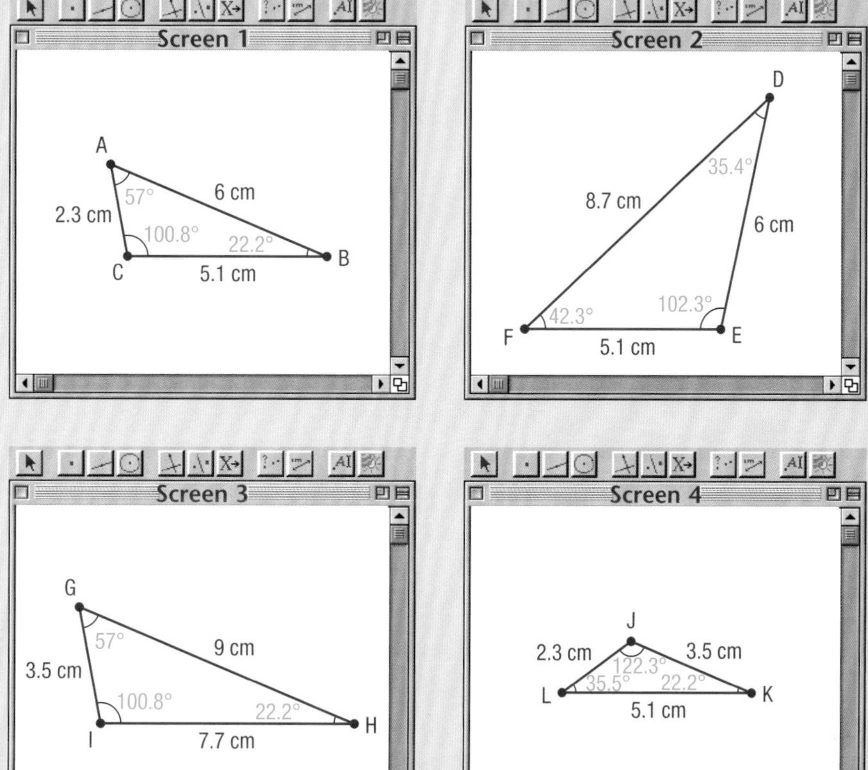

a. By comparing the measurements of angles and sides, what do the triangles in the following screens have in common:

1) Screens **1** and **2**? 2) Screens **1** and **3**? 3) Screens **1** and **4**?

b. Among the triangles shown in Screens **1**, **2**, **3** and **4**, is there a pair of congruent triangles?

c. Can you conclude that two triangles are congruent if:

1) two of the sides of a triangle are congruent to two sides of the other triangle?

2) three angles of a triangle are congruent to three angles of the other triangle?

d. Two sides and an angle of triangle MNO are congruent with two sides and an angle of triangle PQR. With the help of dynamic geometry software, do the following:

1) Construct triangles MNO and PQR.

2) Explore several possible configurations and explain under what condition triangles MNO and PQR are congruent.

knowledge 2.1

MINIMUM CONDITIONS FOR CONGRUENT TRIANGLES

Congruent triangles are triangles whose **corresponding angles** and **corresponding sides** are congruent.

The geometric statements below describe 3 sets of minimum conditions necessary to state that two triangles are congruent.

1. If the corresponding sides of two triangles are congruent, then the triangles are congruent (SSS).

The abbreviation SSS (Side-Side-Side) is used to simplify the written form of this statement.

E.g. $\overline{AC} \cong \overline{DE}$
$\overline{AB} \cong \overline{DF}$
$\overline{BC} \cong \overline{EF}$

Thus, $\Delta ABC \cong \Delta DEF$.

2. If two angles and the contained side of one triangle are congruent to the corresponding two angles and contained side of another triangle, then the triangles are congruent (ASA).

The abbreviation ASA (Angle-Side-Angle) is used to simplify the written form of this statement.

E.g. $\angle B \cong \angle E$
$\overline{BC} \cong \overline{EF}$
$\angle C \cong \angle F$

Thus, $\Delta ABC \cong \Delta DEF$.

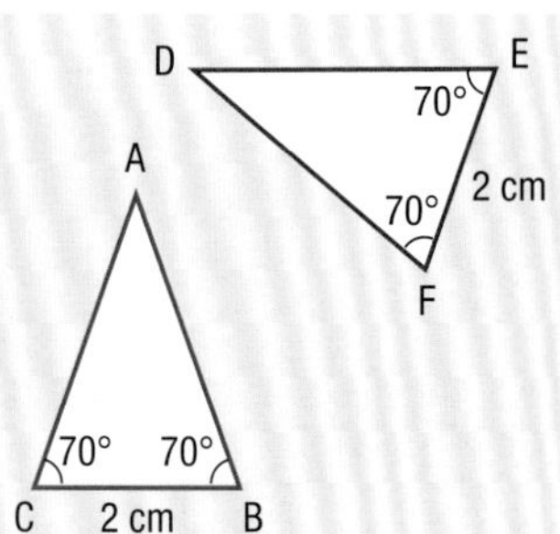

3. If two sides and the contained angle of one triangle are congruent to the corresponding two sides and contained angle of another triangle, then the triangles are congruent (SAS).

The abbreviation SAS (Side-Angle-Side) is used to simplify the written form of this statement.

E.g. $\overline{AC} \cong \overline{DF}$
$\angle C \cong \angle F$
$\overline{BC} \cong \overline{EF}$

Thus, $\Delta ABC \cong \Delta DEF$.

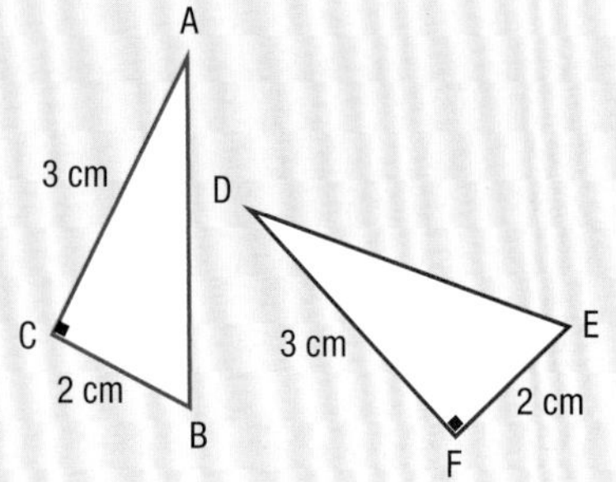

DEDUCTIVE REASONING

In geometry, there are various types of statements that allow you to organize a line of deductive reasoning.

Conjecture

A **conjecture** is a statement that has neither been proved nor refuted.

Theorem

A **theorem** is a conjecture that has been proved.

Counter-example

A **counter-example** is an example that disproves a conjecture.

E.g. In a circle, any triangle formed with the diameter of the circle as one of its sides is a right triangle.

The adjacent illustration is a counter-example. The illustration shows that triangle ABC is not a right triangle although side AC is the diameter of the circle with centre O. Therefore, this conjecture is false.

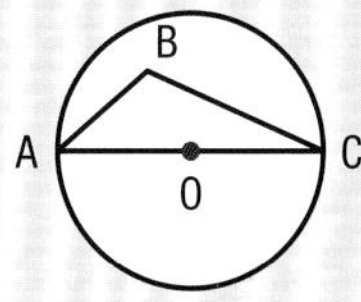

Proof

A **proof** is a logical line of reasoning that allows you to make statements based on previously established or accepted properties.

E.g. Following is a way to prove that $\angle$ BCD $\cong$ $\angle$ CBE shown in the adjacent figure.

Information given. → Hypotheses

The statement that needs to be proved. → Conclusion

Hypotheses:	• $\overline{AB} \cong \overline{AC}$ • $\overline{BD} \cong \overline{CE}$
Conclusion:	$\angle$ BCD $\cong$ $\angle$ CBE

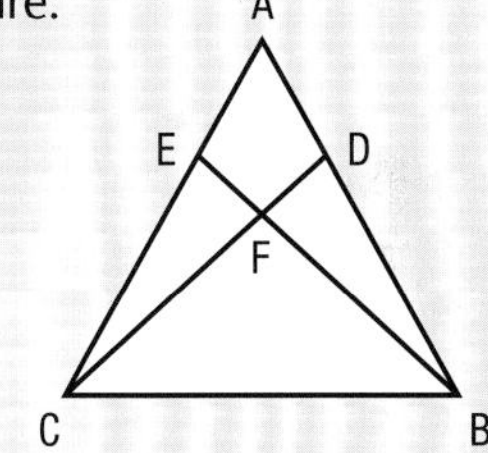

The basis of the proof. ↓ (STATEMENT)

The explanation that supports the statement provided. ↓ (JUSTIFICATION)

STATEMENT	JUSTIFICATION
Triangle ABC is isosceles.	By hypothesis ($\overline{AB} \cong \overline{AC}$).
$\angle$ ACB $\cong$ $\angle$ ABC	Angles opposite the congruent sides of an isosceles triangle are congruent.
$\overline{BD} \cong \overline{CE}$	By hypothesis ($\overline{BD} \cong \overline{CE}$).
$\overline{BC} \cong \overline{BC}$	Common side.
Δ BCE $\cong$ Δ BCD	Two triangles that have one congruent angle contained by corresponding congruent sides are congruent (SAS).
$\angle$ BCD $\cong$ $\angle$ CBE	In congruent triangles, corresponding angles are congruent.

practice 2.1

1 In each of the following cases, what allows you to conclude that the green and yellow triangles are congruent?

a) 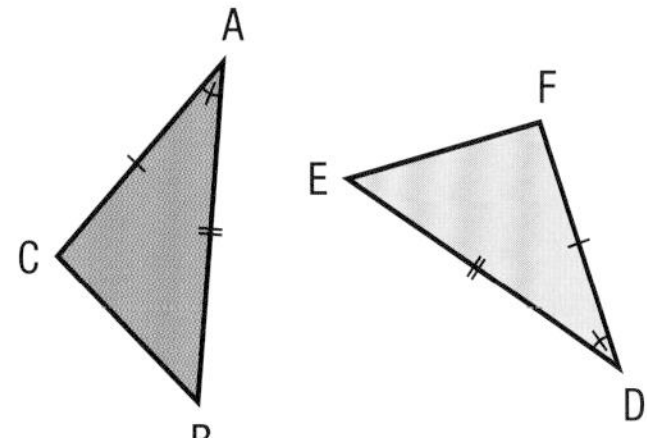

b) 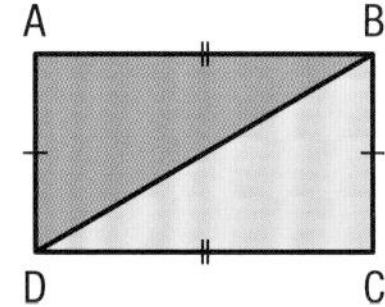

c) 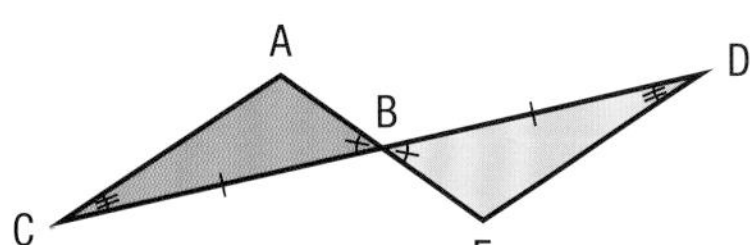

d) 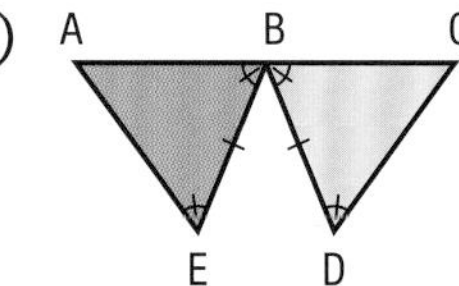

e) 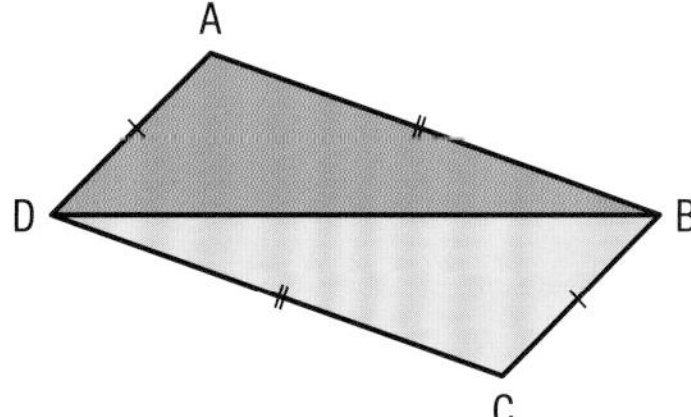

f) 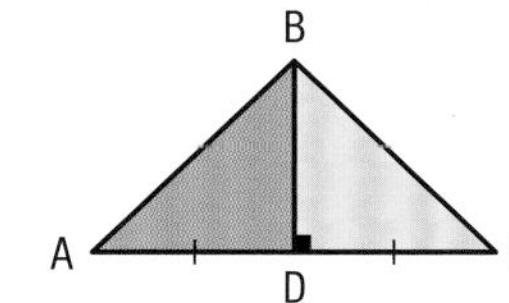

g) 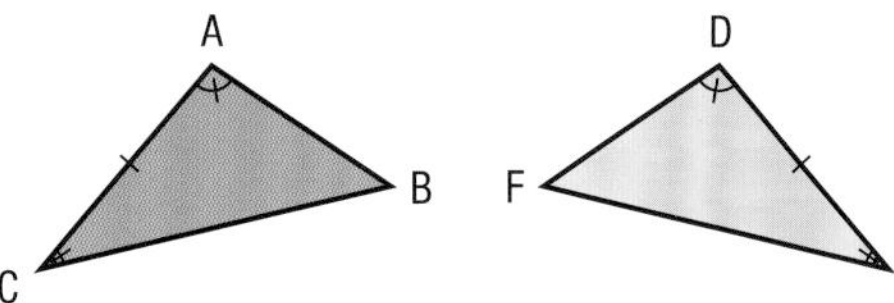

h) 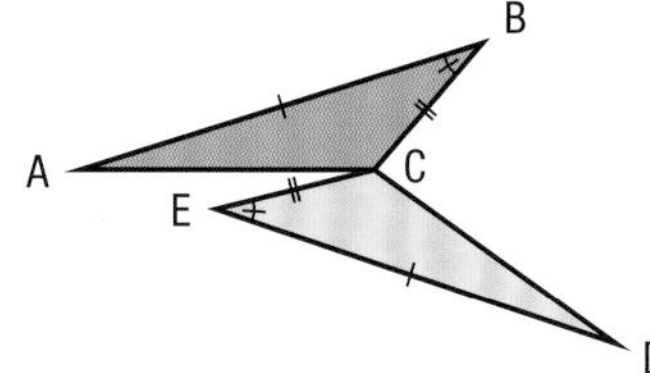

2 For each of the conjectures below, do the following:

1) Determine the hypothesis or hypotheses and the conclusion.
2) Verify whether the conjecture is true or false. If it is false, provide a counter-example.

a) If you double the length of the legs of a right triangle, you also double the length of its hypotenuse.

b) If the height of a right triangle is twice the length of its base, one of the acute angles is twice the size of the other acute angle.

c) In a right triangle, the altitude drawn from the vertex of the right angle divides the triangle into two congruent triangles.

3 a) Determine the length of segment BC as shown in the adjacent illustration.

b) On what geometric statement(s) do you base your reasoning?

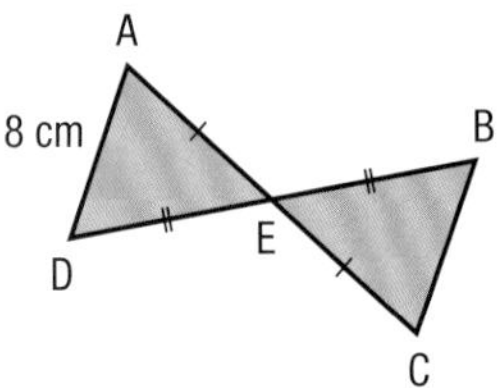

4 The entrance to the big-top tent at a circus is in the shape of a regular pentagon as shown. The curtains open from vertex A, and are attached at the bottom at vertices C and D.

a) Prove that triangles ADE and ABC are congruent.

b) Prove that angle CAD measures 36°.

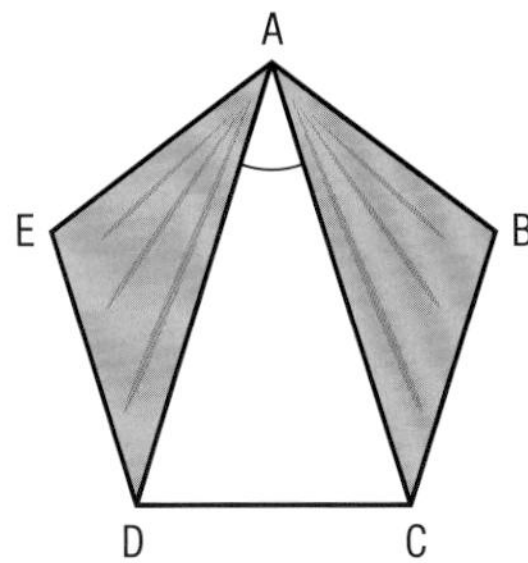

5 The quadrilateral ABCD shown below is a parallelogram.

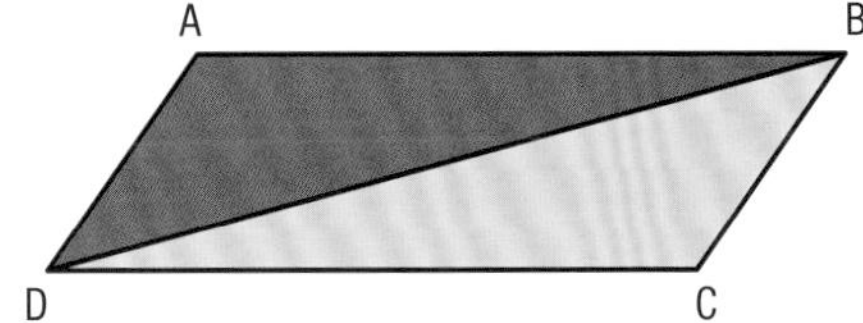

a) What geometric statement allows you to state that:

1) $\overline{AD} \cong \overline{BC}$?

2) $\angle ADB \cong \angle CBD$?

3) $\Delta ABD \cong \Delta CDB$?

b) What geometric transformation would allow you to associate triangle ABD to triangle BCD?

6 The two adjacent triangles are associated by a rotation. Determine the length of segment AC.

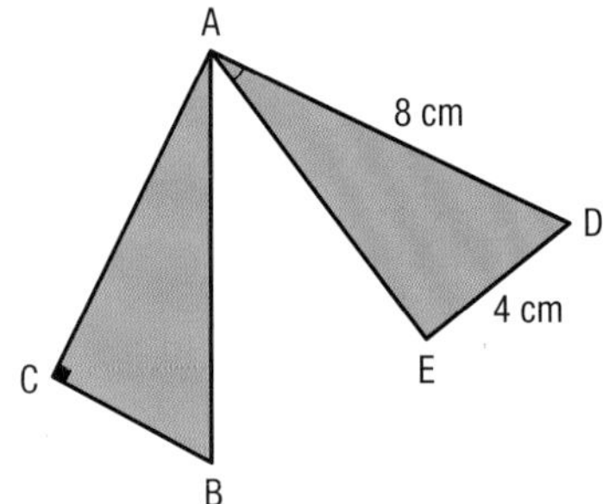

7 In a forest there are two cross-country ski trails. Considering that C is the midpoint of segments AE and BD, show that segments AB and DE are parallel.

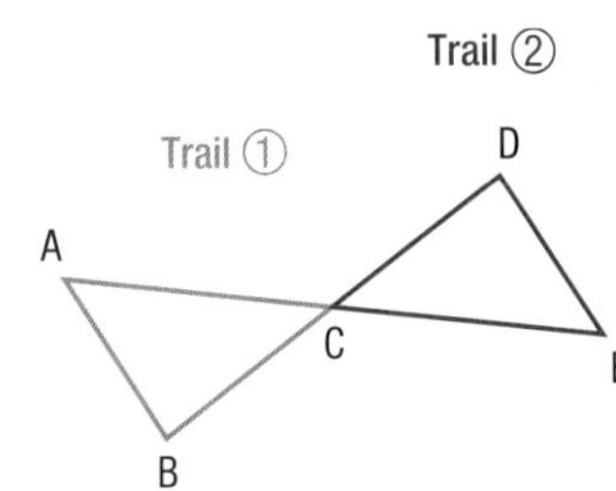

8 In the diagram below, lines MP and NQ are parallel. The midpoint of $\overline{MN}$ is point O. Prove that point O is also the midpoint of $\overline{PQ}$.

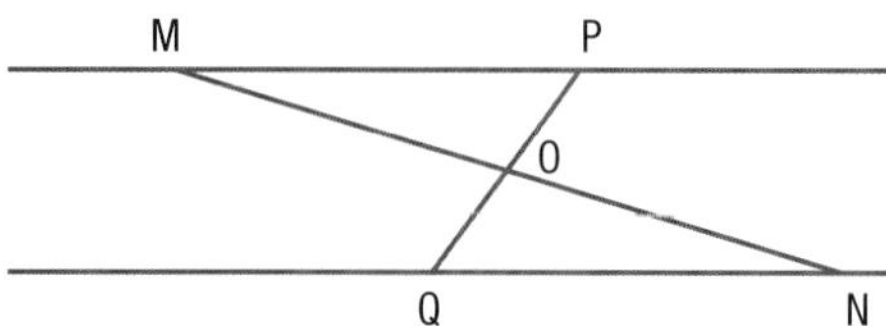

9 The diagram below shows the position of a lamp that lights a painting hanging on a wall. Indicate the location where a second lamp should be placed so that the triangle formed by its light beam is congruent to that of the lamp already in place.

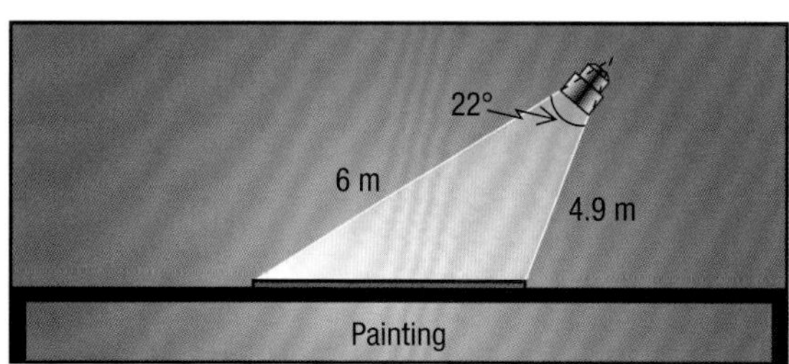

10 A deltoid is a concave quadrilateral with two pairs of congruent adjacent sides. The arrowhead shown has the shape of a deltoid. In this figure, the measure of angle BAD is equal to half of the measure of angle ①.

a) What geometric statement allows you to state that triangles ABC and ADC are congruent?

b) Prove that triangles ABC and ADC are isosceles triangles.

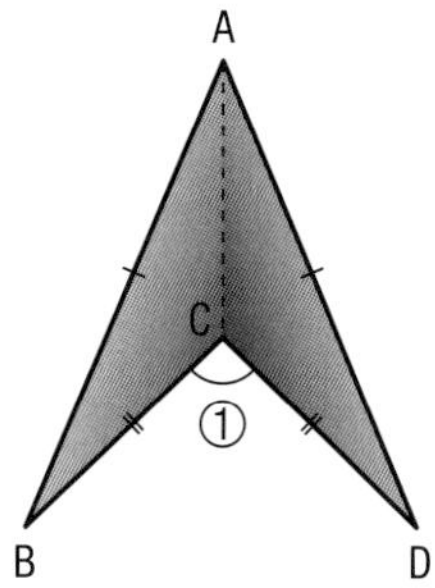

11 Prove the following:

a) The four triangles formed by the diagonals of a rectangle form two pairs of congruent triangles.

b) The six triangles formed by the altitudes of an equilateral triangle are congruent.

12 In the adjacent diagram, segment DB is the bisector of angle ADC. If $\overline{AD} \cong \overline{CD}$, prove that $\overline{AB} \cong \overline{BC}$.

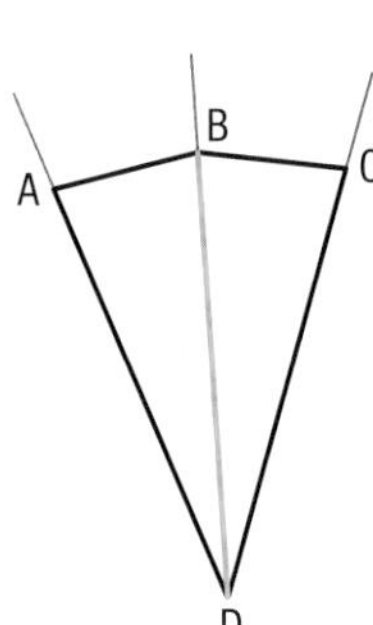

13 **POLARIS** The cross-staff, or Jacob's staff, was one of the first astronomical instruments used to determine the angular height of a star. This device consists of a ruler graduated in degrees and a sliding hammer centred on the ruler. In this particular case involving the Northern Star (Polaris), the angular height reflects the latitude where the observer is located.

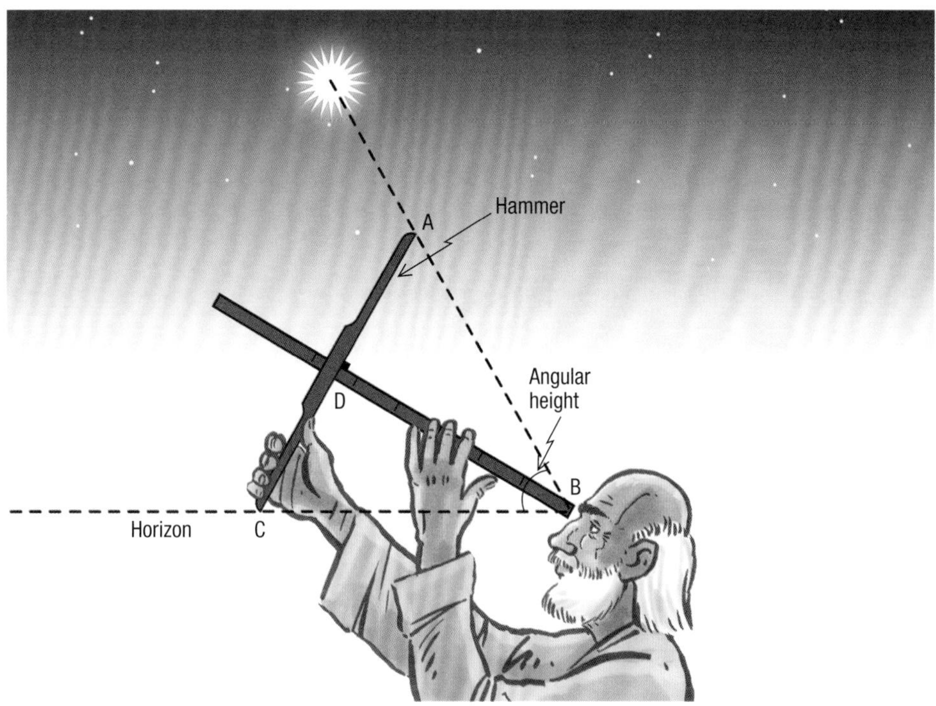

a) Prove that triangles ABD and BCD are congruent.

b) An observer located at a latitude of 60° measures the angular height of Polaris. At what distance from his eye should he place the hammer if the latter measures 76 cm?

14 A graphic artist drew a triangle, then incorporated three congruent rhombuses to create the logo shown below.

a) Prove that the three white sections inside triangle ABC correspond to congruent equilateral triangles.

b) Prove that triangle ABC is equilateral.

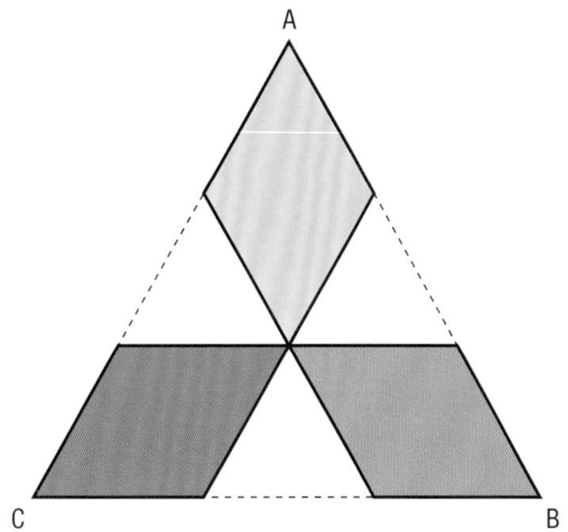

15 Indicate whether the following conjectures are true or false. If they are false, provide a counter-example.

a) An isosceles triangle with one angle measuring 30° is an obtuse triangle.

b) If the diagonals of a quadrilateral are congruent, then the quadrilateral is a rectangle.

c) Two transversals intersected by a third transversal result in congruent corresponding angles.

d) All scalene triangles are obtuse triangles.

e) The diagonals of a rhombus create four right triangles which are all congruent to each other.

f) The opposite angles in a quadrilateral are supplementary.

g) In a trapezoid, the difference between the measures of the bases is less than the sum of the measures of the two non-parallel sides.

16 In the adjacent figure, $\angle ABC \cong \angle ACB$ and $\overline{AD} \cong \overline{AE}$. Prove that $\overline{CD} \cong \overline{BE}$.

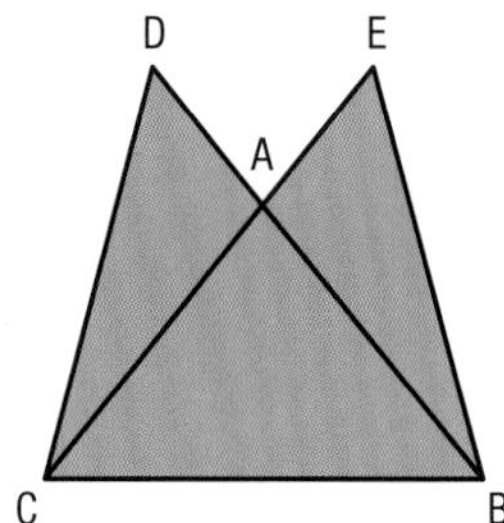

17 A clock was photographed twice during the day. Are the triangles formed by the ends of the hands and the centre of the clock congruent? Explain your answer.

12:15

3:30

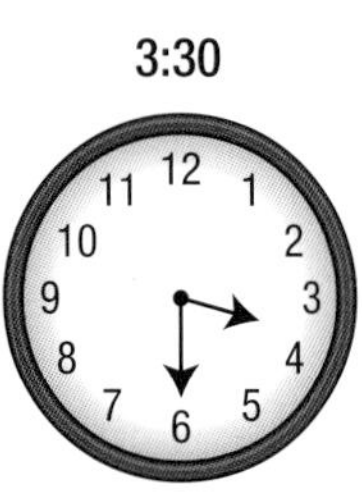

An atomic clock makes it possible to measure time with a high degree of precision. A country's official time is set using atomic clocks.

18 Determine if triangles EFG and LMN are congruent. Justify your answer.

Regular square-based pyramid ①

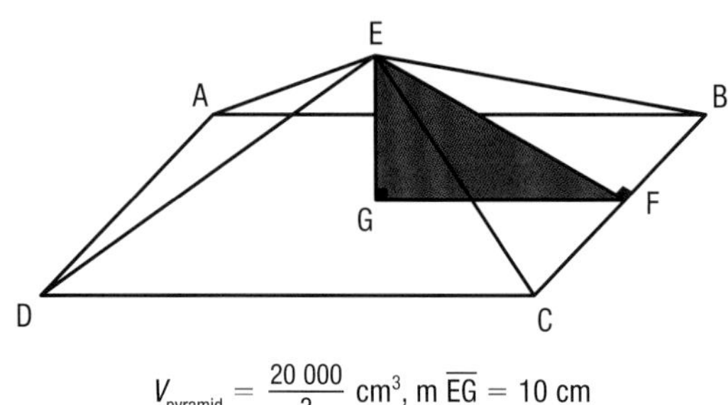

$V_{pyramid} = \frac{20\ 000}{3}$ cm³, m $\overline{EG}$ = 10 cm

Regular square-based pyramid ②

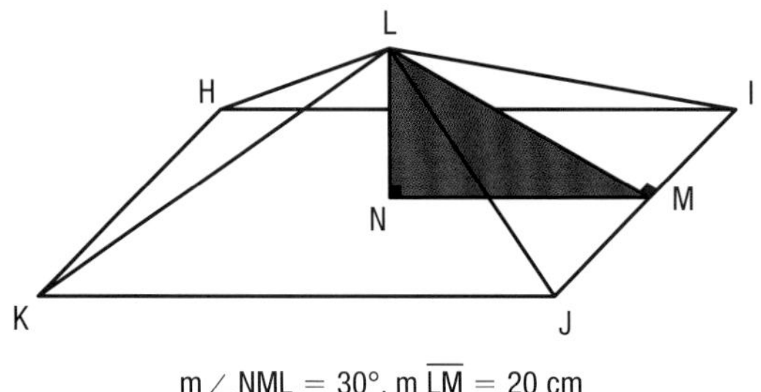

m∠NML = 30°, m $\overline{LM}$ = 20 cm

19 The following are characteristics of a geometric figure:

- P, Q and R are the vertices of a triangle.
- Point S is located at the midpoint of $\overline{QR}$.
- $\overline{PS}$ is a median.
- Points P, S and T form a straight line.
- Segment ST is outside of the triangle.
- $\overline{PS} \cong \overline{ST}$.

Prove that the quadrilateral PQTR is a parallelogram.

20 In the past, some scientists measured time using a metal rod with a spherical mass attached to the end. This pendulum, similar to the one shown below, had to be manually activated and would complete a back and forth motion in 2 s.

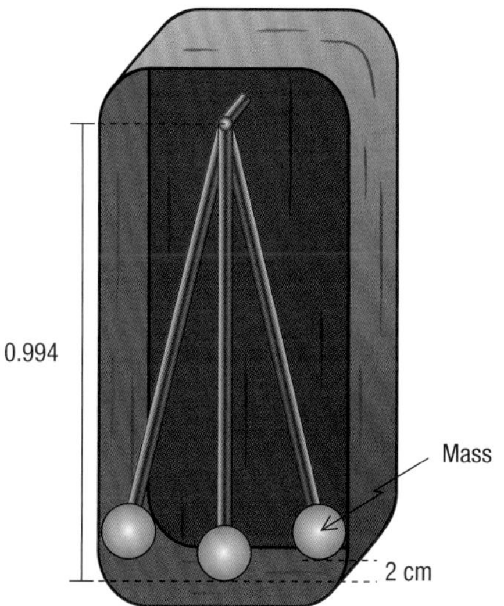

Credit for creating the first accurate pendulum clock goes to Huygens, a 17th century mathematician, astronomer and physicist. Its margin of error did not exceed 10 seconds per day, and it had an autonomous period of 30 days.

What is the smallest possible width of the case if the diameter of the mass is 8 cm?

21 In triangle ABC, point E is the midpoint of $\overline{AC}$ and point D is the midpoint of $\overline{BC}$. Segment DE is extended to point F so that $\overline{ED} \cong \overline{EF}$.

a) Construct this figure.

b) Prove that $\overline{AF} \cong \overline{CD}$.

c) Using this figure, prove that the segment that joins the midpoint of two sides of the triangle is parallel to the third side and that it measures half the length of the third side.

22 In each case, indicate the possible number of triangles ABC that can be constructed.

a) $m\,\overline{AB} = 10$ km, $m\,\overline{BC} = 13$ km, $m\,\overline{AC} = 16$ km

b) $m\,\overline{AB} = 8$ km, $m\angle A = 65°$, $m\angle B = 25°$

SECTION 2.2 Similar triangles

This section is related to LES 3 and 4.

PROBLEM The balalaïka

The raw material, the volume and the shape of a stringed instrument are factors that determine the tone, intensity and duration of the sounds produced by this instrument.

A stringed-instrument maker wants to make a balalaïka with a sound box the top of which:

- would be an enlargement of the triangle illustrated below
- would be made of a single piece cut out of a rectangular sheet of Norway spruce measuring 140 cm by 210 cm

The balalaïka, an easy-to-build instrument, was invented at the end of the 17th century, in Russia, after Tsar Alexis Mikailovitch decreed that possessing or playing a musical instrument was prohibited.

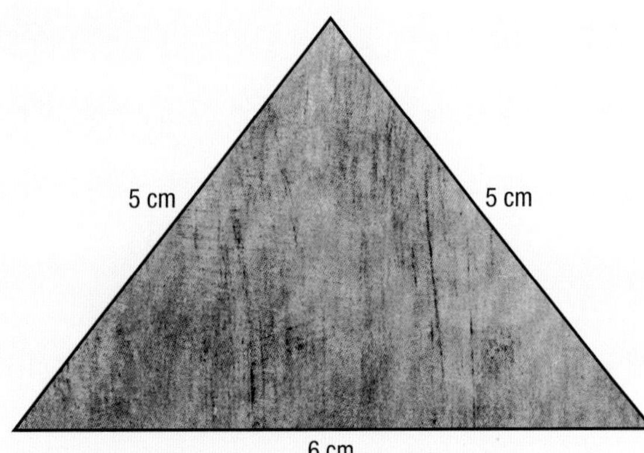

The Norway spruce, simply called spruce in Québec, is a common softwood in Europe and North America.

What are the maximum possible dimensions of the sound box?

ACTIVITY 1 Projecting a logo

A property of light is that it shines in a straight line; this makes it possible to enlarge a small image by projecting it onto a screen or wall.

A company wants to project its logo onto one of the outside walls of the building shown below. A spotlight is set up at C and lights up the company's logo which is represented below by segment BD.

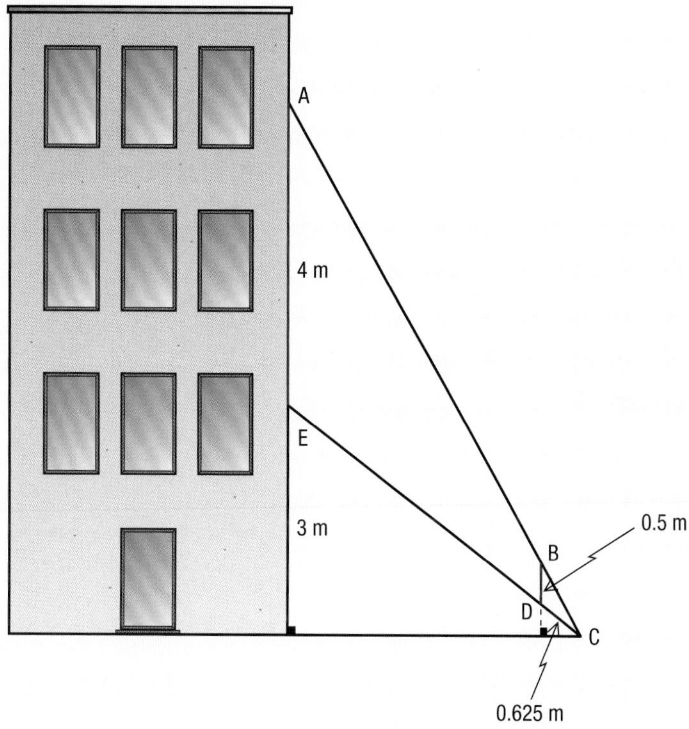

a. What is the relative position of segments AE and BD in relation to each other?

b. What geometric statement allows you to state that $\angle AEC \cong \angle BDC$?

c. What can you state regarding angles ACE and BCD?

d. Triangle ACE is an enlargement of triangle BCD. What is the ratio of similarity between these two triangles?

e. Determine at what distance from the outside wall the spotlight is set up.

Isaac Newton proved that white light passing through a prism would break down into multi-coloured rays called the chromatic spectrum. This discovery led him to devise the circle of chromatic colours which allows us to classify colours.

ACTIVITY 2 The Thales theorem

Thales of Miletus (ca. 625 - ca. 546 BCE) is regarded by many as the first philosopher, scientist and mathematician in Greece. He was particularly interested in geometric figures, and he deduced a number of theorems.

One of Thales' theorems concerns two transversal lines intersected by parallel lines. This theorem, however, was only later proved by Euclide.

Drawing based on a statue of Hermes found in Tivoli and engraved by Ambroise Tardieu.

Born in Miletus, Asia Minor, Thales was the founder of Greek philosophy. He is also known for his knowledge of astronomy and is said to have introduced geometry to the Greeks.

Using dynamic geometry software, transversals t_1 and t_2, as well as parallel lines l_1, l_2 and l_3, were drawn. By changing the position of one or more lines on Screen **1**, changes in the length of certain segments were observed.

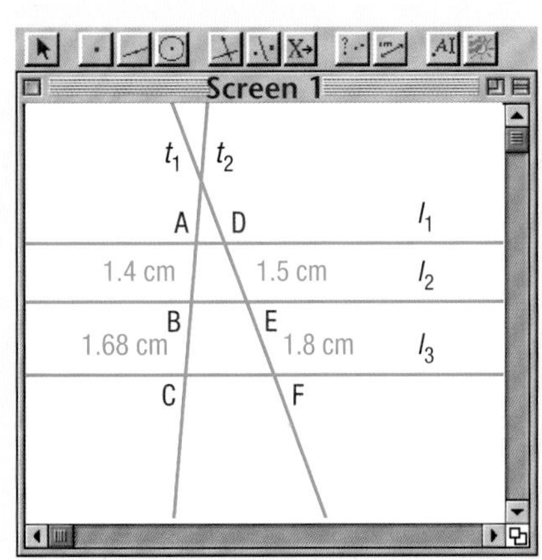

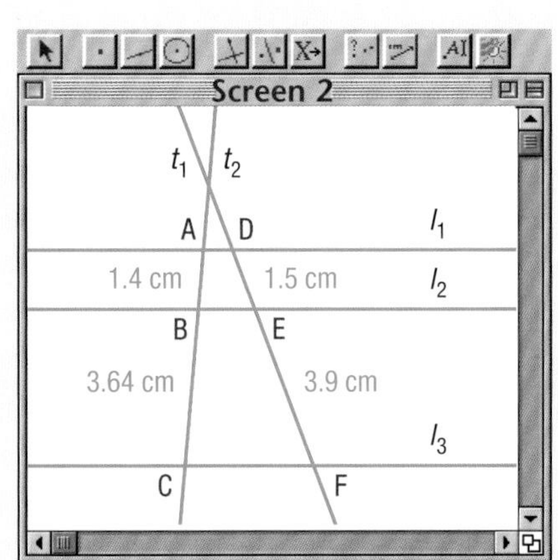

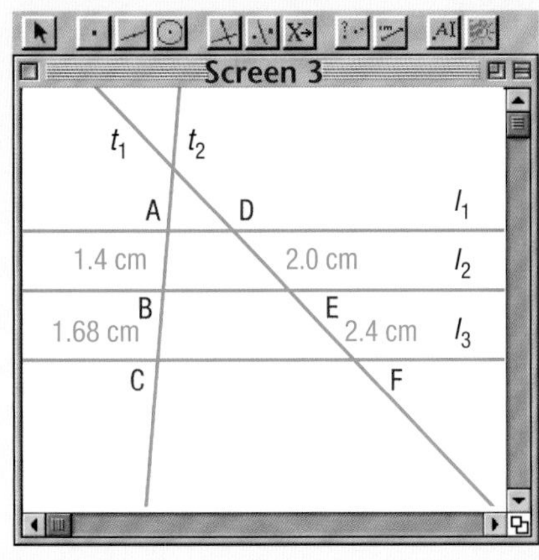

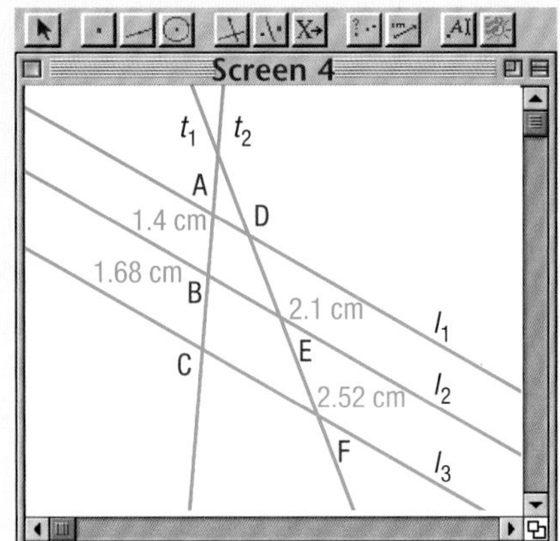

a. Referring to Screen **1**, what changes have been made in:

1) Screen **2**? 2) Screen **3**? 3) Screen **4**?

b. For each of Screens **1**, **2**, **3** and **4**, calculate:

1) $\frac{m\ \overline{DE}}{m\ \overline{AB}}$ 2) $\frac{m\ \overline{EF}}{m\ \overline{BC}}$

c. Based on the results above, what conjecture can you formulate?

Techno math

Dynamic geometry software allows you to compare geometric figures. By using the tools TRIANGLE, DISTANCE, and ANGLE MEASUREMENT, you can draw triangles and check whether they are similar.

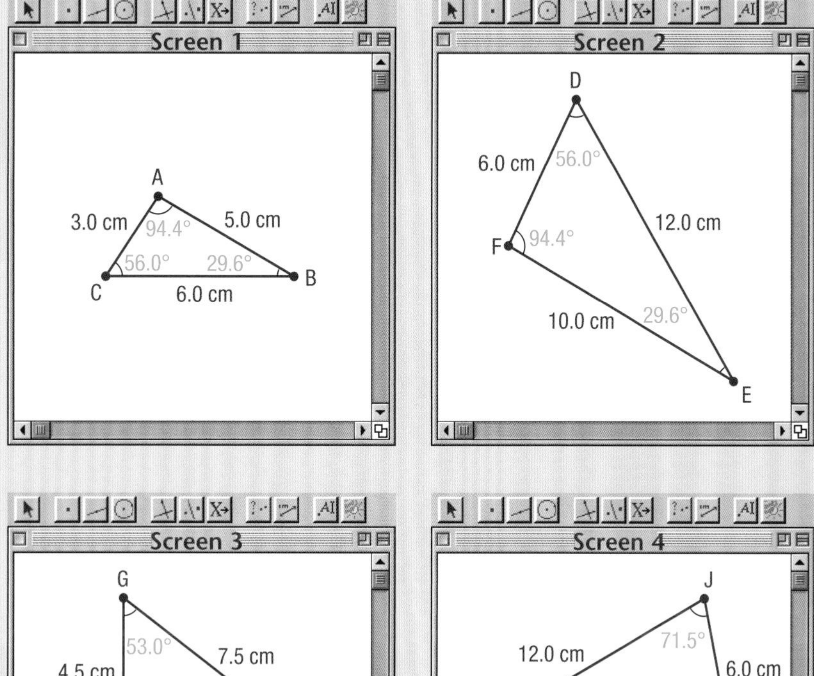

a. How many pairs of congruent angles are there if you compare the triangles in:

1) Screens **1** and **2**? 2) Screens **1** and **3**? 3) Screens **1** and **4**?

b. Verify that:

1) $\frac{m\,\overline{EF}}{m\,\overline{AB}} = \frac{m\,\overline{DE}}{m\,\overline{BC}} = \frac{m\,\overline{DF}}{m\,\overline{AC}}$ 2) $\frac{m\,\overline{GH}}{m\,\overline{AB}} = \frac{m\,\overline{GI}}{m\,\overline{AC}}$ 3) $\frac{m\,\overline{JK}}{m\,\overline{AC}} = \frac{m\,\overline{JL}}{m\,\overline{BC}}$

c. Among the triangles in Screens **1**, **2**, **3** and **4**, is there a pair of similar triangles?

d. Can you state that two triangles are similar if:

1) the lengths of two sides of one triangle are proportional to the lengths of two sides of the other triangle?

2) one angle of a triangle is congruent to one angle of the other triangle?

e. The lengths of two sides of triangle MNO are proportional to the lengths of two sides of triangle PQR. One angle of triangle MNO is congruent to one angle of triangle PQR. Using dynamic geometry software, do the following:

1) Construct triangles MNO and PQR.

2) Explore several possible configurations and explain under what condition triangles MNO and PQR are similar.

knowledge 2.2

MINIMUM CONDITIONS FOR SIMILAR TRIANGLES

Similar triangles are triangles whose corresponding angles are congruent and whose lengths of corresponding sides are proportional.

The geometric statements below describe 3 sets of minimum conditions necessary to state that two triangles are similar.

1. Two triangles that have two congruent corresponding angles are similar (AA).

The abbreviation AA (Angle-Angle) is used to simplify the written form of this statement.

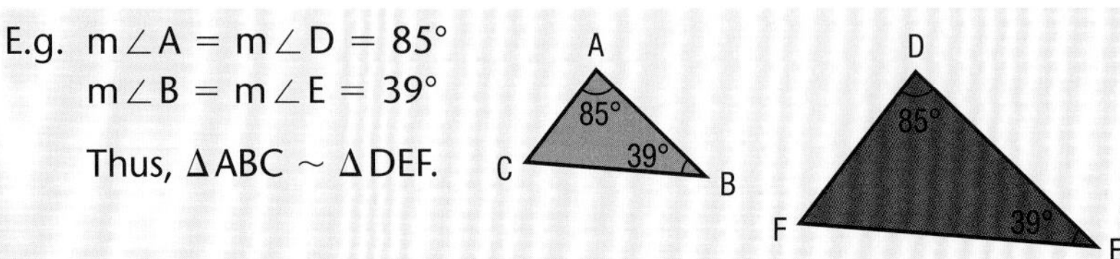

2. Two triangles that have one congruent angle contained between corresponding sides of proportional length are similar (SAS).

The abbreviation SAS (Side-Angle-Side) is used to simplify the written form of this statement.

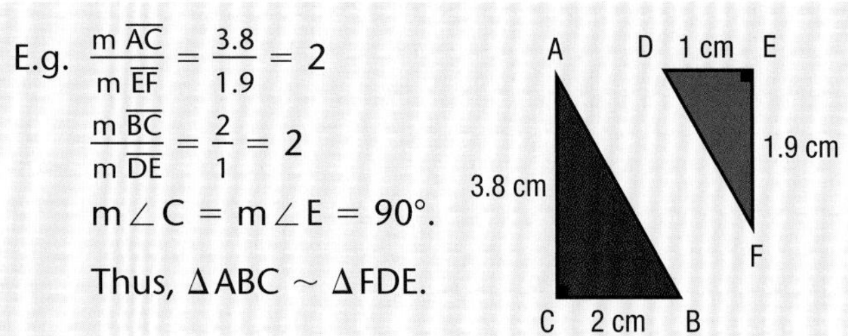

3. Two triangles whose corresponding sides are proportional are similar (SSS).

The abbreviation SSS (Side-Side-Side) is used to simplify the written form of this statement.

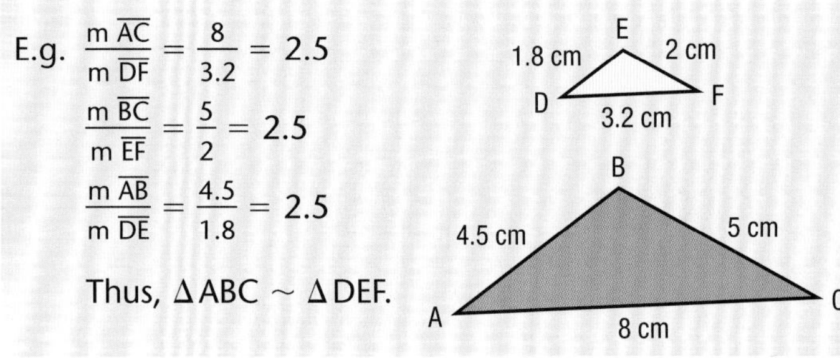

When two triangles are similar, it is possible to find missing measurements.

E.g. Considering that the two adjacent triangles are similar, the length of side AB can be calculated as follows.

$$\frac{m\,\overline{AB}}{m\,\overline{DE}} = \frac{m\,\overline{BC}}{m\,\overline{EF}} = \frac{m\,\overline{AC}}{m\,\overline{DF}}$$

Substitute the known measurements $\frac{m\,\overline{AB}}{52.5} = \frac{9}{31.5} = \frac{12}{42}$.

Therefore, m $\overline{AB}$ = 15 cm.

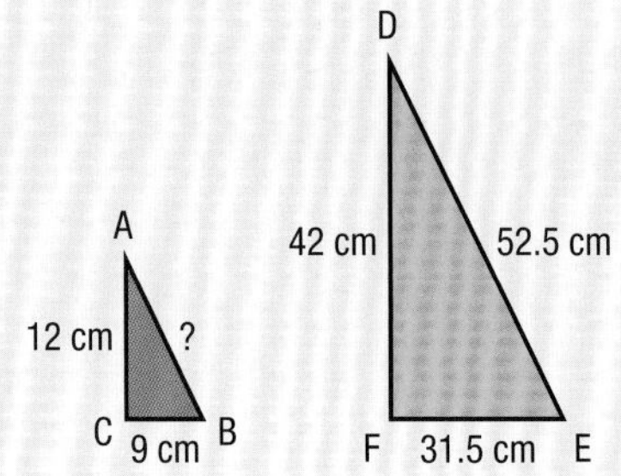

practice 2.2

1 Consider the following information about three triangles.

What geometric statement allows you to state that:

a) Triangles ① and ② are similar?

b) Triangles ② and ③ are similar?

c) Triangles ① and ③ are similar?

	m∠A	m∠B	m∠C	m $\overline{AB}$	m $\overline{BC}$	m $\overline{AC}$
Triangle ①	37°	90°		4 cm	3 cm	5 cm
Triangle ②			53°	16 cm	12 cm	20 cm
Triangle ③	37°	90°	53°		9 cm	15 cm

2 In each case, identify the geometric statement that allows you to state that the pairs of triangles are similar.

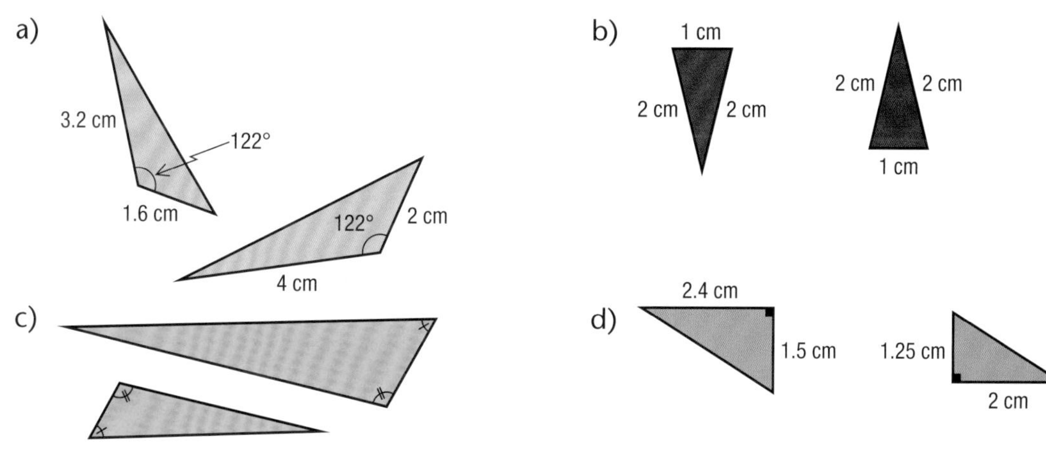

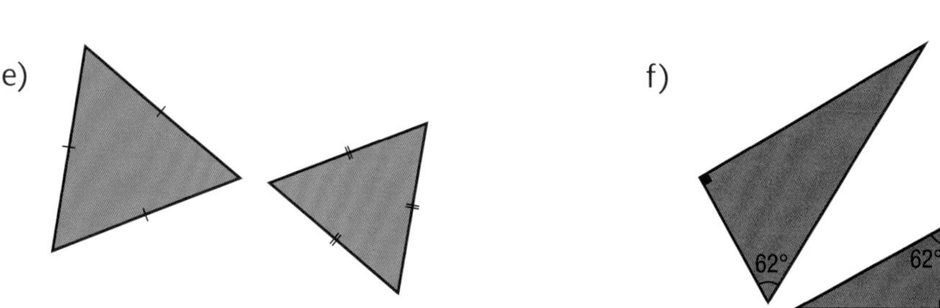

3 Identify the geometric statement that allows you to state that $\Delta ABC \sim \Delta AED$ in the figure below.

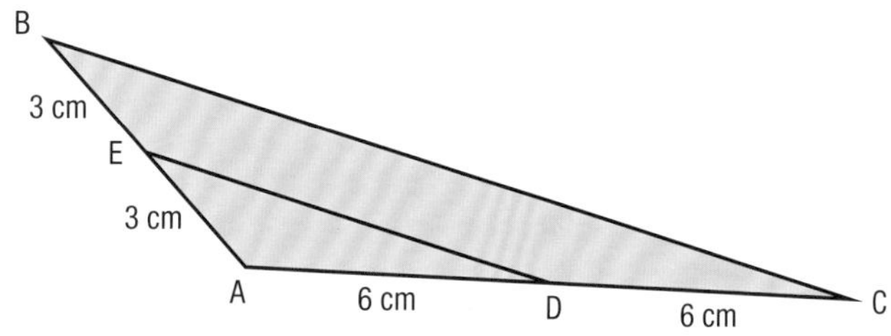

4 Consider the following representation of a jump ramp used during a water-skiing competition:

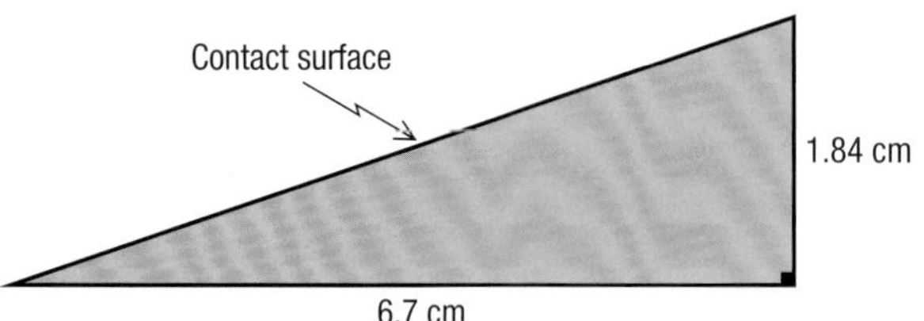

In water-skiing competitions, only the length of the jump is evaluated.

Depending on the ability of the competitors, other types of ramps similar to the one shown above can be used. What is the length of the contact surface of a ramp with a base measuring 6.4 m?

5 The sides of a triangle ABC respectively measure 2.5 cm, 4 cm and 5 cm. What is the perimeter of triangle DEF, similar to triangle ABC, whose side opposite the obtuse angle measures 12 cm?

6 In each of the following, determine the lengths represented by x and y.

a)

b)

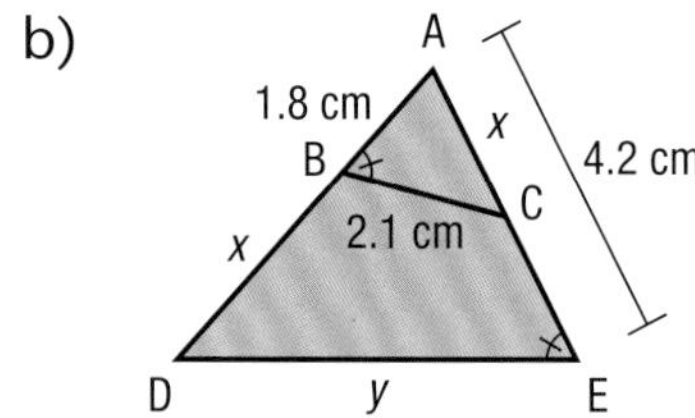

c)

d)

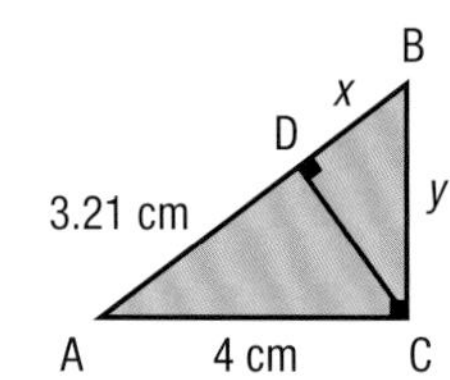

e)

f)

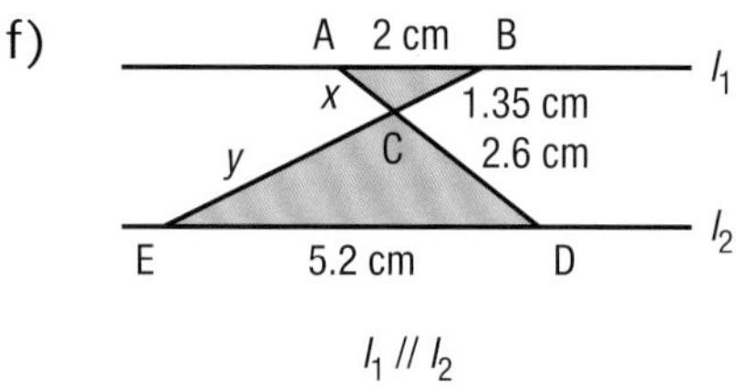

7 a) Billy states that two isosceles triangles each having an angle measuring 50° are necessarily similar. Is his statement correct? Justify your answer.

b) Would your answer be the same if the measure of the chosen angle were different? Justify your answer.

8 The adjacent diagram shows one of the lateral faces of a telecommunications tower. On this face, how many triangles are there that are similar to:

a) triangle ABC? Explain your answer.

b) triangle ADC? Explain your answer.

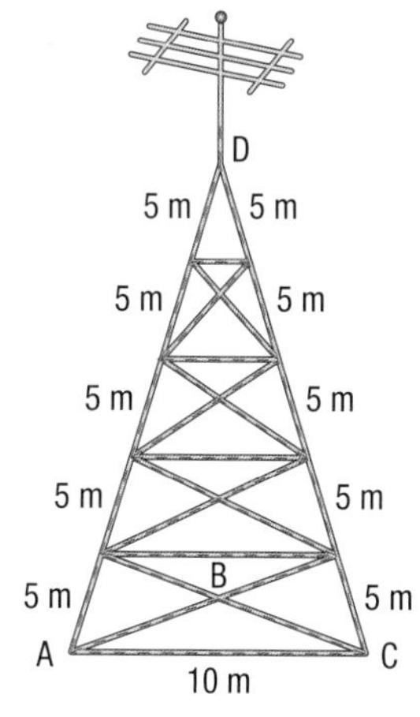

9 A square object with an area of 20 cm^2 is placed so that it is parallel to a screen. The distance between the object and the screen is 2.1 m. Determine the distance of the light source from the object so that the shadow projected on the screen is three times greater than the object.

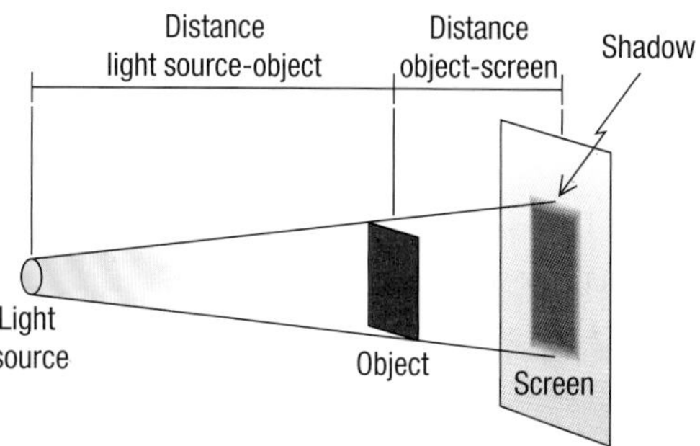

10 The diagram below represents the side view of a municipal building. Determine:

a) m $\overline{AB}$ b) m $\overline{CF}$

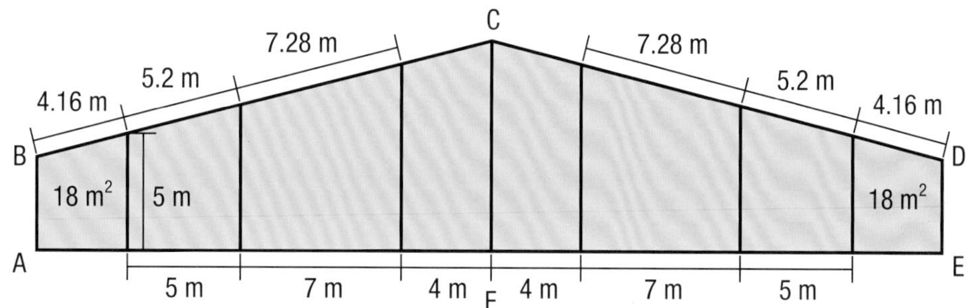

11 Based on the adjacent triangle, do the following:

a) Prove that $\Delta ABD \sim \Delta BCD$.

b) Determine the length:

1) of segment CD

2) of segment BC

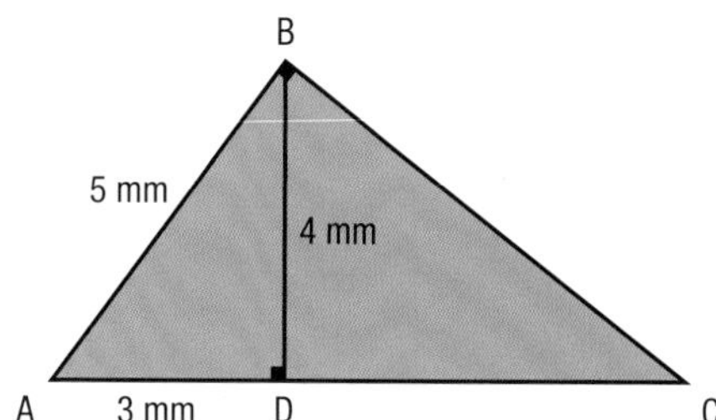

12 Consider the adjacent trapezoid.

Prove that $\frac{m\ \overline{BE}}{m\ \overline{ED}} = \frac{m\ \overline{AE}}{m\ \overline{CE}}$.

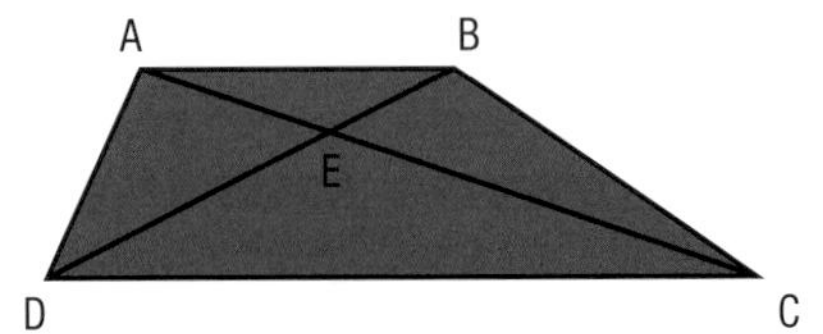

13 **THALES OF MILETUS** It is said that one day during a trip to Egypt, Thales decided to calculate the height of the Cheops pyramid. As reference points, he used his body, his shadow and the shadow of this regular pyramid that has one side of its base measuring 230 m. He noted the following:

"When the length of my shadow is equal to my height, the height of the pyramid is equal to the length of its shadow added to half the length of base of the pyramid."

In this situation:

a) If the shadow of the pyramid on the ground is 22 m, what is the height of the pyramid?

b) If Thales had instead used a 1-m stick to project a shadow of 0.78 m to calculate the height of the pyramid, what would have been the length of the shadow of the pyramid?

14 The altitudes in the adjacent isosceles triangle ABC have been drawn. Using geometric statements, prove that $\frac{m\ \overline{AE}}{m\ \overline{CD}} = \frac{m\ \overline{BE}}{m\ \overline{BD}}$.

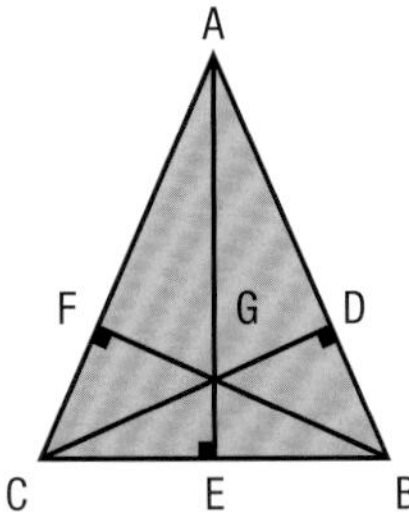

15 It is possible to estimate the height of a building using a mirror. Place the mirror on the ground at a certain distance from the building. Stand next to the mirror and then move away from the building until the top of the building is visible in the mirror as shown in the illustration below. What is the height of the building represented by this diagram?

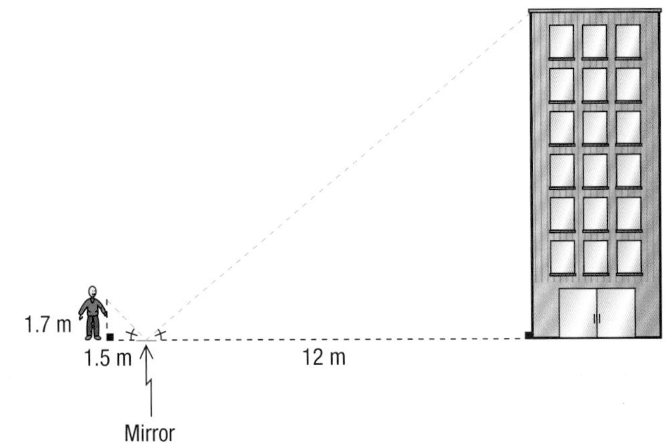

SECTION 2.3 Manipulating algebraic expressions

This section is related to LES 4.

PROBLEM Solar panels

Solar panels are devices that use energy transmitted by the sun and convert it into heat or electricity. Thermal solar panels convert light into heat, and photovoltaic solar panels transform light into electricity.

A horticulturist wants to use the solar panel illustrated below to supply electricity to his greenhouse. The area of this panel, in decimetres squared, corresponds to the expression $4n(n + 3) - (n + 3)$.

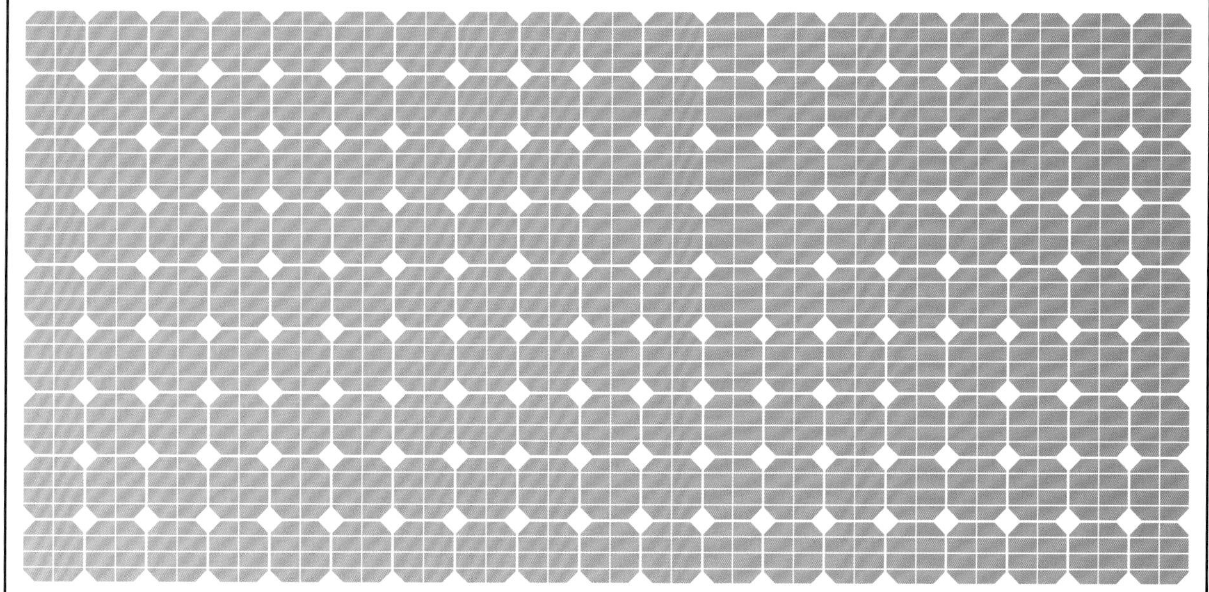

Following are the horticulturist's needs:

- The perimeter of the panel must be less than 88 dm.
- The panel must produce a minimum of 0.003 kW of electricity.
- The length of the base and the height must be whole numbers.

Determine the possible dimensions of this panel considering that the mean output of this model is 0.013 W/dm².

ACTIVITY 1 A useful language

In the 16th century, François Viète introduced letters to certain algebraic calculations. Drawing inspiration from Viète's work, the French mathematician René Descartes wrote essays about algebra and geometry. In his study of arithmetic and algebraic operations, he established a relationship between certain algebraic operations and geometric constructions.

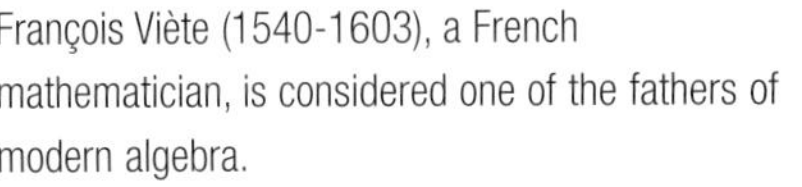

François Viète (1540-1603), a French mathematician, is considered one of the fathers of modern algebra.

For example, the product of $(2x + 3)$ and $(3y + 1)$ can be represented by the adjacent rectangle. By adding the area of each shaded section, you can deduce that
$(2x + 3)(3y + 1) = 6xy + 2x + 9y + 3$.

Consider the following situations that involve algebraic operations and geometric constructions.

a. Express the sum of the areas of rectangles ABCD and EFGH using:

1) a polynomial with four terms
2) a sum of the product of two factors
3) a product of two factors

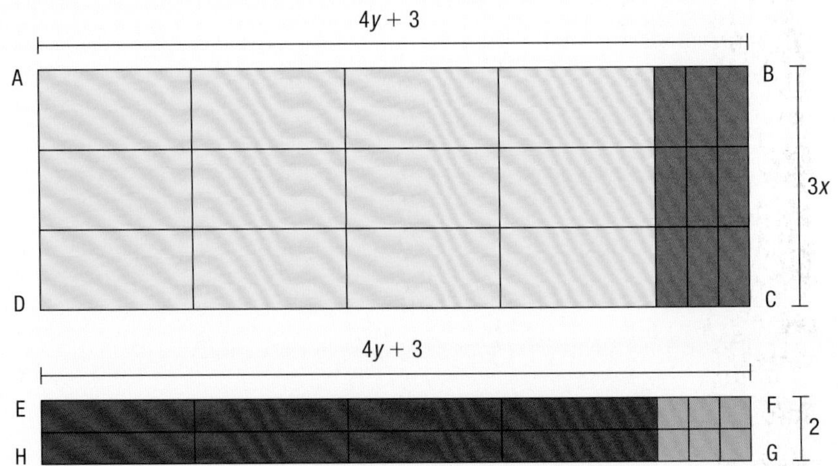

b. Can you state that the result of $(12xy + 9x + 8y + 6) \div (4y + 3)$ is $(3x + 2)$? Explain your answer.

c. Remove rectangle MJKN from rectangle IJKL, and SPQT from rectangle OPQR. Express the sum of the areas of rectangles IMNL and OSTR using:

1) a polynomial with four terms
2) a sum of the product of two factors
3) a product of two factors

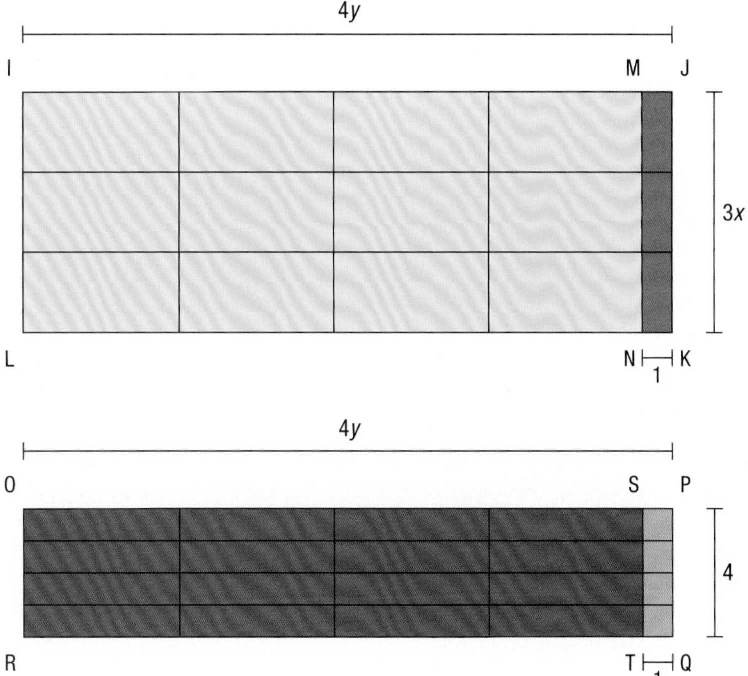

d. 1) Represent the algebraic expression $6xy + 4x + 9y + 6$ using a rectangle such as the one shown above.

2) Express the area of this rectangle with a product of two factors.
3) What do you notice about the product of these two algebraic expressions?
4) What algebraic expression corresponds to the quotient $(6xy + 4x + 9y + 6) \div (3y + 2)$?

e. Express the following as a product of two factors:

1) $3xy + 9x + 2y + 6$
2) $2xy + 4x + 3y + 6$

The word "algebra" is derived from the title of a mathematical essay by Iranian mathematician and astronomer Al-Khawarizmi that dates back to the 9th century.

ACTIVITY 2 A question of squares

For centuries, many mathematicians, including the Pythagoreans, have used geometry as a tool to better understand algebraic operations. Among other shapes, the square has been used to complete various calculations.

As shown in the adjacent illustration, square ABHG is removed from square ACDF.

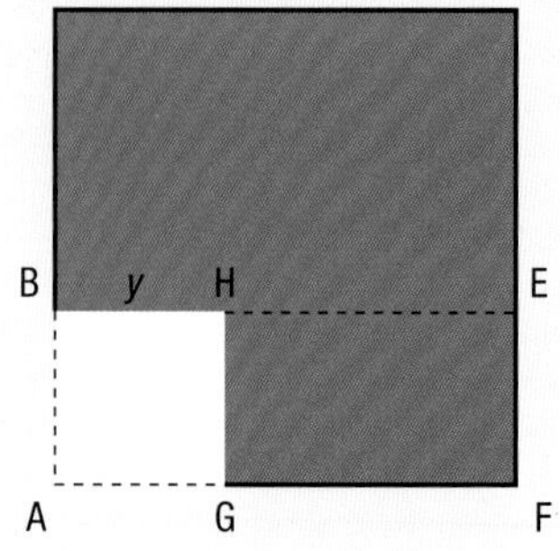

a. Express the area of the blue shape using a difference of two terms.

b. Is it possible to factor the algebraic expression obtained in **a.** by removing a common factor? Explain your answer.

c. Express the area:

1) of rectangle BCDE using a product of two factors
2) of rectangle EFGH using a product of two factors
3) of the blue shape using a sum of the product of two factors
4) of the blue shape using a product of two factors

d. Show that the algebraic expression found in **c. 4)** is equivalent to the one found in **a.**

Shape ABCD, shown on the right, is represented by the algebraic expression $4x^2 + 12x + 9$.

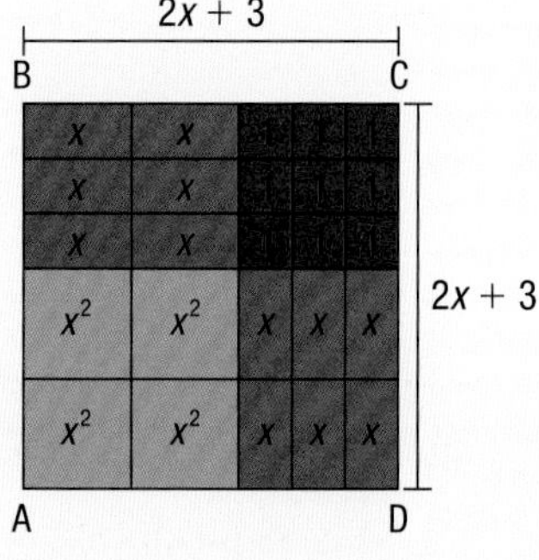

e. What type of quadrilateral is represented by:

1) the green section?
2) the red section?
3) shape ABCD?

f. Express the area of quadrilateral ABCD using a product of two factors.

g. Show that the algebraic expression found in **f.** is equivalent to the expression $4x^2 + 12x + 9$.

The Pythagorean school was founded by Pythagoras (580 ca. – 490 ca. BCE). The Pythagoreans were interested in astronomy, politics, philosophy and mathematics. Their teachings were mainly oral and were kept as secret. This school also welcomed women.

ACTIVITY 3 From fractions to rational expressions

Following are various situations that require algebraic fractions.

Situation ①

While training, a person walks a distance of 6 km, turns around and returns running to her point of departure. The speed s of the return trip corresponds to double the speed of her departure.

a. Write the algebraic expression that represents:

1) the time that this person spent walking 2) the time that this person spent running

b. Using operations on fractions, write the simplified algebraic expression that represents the total time that this person spent training.

c. Without taking the context into account, are the operations that were completed in **b.** valid for any value of s? Explain your answer.

Situation ②

During an experiment, an object moves from point A to point B at a speed of $\left(\frac{3n}{n+3}\right)$ m/s in $\left(\frac{4n+12}{4n}\right)$ s.

d. Show that the expression $\frac{4n+12}{4n}$ is equivalent to the expression $\frac{n+3}{n}$.

e. Using operations on fractions, write the simplified algebraic expression that represents the total distance travelled by this object.

f. Without taking the context into account, are the operations that were completed in **e.** valid for any value of n? Explain your answer.

Situation ③

During a flight, an airplane travels $\left(\frac{5t^2}{2t+1}\right)$ km in $\left(\frac{5t}{6t^2+3t}\right)$ h.

g. Is the expression $\frac{5t}{6t^2+3t}$ equivalent to the expression $\frac{5}{3(2t+1)}$?
Explain your answer.

h. Using operations on fractions, write the simplified algebraic expression that represents the mean speed of the airplane.

i. Without taking the context into account, are the operations that were completed in **h.** valid for any value of t? Explain your answer.

knowledge 2.3

FACTORING

Factoring an algebraic expression results in expressing it as a product of factors.

E.g.

	Expanded form	Factored form	Factors
1)	$3xy + 6x + 2y + 4$	$(3x + 2)(y + 2)$	$3x + 2$ and $y + 2$
2)	$ax + 3x + ay + 3y$	$(x + y)(a + 3)$	$x + y$ and $a + 3$
3)	$8mn - 10m + 12n - 15$	$(2m + 3)(4n - 5)$	$2m + 3$ and $4n - 5$

There are various methods used to factor an algebraic expression.

Grouping

Factoring an algebraic expression using this method consists of doing the following:

1. Reorder the terms that have a common factor.	E.g. In the expression $ab + 6 + 3b + 2a$, ab and $2a$ have a as a common factor and $3b$ and 6 have 3 as a common factor. Write the expression as: $ab + 2a + 3b + 6$.
2. Remove the common factor from each of the groups.	$a(b + 2) + 3(b + 2)$
3. Remove the common factor from both terms.	$a(b + 2) + 3(b + 2)$ → $(b + 2)(a + 3)$ Factor $b + 2$ is common to both terms.
The result can be validated by expanding the factored form using the distributive property over addition or subtraction.	$(b + 2)(a + 3) = b \times a + b \times 3 + 2 \times a + 2 \times 3$ $= ab + 2a + 3b + 6$

E.g. 1) $2xy + 4x + 3y + 6 = 2x(y + 2) + 3(y + 2)$
$= (y + 2)(2x + 3)$

2) $4a^2b - 8ab + 6a - 12 = 4ab(a - 2) + 6(a - 2)$
$= (a - 2)(4ab + 6)$

Difference of squares

This method allows you to factor an algebraic expression of the form $a^2 - b^2$. This type of polynomial can be factored by applying the following model:

$$a^2 - b^2 = (a + b)(a - b)$$

E.g. 1) Since $36x^2 - y^2 = (6x)^2 - y^2$, the factors of the algebraic expression $36x^2 - y^2$ are $6x + y$ and $6x - y$.

2) Since $4a^2 - 9b^6 = (2a)^2 - (3b^3)^2$, the factors of the algebraic expression $4a^2 - 9b^6$ are $2a + 3b^3$ and $2a - 3b^3$.

3) It is possible to determine the algebraic expression that corresponds to the length of each of the diagonals in the adjacent rhombus, as follows.

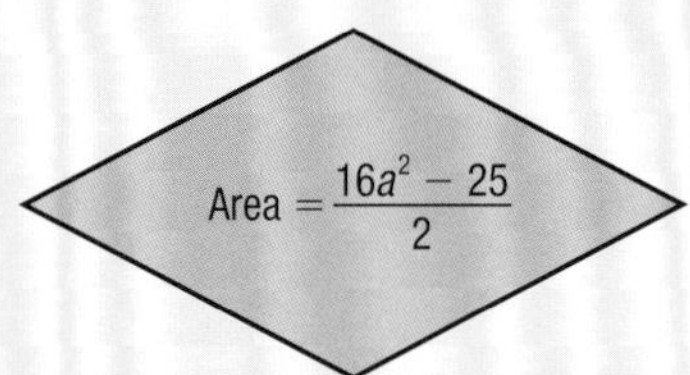

Since $16a^2 - 25$ corresponds to a difference of squares:

Area $= \frac{D \times d}{2} = \frac{(4a - 5)(4a + 5)}{2}$.

The algebraic expressions that correspond to the length of the diagonals of the rhombus are $4a - 5$ and $4a + 5$.

Perfect square trinomials

This method allows you to factor an algebraic expression of the form $ax^2 + bx + c$, where $a > 0$, $c > 0$ and $b = 2 \times \sqrt{a} \times \sqrt{c}$. This type of polynomial can be factored using the following method.

1. Verify that the trinomial has the characteristics of a perfect square.	E.g. In the expression $x^2 + 8x + 16$: $1 > 0$, $16 > 0$ and $8 = 2 \times \sqrt{1} \times \sqrt{16}$.
2. Determine if the factors are sums or differences based on the sign of the middle term.	Since the middle term is positive, each factor corresponds to a sum. $x^2 + 8x + 16 = (... + ...)^2$
3. Determine the factors.	• x^2 is the square of x. • 16 is the square of 4. Therefore, the factors are $x + 4$ and $x + 4$ or $(x + 4)^2$.

Ex.: 1) The expression $9a^6 - 12a^3 + 4$ is a perfect square since $9 > 0$, $4 > 0$ and $12 = 2 \times \sqrt{9} \times \sqrt{4}$. Since the middle term is negative, each of the factors correspond to a difference. Since $9a^6 = (3a^3)^2$ and that $4 = 2^2$, the factors are $3a^3 - 2$ and $3a^3 - 2$ or $(3a^3 - 2)^2$.

2) It is possible to determine the algebraic expression that corresponds to the length of each side of the adjacent square as follows.

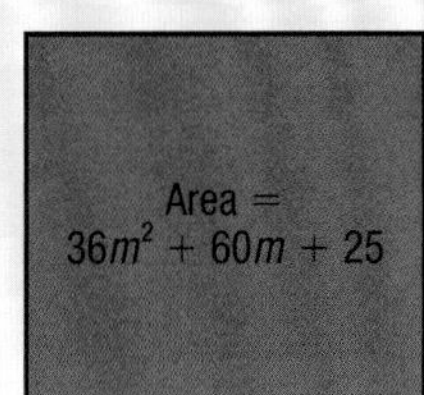

Since $36m^2 + 60m + 25$ is a perfect square trinomial:

Area $= c^2 = (6m + 5)^2$.

The algebraic expression that corresponds to the side length of this square is $6m + 5$.

MANIPULATING ALGEBRAIC EXPRESSIONS

Dividing a polynomial by a binomial

The division of a polynomial by a binomial can be completed by dividing the polynomial by the binomial in successive steps. At each step, you must chose a term for the quotient that cancels the term with the greatest degree in the polynomial that is to be divided.

E.g. 1) $(3x^2 + 5x - 2) \div (x + 2)$

Dividend: $3x^2 + 5x - 2$; Divisor: $x + 2$; Quotient: $3x - 1$

$$\begin{array}{r|l} 3x^2 + 5x - 2 & x + 2 \\ -\ (3x^2 + 6x) & 3x - 1 \\ \hline -x - 2 & \\ -\ (-x - 2) & \\ \hline 0 & \end{array}$$

Therefore:
$(3x^2 + 5x - 2) \div (x + 2) = 3x - 1$

2) $(5x^2 - 15x + 5) \div (x - 2)$

Dividend: $5x^2 - 15x + 5$; Divisor: $x - 2$; Quotient: $5x - 5$

$$\begin{array}{r|l} 5x^2 - 15x + 5 & x - 2 \\ -\ (5x^2 - 10x) & 5x - 5 \\ \hline -5x + 5 & \\ -\ (-5x + 10) & \\ \hline -5 & \end{array}$$

Remainder → -5

If there is a remainder, show this by placing it on the divisor, that is:
$(5x^2 - 15x + 5) \div (x - 2) = 5x - 5 - \frac{5}{x - 2}$

Rational expressions

A rational expression is an expression of the form $\frac{P}{Q}$ where P and Q are polynomials and $Q \neq 0$.

E.g. The expressions $\frac{1}{x}$, $\frac{3x - 1}{2}$ and $\frac{x + y}{x - 1}$ are rational expressions. The expression $\frac{1}{x}$ is only defined when $x \neq 0$, just as the expression $\frac{x + y}{x - 1}$ is only defined when $x \neq 1$.

Equivalent algebraic expressions can be generated by multiplying both the numerator and denominator of an expression by the same number that is not equal to 0.

E.g. The expressions $\frac{-4}{x + 3}$ and $\frac{-4x}{x^2 + 3x}$ are equivalent expressions since $\frac{-4}{x + 3} \times \frac{x}{x} = \frac{-4x}{x^2 + 3x}$.
These expressions are only defined when $x \neq -3$ and $x \neq 0$.

It is possible to simplify a rational expression when the numerator and denominator have at least one common factor. Cancel the common factors of the numerator and denominator presuming that they are both not equal to 0.

E.g. $\frac{6x + 3}{2x^2 + x} = \frac{3\cancel{(2x + 1)}}{x\cancel{(2x + 1)}} = \frac{3}{x}$. These equalities are true if $x \neq 0$ and $x \neq -\frac{1}{2}$.

Operations on rational expressions

Operations on rational expressions are completed by applying the same rules that are used on numbers that are written in fraction form.

Addition and **subtraction** of rational expressions require finding equivalent expressions that have the same denominator.

E.g. 1) $\frac{1}{2x} + \frac{4}{x} = \frac{1}{2x} + \frac{8}{2x} = \frac{9}{2x}$.
These equalities are true if $x \neq 0$.

2) $\frac{3}{x} - \frac{4}{xy} = \frac{3y}{xy} - \frac{4}{xy} = \frac{3y - 4}{xy}$.
These equalities are true if $x \neq 0$ and $y \neq 0$.

To **multiply** rational expressions, multiply the numerators together and the denominators together.

E.g. $\frac{x^2 + x}{2x} \times \frac{3}{x} = \frac{3(x^2 + x)}{2x^2} = \frac{3\cancel{x}(x + 1)}{2x^{\cancel{2}}} = \frac{3(x + 1)}{2}$. These equalities are true if $x \neq 0$.

To **divide** rational expressions, multiply the first fraction by the reciprocal of the second fraction.

E.g. $\frac{7}{x} \div \frac{y - 3}{5x} = \frac{7}{x} \times \frac{5x}{y - 3} = \frac{35\cancel{x}}{\cancel{x}(y - 3)} = \frac{35}{(y - 3)}$. These equalities are true if $x \neq 0$ and $y \neq 3$.

Following is a situation that involves operations on rational expressions and the adjacent rectangle.

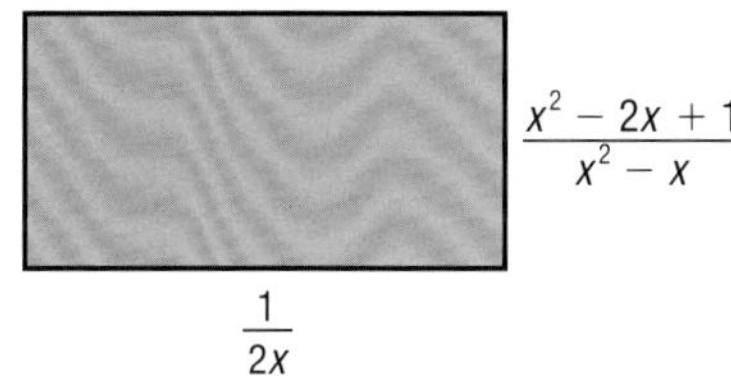

E.g. 1) It is possible to determine the simplified algebraic expression that corresponds to the perimeter of the rectangle shown above.

Perimeter $= 2h + 2b$

$= 2 \times \frac{x^2 - 2x + 1}{x^2 + x} + 2 \times \frac{1}{2x}$

$= 2 \times \frac{\cancel{(x - 1)}(x - 1)}{x\cancel{(x - 1)}} + \cancel{2} \times \frac{1}{\cancel{2}x}$

$= \frac{2(x - 1)}{x} + \frac{1}{x}$

$= \frac{2x - 2 + 1}{x}$

$= \frac{2x - 1}{x}$

The simplified algebraic expression that corresponds to the perimeter of this rectangle is $\frac{2x - 1}{x}$, if $x \neq 0$ and $x \neq 1$.

2) It is possible to determine the simplified algebraic expression that corresponds to the area of the rectangle shown above.

Area $= b \times h$

$= \frac{x^2 - 2x + 1}{x^2 - x} \times \frac{1}{2x}$

$= \frac{\cancel{(x - 1)}(x - 1)}{x\cancel{(x - 1)}} \times \frac{1}{2x}$

$= \frac{x - 1}{x} + \frac{1}{2x}$

$= \frac{x - 1}{2x^2}$

The simplified algebraic expression that corresponds to the area of this rectangle is $\frac{x - 1}{2x^2}$, if $x \neq 0$ and $x \neq 1$.

practice 2.3

1 In each case, indicate if the trinomial has the characteristics of a perfect square. Explain each of your answers.

a) $x^2 + 4x + 4$ b) $x^2 - 24x - 144$ c) $-9x^2 + 42x + 49$

d) $64x^2 - 144xy + 81y^2$ e) $3x^2 + 15x + 5$ f) $\frac{4}{9}x^2 - \frac{4}{3}x + 1$

2 Factor the following polynomials.

a) $xy + 4x + 2y + 8$ b) $169x^2 - 26x + 1$

c) $xy - 2x + 4y - 8$ d) $-2xy - 2y + 4x + 4$

e) $4x^2 - 36y^2$ f) $21xy - 12 - 18x + 14y$

g) $5x^2 - 5xy + 3xy - 3y^2$ h) $3xy - 1 - 3x + y$

i) $49 - 14x + x^2$ j) $x^2 - x - xy + y$

k) $\frac{xy}{3} + \frac{8x}{3} + \frac{y}{2} + 4$ l) $2x^2 - (2x + 3)^2$

3 Divide the following.

a) $(x^2 - x - 2) \div (x - 2)$ b) $(6x^2 - 5x - 4) \div (2x + 1)$

c) $(3x^2 + 2x - 4) \div (x + 3)$ d) $(10x^2 - 23x + 12) \div (2x - 3)$

e) $(x^2 + 4x + 4) \div (x + 2)$ f) $(x^3 - 3x^2 + x + 1) \div (x - 1)$

4 Simplify each of the following rational expressions.

a) $\frac{16x^4y}{6x^2y}$ b) $\frac{7xy}{7x - 21}$ c) $\frac{x - 3}{x^2 - 6x + 9}$

d) $\frac{x^2 - y^2}{3(x - y)}$ e) $\frac{8x^2 - 8}{6x + 6}$ f) $\frac{-32x^2}{16x^2 - 64x^3}$

5 In each case, complete the indicated operation and, if possible, simplfy the result to its simplest form.

a) $\left(\frac{-4y}{5x^3}\right)\left(\frac{x^2}{-6y}\right)$ b) $\frac{1}{x} + \frac{x}{2x^2}$ c) $\left(\frac{-15x}{2x + 6}\right) \div \left(\frac{10x}{7x + 21}\right)$

d) $\frac{-x}{xy} - \frac{x}{y}$ e) $\frac{3x}{y^2} + \frac{2x^2 - x}{xy^2}$ f) $\left(\frac{x^2 - 49}{2x - 6}\right)\left(\frac{(x - 3)^2}{x + 7}\right)$

g) $\left(\frac{12xy^2}{9xy}\right) \div \left(\frac{(3xy)^2}{6xy^2}\right)$ h) $\frac{5x + 2}{x^2 + x} + \frac{3}{x + 1}$ i) $\frac{x}{x^2 + 8x + 16} - \frac{x - 2}{x + 4}$

6 A carpenter cuts a rectangular wooden board using the measurements indicated below.

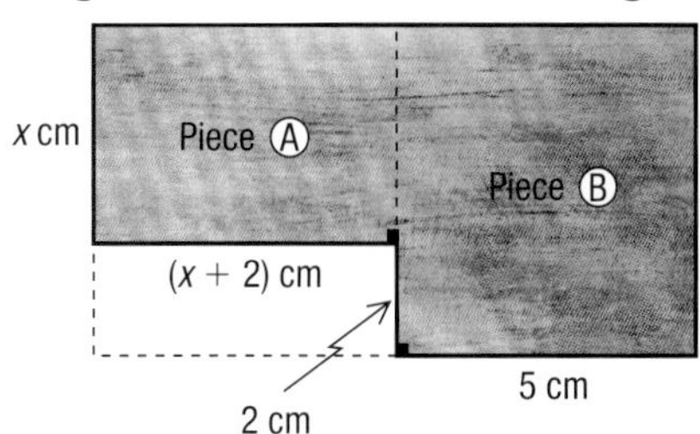

a) What algebraic expression represents the area of:

1) Piece Ⓐ?
2) Piece Ⓑ?
3) Piece Ⓐ and Ⓑ?

b) Express the total area of the board in the form of:

1) a product of factors
2) the difference of the product of two factors

c) Factor the algebraic expression found in **b) 2)**.
What do you notice?

7 For what values of m and n are the following equations true?

a) $mn - 5n - m + 5 = 0$

b) $mn + 3n + 2m + 6 = 0$

c) $24mn + 1 + 8m + 3n = 0$

d) $3mn - 9m + n = 3$

e) $2mn + 2 = n + 4m$

f) $2mn - 3n = {}^{-}2m + 3$

8 Around 1924, the invention of the rotary phone made it possible to make calls without an operator. This device was made up of a handset and a plastic circular dial containing 10 holes. Determine the radius of the dial on the telephone shown in the adjacent picture considering that the holes have a diameter of 0.5 cm each and that the area of the dial is 13.5π cm^2.

9 Judo is a combat sport practised on a tatami. The combat surface is delineated by a red line marked on the ground. The total area (in m^2) of the tatami shown in the adjacent diagram is represented by the algebraic expression $4x^2 + 40x + 100$:

a) Determine the length of the red boundary line of the combat area.

b) Determine the area of the combat surface.

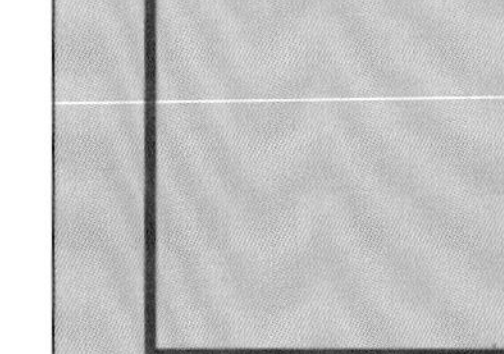

10 In each case, determine the simplified algebraic expression that represents the missing length.

a)

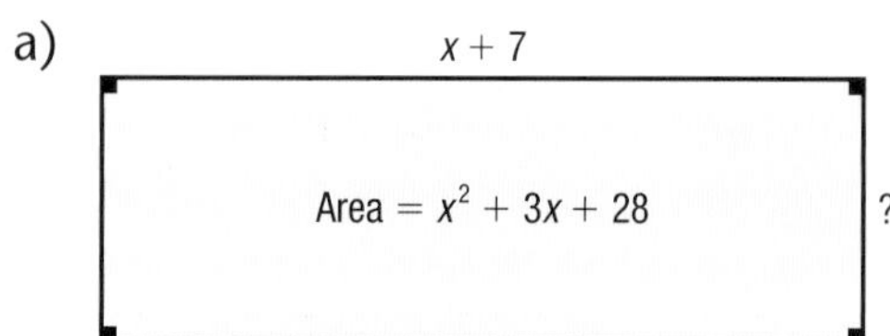

b)

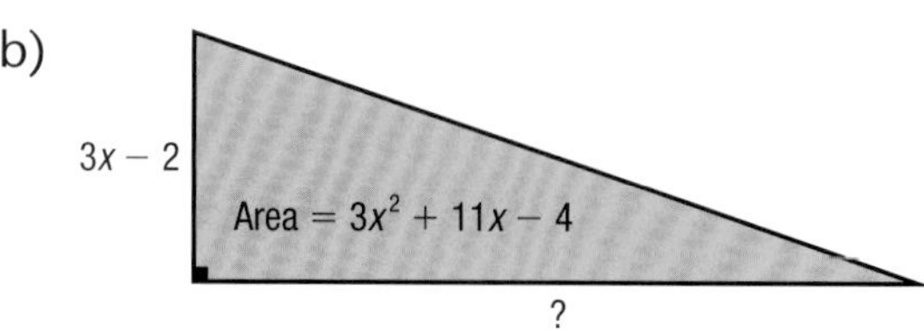

c)

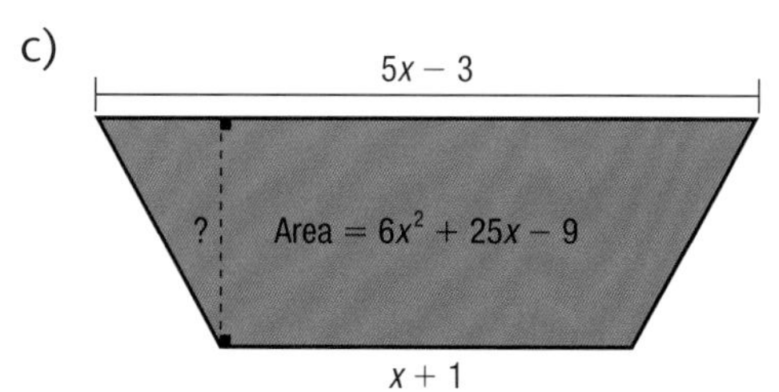

d)

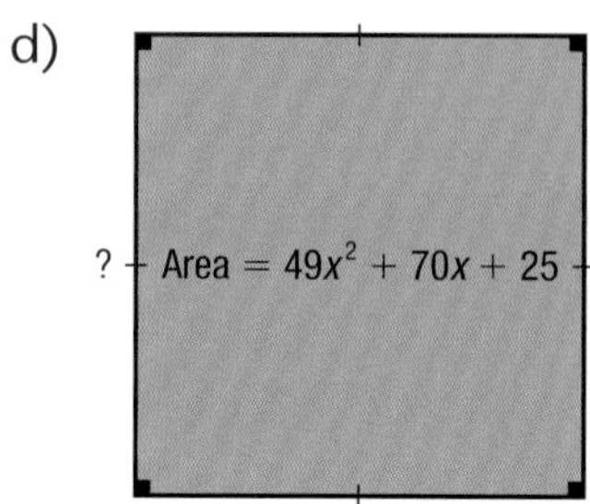

11 In the illustration shown below, segments AE and BC are parallel, and all the lengths are expressed in centimetres.

a) Which geometric statement allows you to conclude that triangles ADE and CDB are similar?

b) Determine an algebraic expression to represent triangle CDB considering that the area of triangle ADE corresponds to the algebraic expression $\frac{x^2}{2} + x + \frac{x}{2} + 1$.

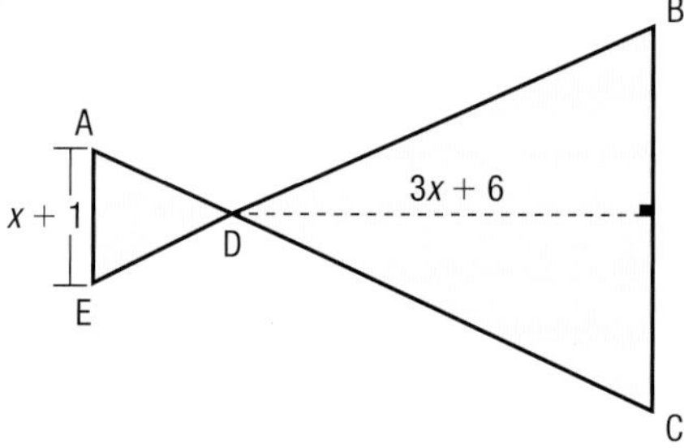

12 The diagram below shows a rectangular gorilla habitat in a zoo. This habitat consists of a dry area and a water-filled trench of consistent width. If the area of the water trench is $(10x + 12y + 120)$ m^2, do the dollowing:

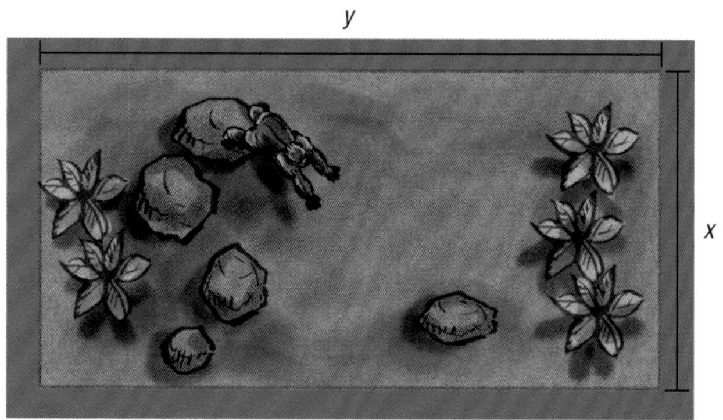

a) Express the area of the gorilla habitat as a product of factors.

b) Find the width of the water trench.

13 Determine the simplified algebraic expression that represents the volume of the adjacent right triangular-base prism.

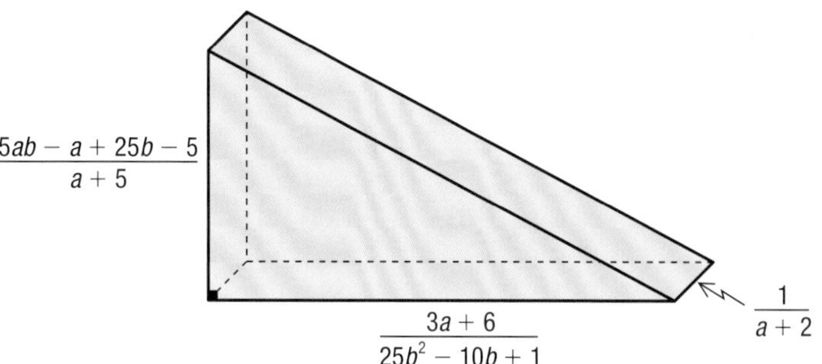

14 Determine the simplified algebraic expression that represents the area of the figure illustrated below.

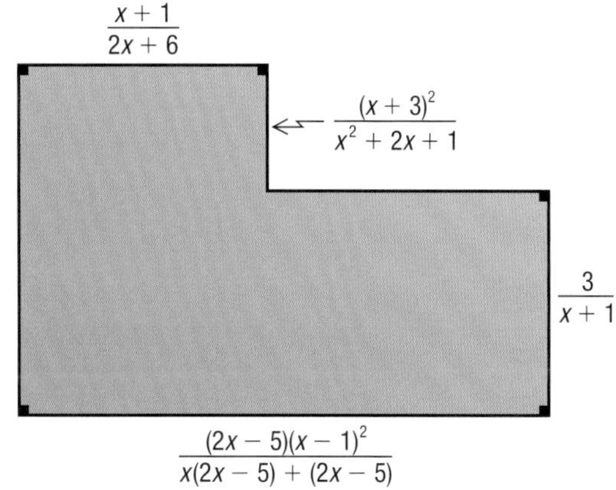

15 A triathlon is a race that consists of successively covering a certain distance by swimming, cycling and running. During a triathlon, Matty swam for 20 min, cycled 40 km and ran 10 km. The mean speed of the cycling corresponds to quadruple of the speed *s* of running.

a) Determine an algebraic expression that represents:
   1) the time spent cycling
   2) the time spent running
   3) the total time

b) If Matty completed the triathlon in 2 h and 20 min, how fast was he travelling by bicycle?

16 The division of a polynomial by $(x - 2)$ has a quotient of $(7x + 19)$ and a remainder of 39. What is this polynomial?

SECTION 2.4 Optimizing a distance

This section is related to LES 4.

PROBLEM The Snell-Descartes law

Optics is a branch of physics that focuses on the characteristics of light dispersion. According to the Snell-Descartes law of reflection, when a ray of light hits a reflecting surface, the angle of incidence is congruent to the angle of reflection. Examples of this occur with mirrors and windows.

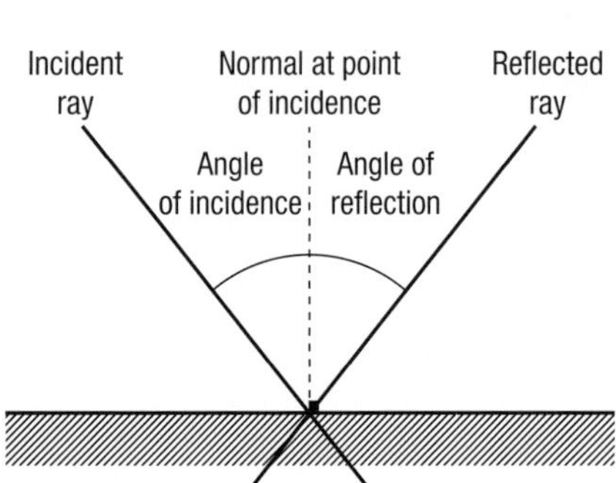

Two people assume the following positions in front of a mirror. Looking into the mirror, Person **A** looks at Person **B**.

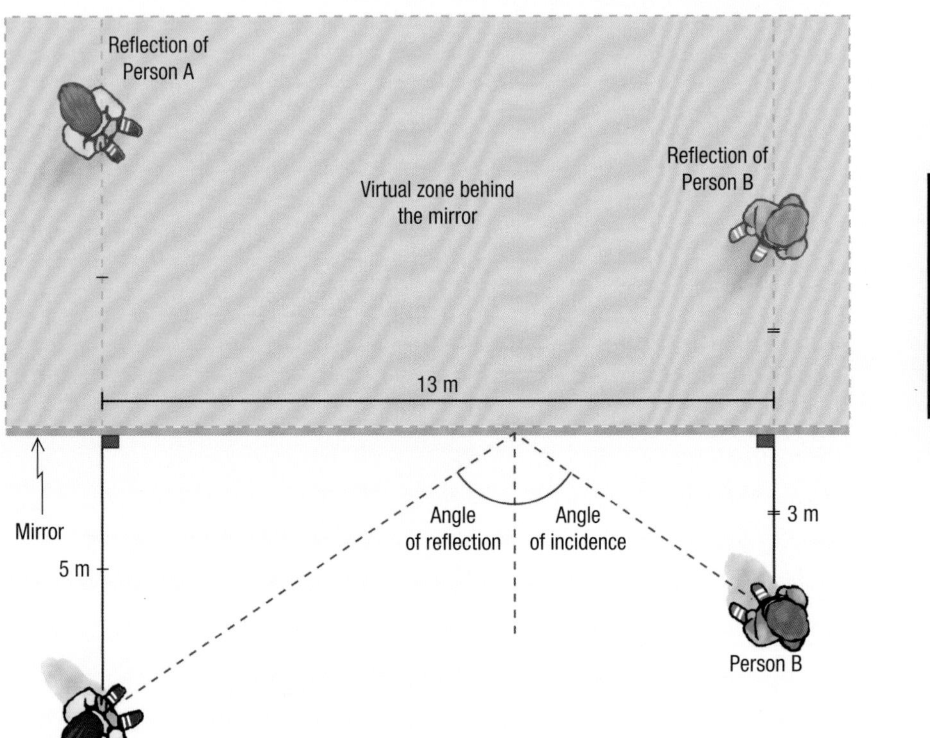

Fibre optics make it possible to transmit information over very long distances in the form of light signals.

What distance separates Person A from the reflection of Person B?

ACTIVITY 1 The *Phoenix* Lander

Following a ten month journey through space, the Phoenix Lander touched down on Martian soil in May 2008 to search for traces of water and life. During its descent, the probe captured images that made it possible to measure the slope of the ground and the presence of boulders that might impede landing.

Below is an aerial view of the landing site of a probe:

The cost of the Phoenix Lander mission to Mars is estimated to have been $420 million.

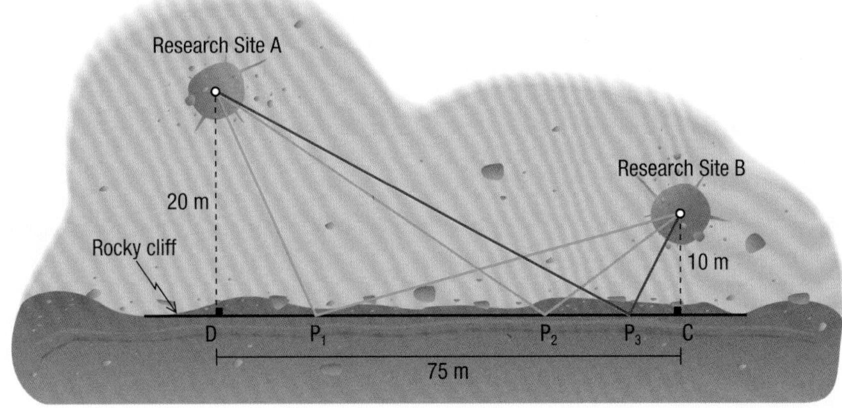

The probe is located at Research Site **A** and has to move to Research Site **B**.

a. If the probe has to stop at one of the three recharging stations $\mathbf{P_1}$, $\mathbf{P_2}$ or $\mathbf{P_3}$, what would be the shortest trajectory?

Scientists must determine the location for a Recharge Station **P** along the rocky cliff so that the trajectory followed by the probe is the shortest possible.

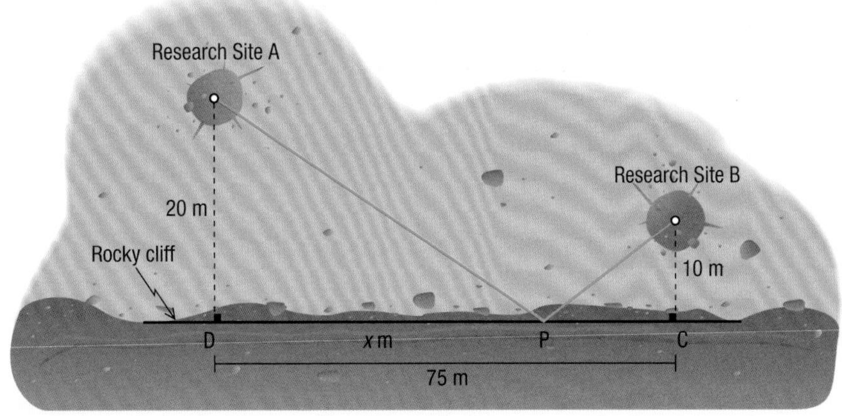

b. Determine an algebraic expression that contains variable x and that represents the distance:

1) between point C and the probe's Recharge Station
2) between Research Site **A** and the probe's Recharge Station
3) between Research Site **B** and the probe's Recharge Station

c. Complete the following table.

Distance from D to P (m)	Distance from C to P (m)	Distance from A to P (m)	Distance from B to P (m)	Length of trajectory (m)
10				
25				
40				
45				
55				
60				
70				

d. Based on this table, where should the scientists locate the probe's Recharge Station?

In order to precisely determine the location of the Recharge Station, scientists illustrate the landing site with the diagram shown below. In this diagram, point B is a reflection of point B over segment CD.

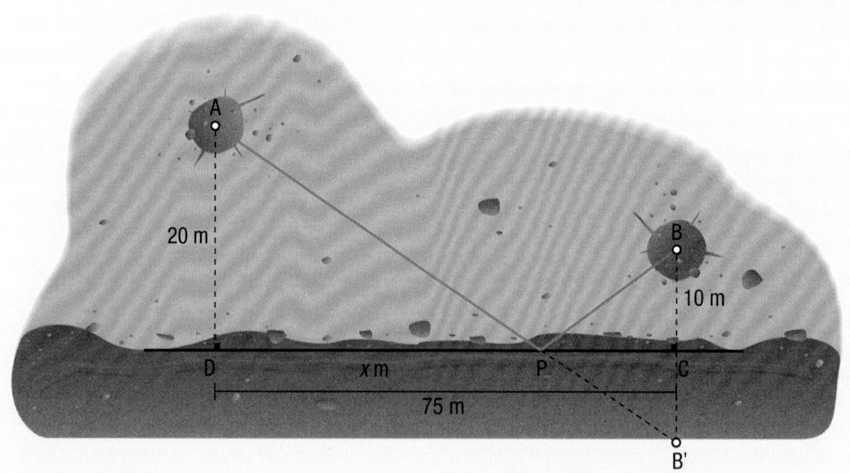

e. Why can it be stated that:

1) triangles ADP and B′CP are similar?
2) triangles BCP and B′CP are congruent?
3) triangles ADP and BCP are similar?

f. Explain why the following proportion can be established:

$$\frac{m\ \overline{DP}}{m\ \overline{CP}} = \frac{m\ \overline{AD}}{m\ \overline{BC}}$$

g. Using the proportion shown above, determine the minimum distance that the probe will have to travel from Research Site **A** to reach Research Site **B**.

Techno math

A graphing calculator allows you to solve problems involving distance. Consider the approach below that allows you to determine the distance at which point B should be placed from point E on segment DE in the adjacent diagram, so the distance between points A and B and points B and C is minimal.

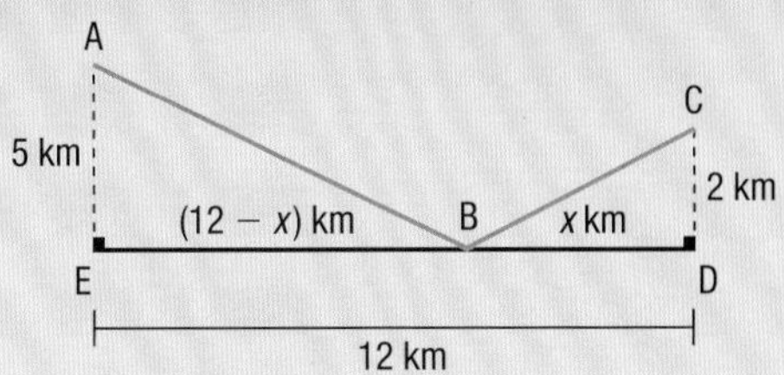

These screens allow you to enter and edit algebraic expressions that represent information given in a problem.

Screen 1

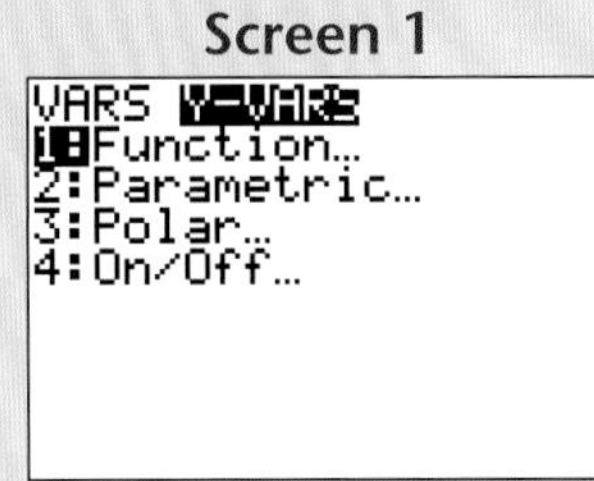

Screen 2

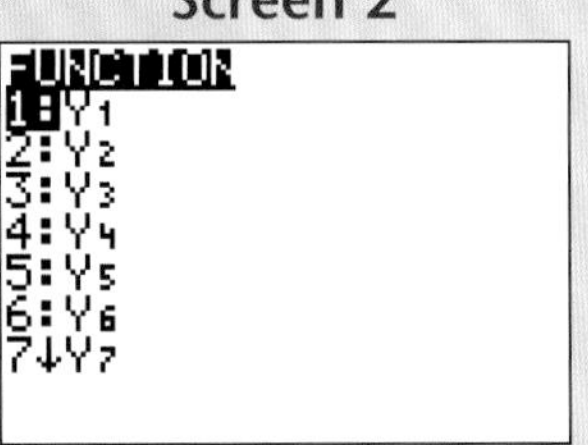

Screen 3

```
Plot1  Plot2  Plot3
\Y1=√((12-X)²+5²
)
\Y2=√(X²+2²)
\Y3=Y1+Y2
\Y4=
\Y5=
\Y6=
```

This screen allows you to visualize various distances associated with the problem.

Screen 4

X	Y1	Y2	Y3
1.5	11.63	2.5	14.13
2	11.18	2.8284	14.009
2.5	10.735	3.2016	13.937
3	10.296	3.6056	13.901
3.5	9.8615	4.0311	13.893
4	9.434	4.4721	13.906
4.5	9.0139	4.9244	13.938

Y1=√((12-X)²+5²)

a. Based on Screen **3**, what algebraic expression represents the distance between points:

1) A and B?

2) B and C?

b. Based on Screen **4**, where should you place point B on segment DE so that the distance between points A and B and points B and C is minimal?

c. Using a graphing calculator, determine the distance at which point B should be placed from point E on segment DE so that the distance between points A and B and points B and C is minimal.

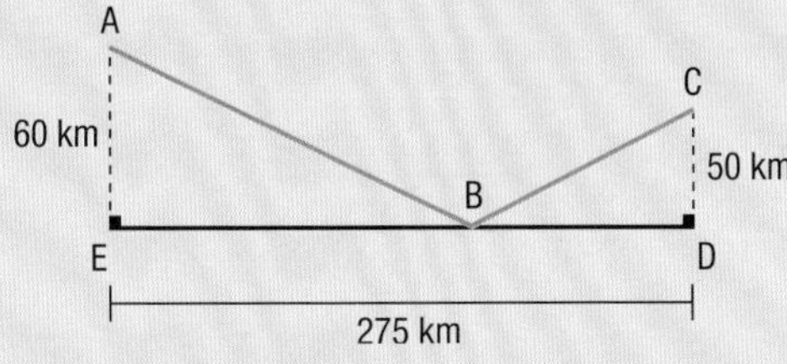

knowledge 2.4

OPTIMIZING A DISTANCE

Geometric statements are useful for solving problems involving distance. Algebra is combined with geometry to solve this type of problem.

E.g. Determine where point B should be located along $\overline{DE}$ so that the sum of the distances between points A and B and points B and C is minimal.

In order for this distance to be minimal, point B must be located so that the two triangles ABE and CBD are similar.

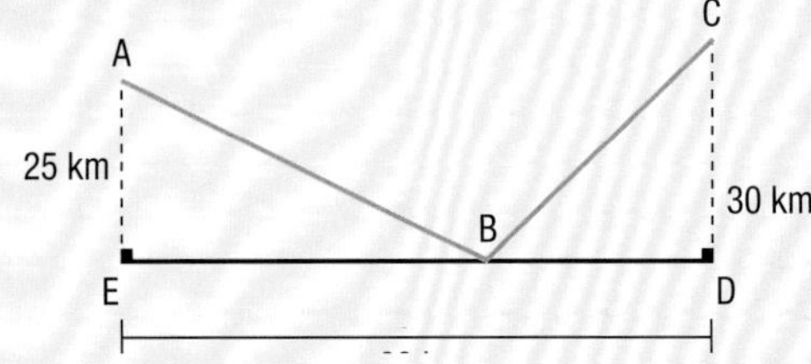

If m $\overline{BE} = x$, then m $\overline{BD} = 60 - x$.

If triangles ABE and CBD are similar, you get:

$$\frac{m\ \overline{BE}}{m\ \overline{BD}} = \frac{m\ \overline{AE}}{m\ \overline{CD}}$$

$$\frac{x}{60 - x} = \frac{25}{30}, \text{ where } x \neq 60$$

$$30x = 25(60 - x)$$

$$30x = 1500 - 25x$$

$$55x = 1500$$

$$x = \frac{300}{11}$$

$$x \approx 27.27 \text{ km}$$

Point B must be located on segment DE at approximately 27.27 km from point E.

It is also possible to determine the distance between points A and B and points B and C using the following approach.

1) $(m\ \overline{AB})^2 = (m\ \overline{AE})^2 + (m\ \overline{BE})^2$

$(m\ \overline{AB})^2 = 25^2 + \left(\frac{300}{11}\right)^2$

$(m\ \overline{AB})^2 \approx 1368.8$

$m\ \overline{AB} \approx 37$ km

2) $(m\ \overline{BC})^2 = (m\ \overline{CD})^2 + (m\ \overline{BD})^2$

$(m\ \overline{BC})^2 = 30^2 + \left(\frac{360}{11}\right)^2$

$(m\ \overline{BC})^2 \approx 1971.07$

$m\ \overline{BC} \approx 44.4$ km

3) $m\ \overline{AB} + m\ \overline{BC} \approx 81.4$ km

The minimal distance between points A and B and points B and C is approximately 81.4 km.

practice 2.4

1 For each case, determine whether the triangles are similar. If the triangles are similar, justify your claim with a geometric statement.

a)

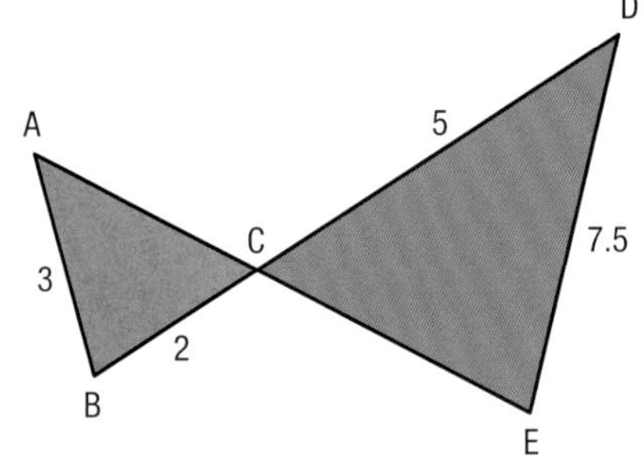

b)

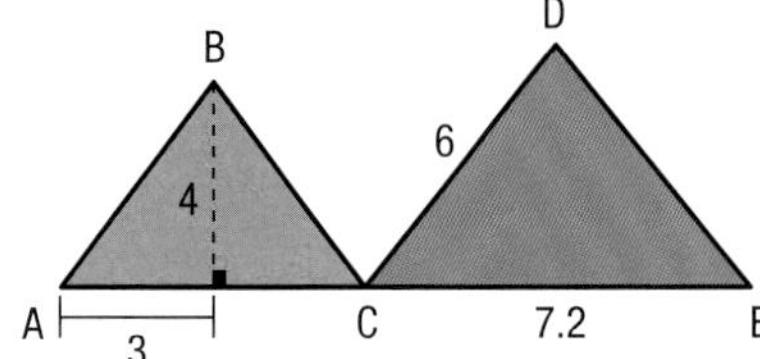

c)

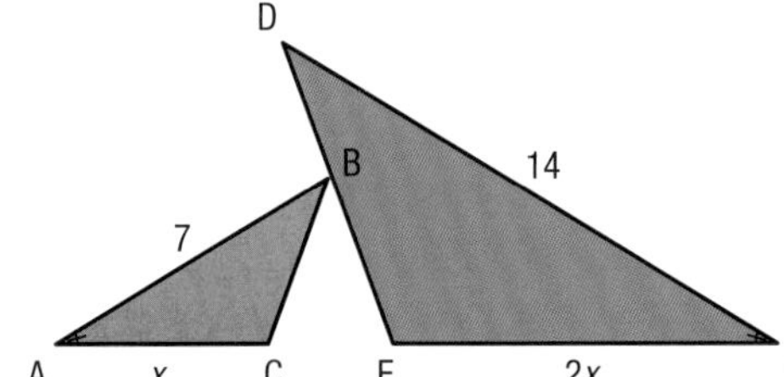

d)

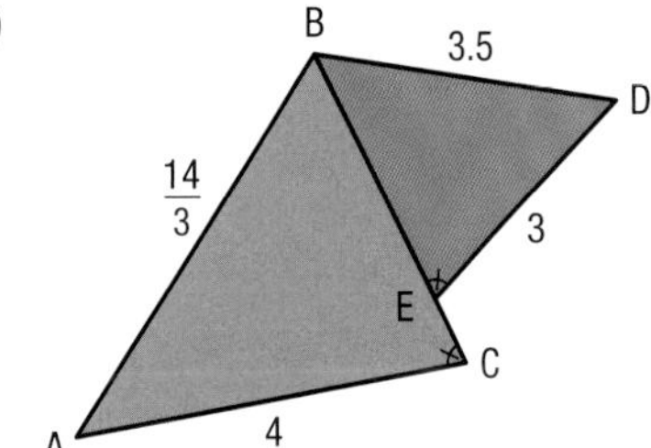

2 In each of the following diagrams, point S is located on $\overline{UV}$ so that the sum of the distances between points R and S and points S and T is minimal. Considering that the lengths are expressed in kilometres, determine:

The distance between points A and B can be written as d(A, B).

1) $d(\text{V, S})$ 2) $d(\text{R, S}) + d(\text{S, T})$

a)

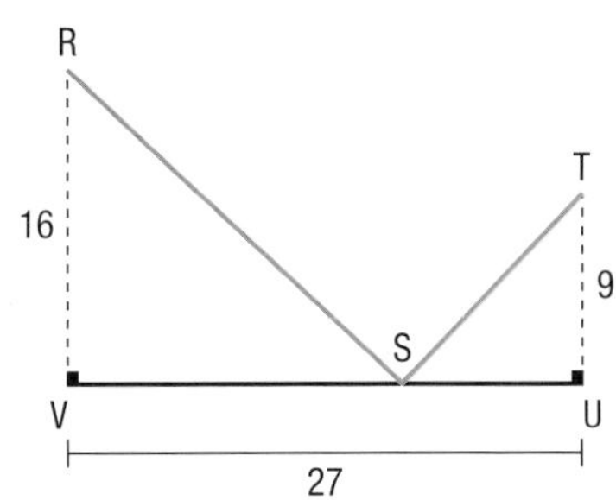

b)

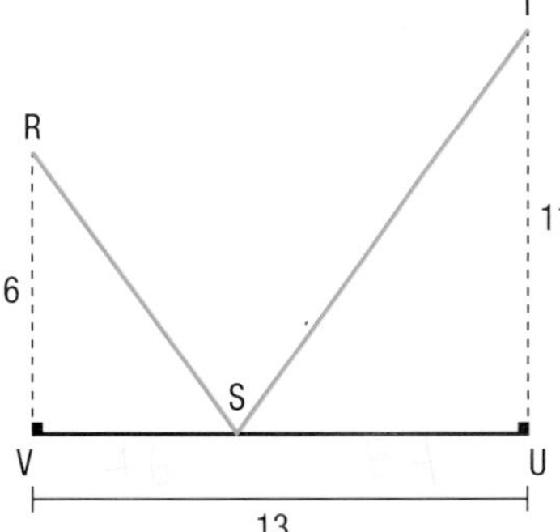

3 Points A(3, 7), B(8, 2) and P(x, y) are plotted in a Cartesian plane so that the sum of the distances between points A and P and points P and B is minimal. Determine the coordinates of point P if it is located on:

a) the x-axis

b) the y-axis

4 In the diagram below, point T is located on $\overline{CD}$ so that the sum of the lengths of the hypotenuses of triangles ADT and BCT is as small as possible. If the ratio of the areas of these two triangles is 16:1, determine the sum of the lengths of the hypotenuses.

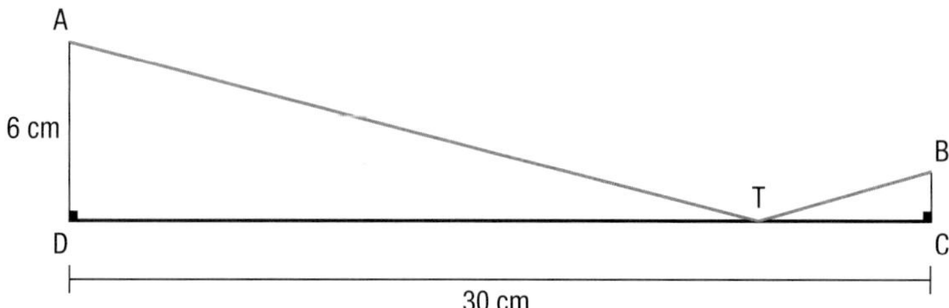

5 The adjacent illustration represents a decorative hourglass comprised of two similar right circular cones. Determine:

a) the height of Part Ⓑ

b) the lateral area of Parts Ⓐ and Ⓑ of this hourglass

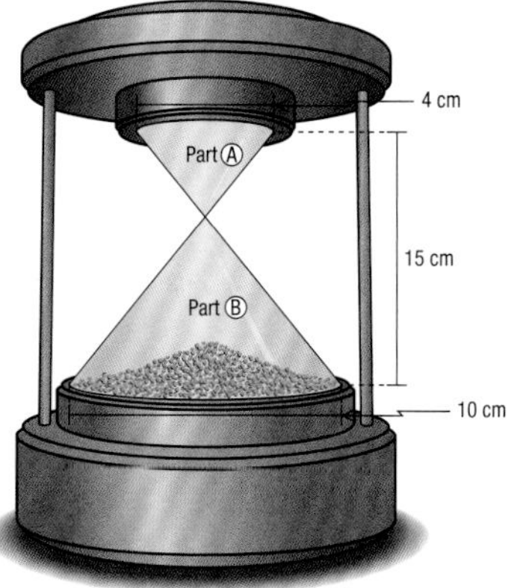

6 An oil platform is an offshore facility that makes it possible to tap into oil deposits located under the ocean floor. In order to minimize construction costs, an oil company decides to build a single platform capable of tapping two oil deposits. At what distance from point A will they have to build the platform so that the pipelines are as short as possible?

Oil platforms were originally set up in very shallow waters, but they can now be built in deeper waters far offshore.

7 In preparation for an important competition, the bottom of the pool shown below is scheduled to be repainted.

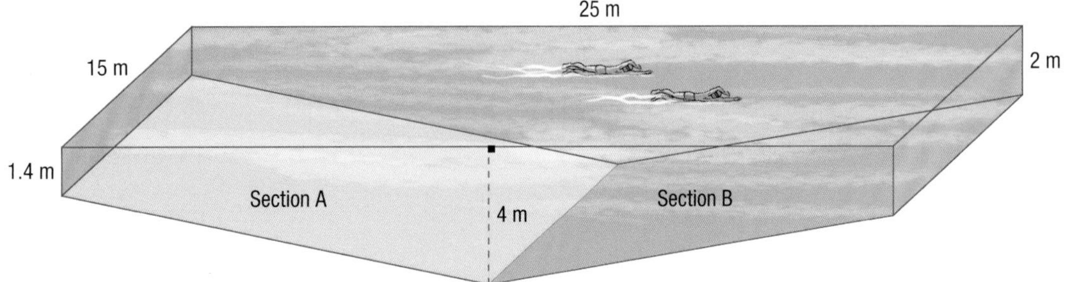

Determine the area of the surface to be painted, considering that the slope of Section **A** is the same as that of Section **B**.

8 Determine the area of the purple sections in the adjacent puzzle considering the following:

- The polygons of the same colour are congruent.
- All of the horizontal and vertical segments are respectively parallel to each other.
- Point C is located on segment BD so that the sum of the lengths of segments AC and CE is minimal.

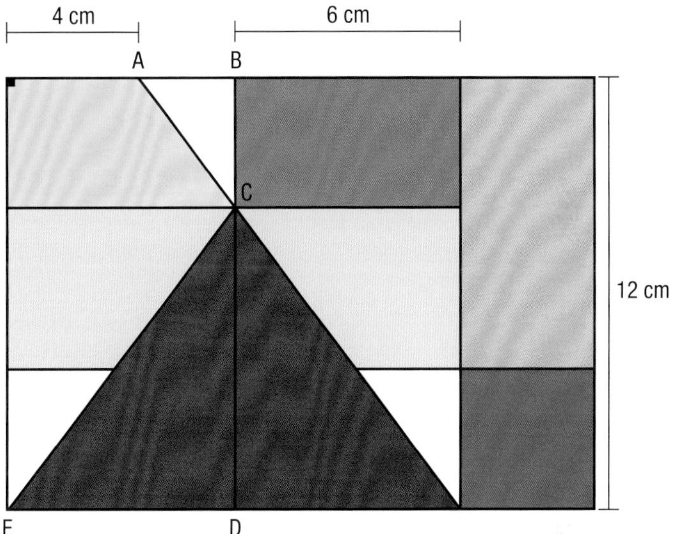

9 In the figure below, parallelogram ABCD is inscribed in rectangle GHIJ. Determine the perimeter of the parallelogram; considering that the dimensions of the rectangle are 10 cm by 20 cm and that point C is located on segment IJ so that the sum of the lengths of segments CD and BC is minimal.

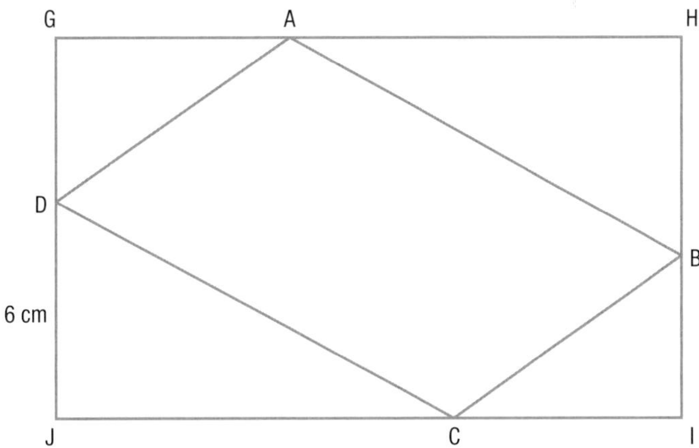

10 What is the amount of lumber required to build the adjacent roof trusses, considering that:

- point G is located on segment HF so that the sum of the lengths of segments AG and GB is minimal?
- point E is located on segment FD so that the sum of the lengths of segments BE and EC is minimal?

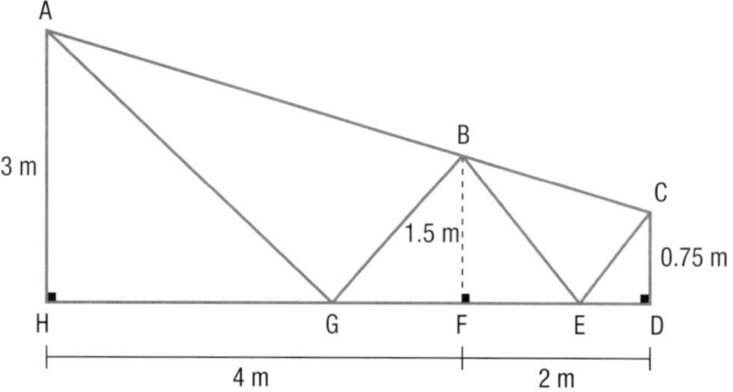

Triangles are the geometric shape most often used in the construction of roof supports because they provide a better load distribution.

11 Inaugurated in 1990 and exclusively for women contenders, the Aïcha des Gazelles Rally takes place in the Sahara desert. During each stage, the teams must travel along a set itinerary with the help of a map and compass. Consider the information below concerning two legs of this race:

Leg ①
Leaving from Town **A**, get to Town **B** passing through the Checkpoint.

Leg ②
Leaving from Town **C**, get to Town **D** passing through the Checkpoint.

Considering that the Checkpoint is located so as to minimize the distance travelled, what is the minimal distance covered during each leg?

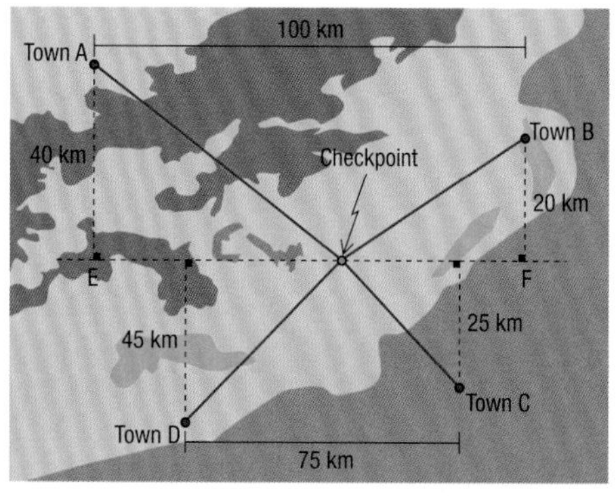

12 For the construction of a new airport, an architect must decide where to locate the main building. The latter has to be above the subway line and allow people catching a connecting flight to walk the shortest distance possible from one building to another. Determine the minimal length of each wing joining Buildings **A**, **B**, and **C** to the main building.

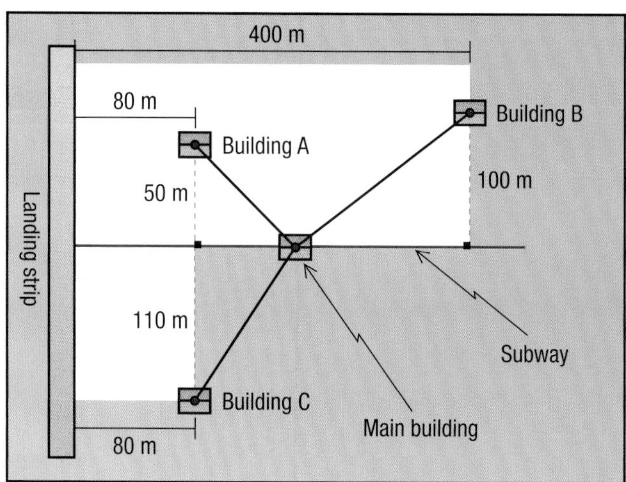

SECTION 2.5 Metric relations

This section is related to LES 3.

PROBLEM An architectural wonder

The Dubai Tower, a skyscraper located in the United Arab Emirates, has a height of approximately 900 m including its communications antennas. Its unprecedented construction scale will require the equivalent of 330 000 m^3 of reinforced concrete, 39 000 tons of steel beams, 142 000 m^2 of glass and 20 million hours of labour.

As shown in the diagram below, during construction, two tourists observe the top of the tower from two different locations.

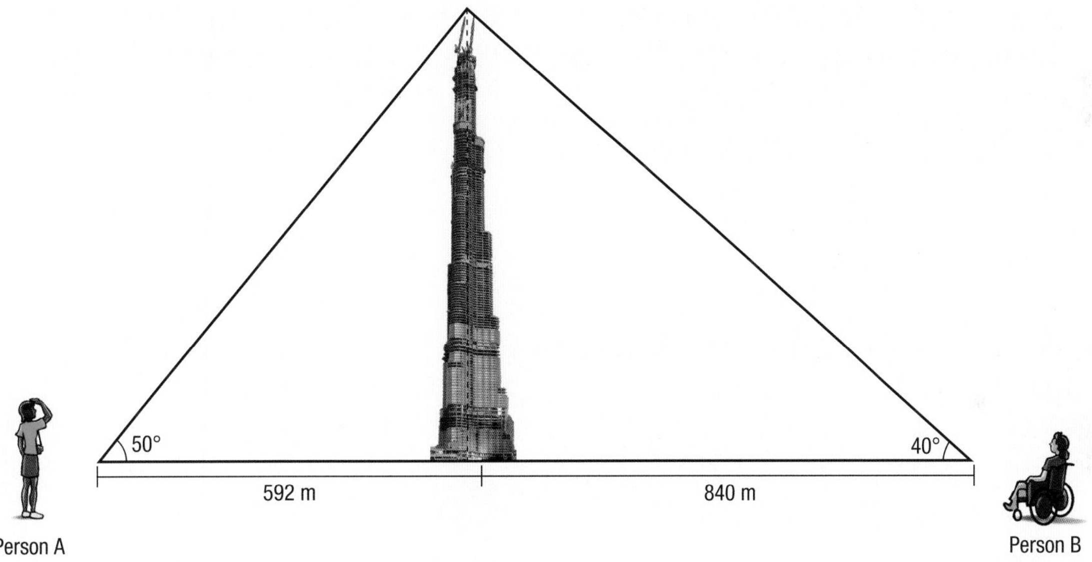

What is the height of the tower at this moment?

With its hotels, housing units, offices and retail outlets, the Dubai Tower has been designed as a real indoor city that can be lived in year round.

ACTIVITY 1 Birds of a feather stick together!

By drawing the altitude from the right angle in a right triangle, two other triangles are formed. These three triangles are arranged in the same orientation, as shown below.

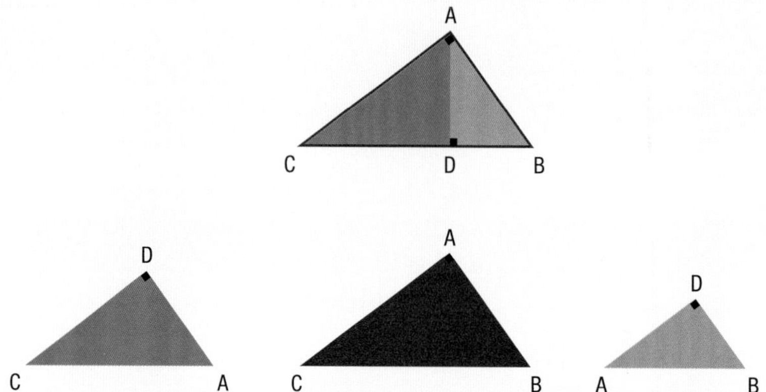

a. What geometrical statement allows you to state that:

1) Δ blue ~ Δ pink? 2) Δ green ~ Δ pink? 3) Δ blue ~ Δ green?

b. Complete the following ratios:

1) using the blue and pink triangles

$$\frac{\Delta \text{ blue: } m\overline{CD}}{\Delta \text{ pink: } m\overline{AC}} = \frac{m\overline{AC}}{BC} = \frac{DA}{m\overline{AB}}$$

2) using the pink and green triangles

$$\frac{\Delta \text{ pink: } m\overline{AC}}{\Delta \text{ green: } m\overline{AD}} = \frac{m\overline{BC}}{BA} = \frac{AB}{m\overline{DB}}$$

3) using the blue and green triangles

$$\frac{\Delta \text{ blue: } CA}{\Delta \text{ green: } m\overline{AB}} = \frac{m\overline{AD}}{BD} = \frac{CD}{m\overline{AD}}$$

c. Determine the missing length in each case.

1) m $\overline{BD}$ 2) m $\overline{FH}$ 3) m $\overline{MO}$

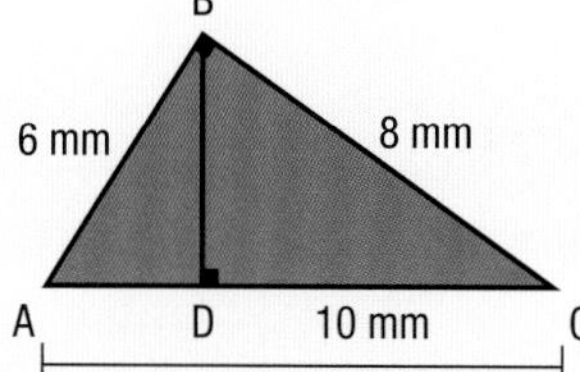

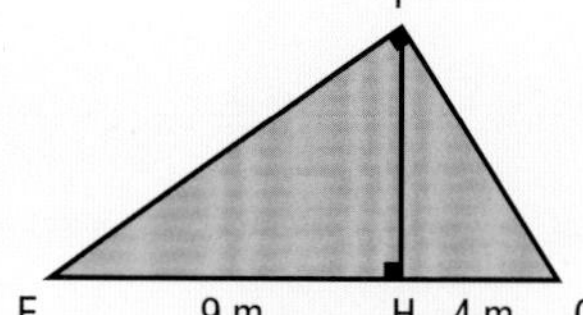

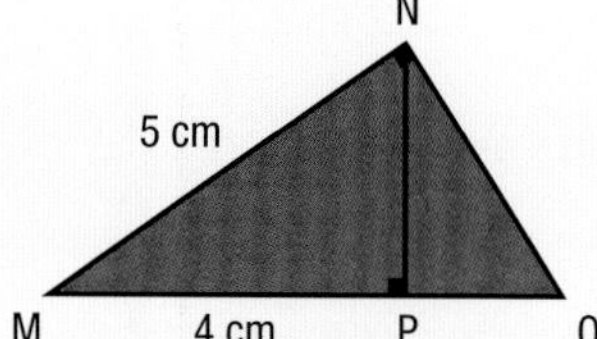

Techno math

Dynamic geometry software program allows you to explore and verify metric relations in right triangles. By using the tools LINE, PERPENDICULAR LINE, TRIANGLE and DISTANCE, you can construct a right triangle and the altitude from the vertex of the right angle.

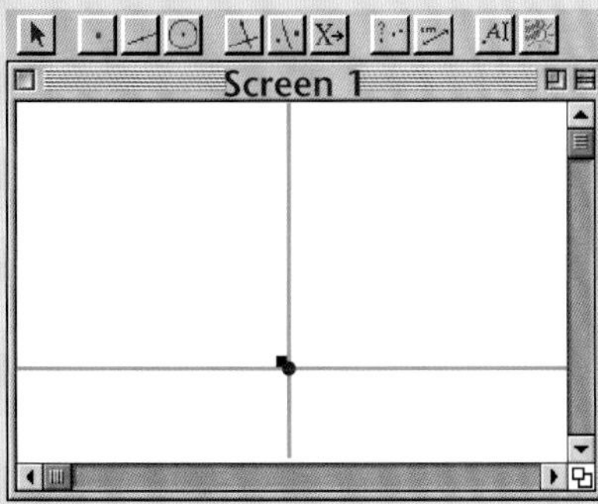

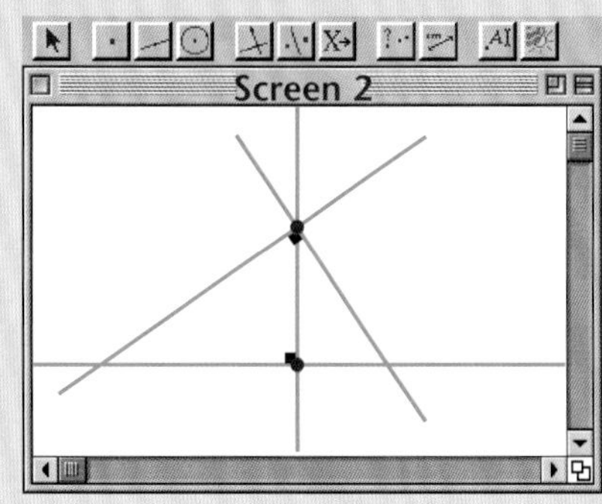

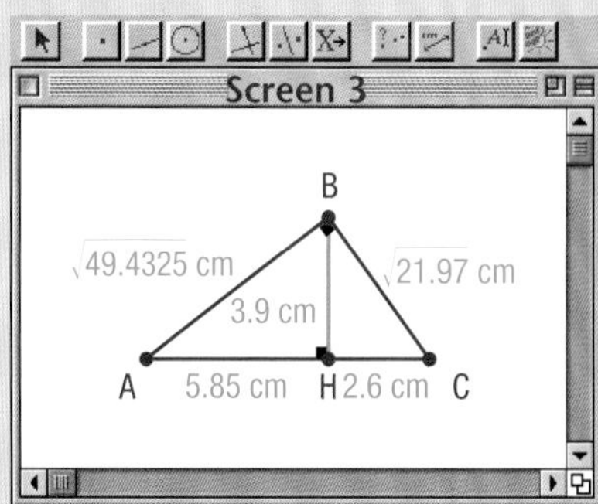

By changing the position of the vertices of the triangle, you can observe certain effects related to the lengths of the sides and altitude of the triangle.

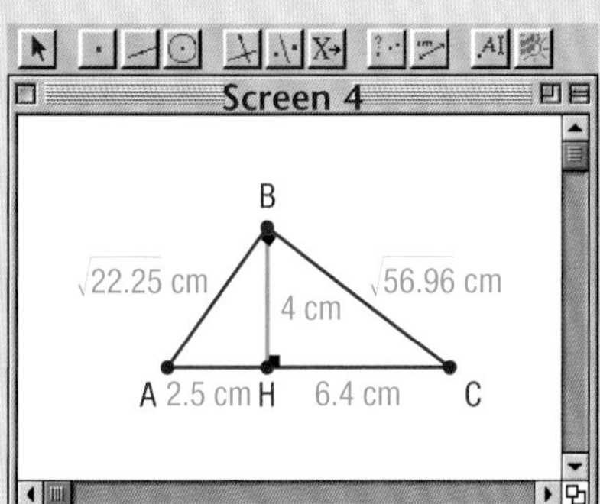

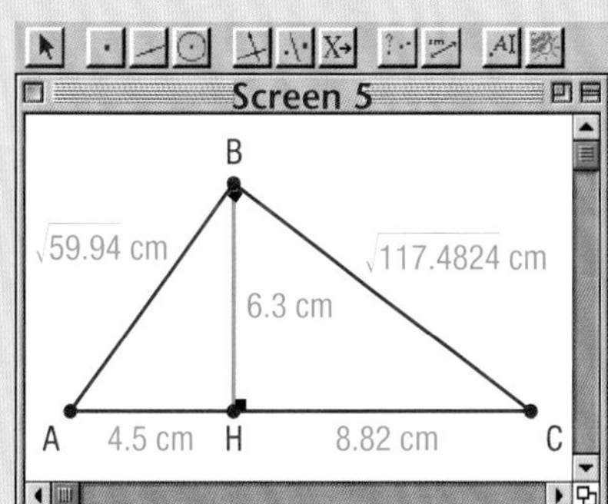

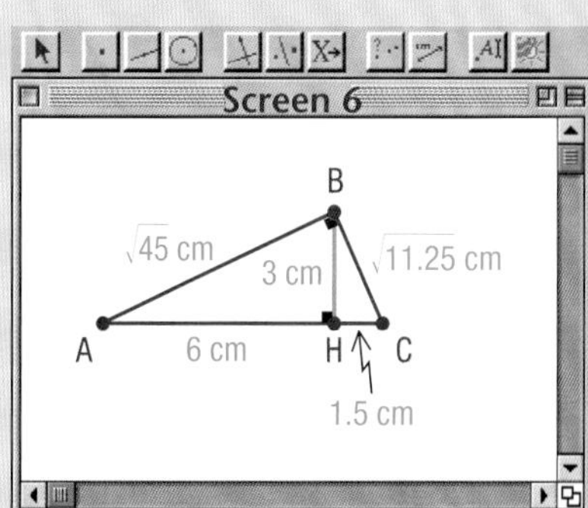

a. Using calculations and a geometric statement, explain why:

1) $\Delta ABH \sim \Delta BCH$
2) $\Delta ABC \sim \Delta AHB$
3) $\Delta ABC \sim \Delta BHC$

b. For each of the Screens **3**, **4**, **5** and **6**, calculate:

1) $m\ \overline{AC} \times m\ \overline{BH}$
2) $m\ \overline{AB} \times m\ \overline{BC}$

c. Based on the results obtained, what conjecture can you formulate?

d. Using dynamic geometry software, verify whether this conjecture can be applied:

1) to isosceles triangles
2) to equilateral triangles

knowledge 2.5

METRIC RELATIONS IN A RIGHT TRIANGLE

By drawing the altitude from the vertex of the right angle in a right triangle, three similar right triangles are formed. Using the lengths of the corresponding sides of the triangles formed, you can establish proportions and deduce various geometric statements.

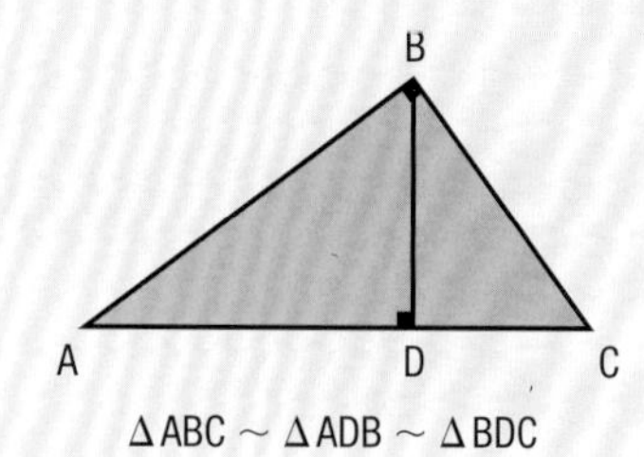

$\Delta ABC \sim \Delta ADB \sim \Delta BDC$

1. In a right triangle, the length of a leg of a right triangle is the geometric mean of the length of its projection on the hypotenuse and the length of the hypotenuse, meaning:

$$\frac{m\overline{AD}}{m\overline{AB}} = \frac{m\overline{AB}}{m\overline{AC}} \text{ or } (m\overline{AB})^2 = m\overline{AD} \times m\overline{AC}$$

$$\frac{m\overline{CD}}{m\overline{BC}} = \frac{m\overline{BC}}{m\overline{AC}} \text{ or } (m\overline{BC})^2 = m\overline{CD} \times m\overline{AC}$$

E.g.

1)

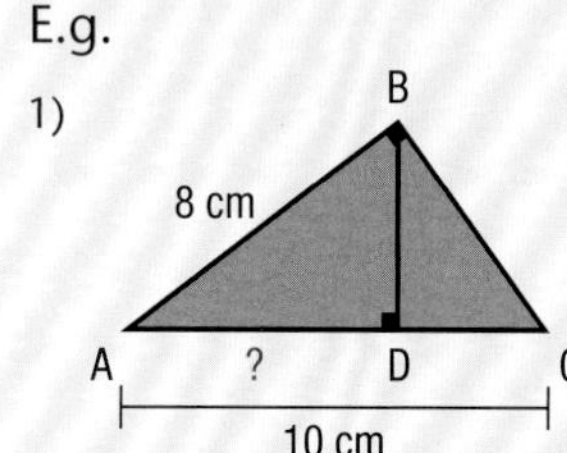

$$\frac{m\overline{AD}}{8} = \frac{8}{10}$$

$$64 = 10 \times m\overline{AD}$$

$$m\overline{AD} = 6.4 \text{ cm}$$

2)

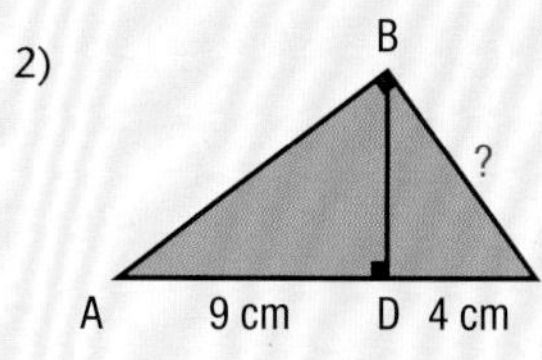

$$\frac{4}{m\overline{BC}} = \frac{m\overline{BC}}{13}$$

$$(m\overline{BC})^2 = 52$$

$$m\overline{BC} = \sqrt{52} \text{ cm or } \approx 7.21 \text{ cm}$$

2. In a right triangle, the length of the altitude drawn from the right angle is the geometric mean of the length of the two segments that determine the hypotenuse, meaning:

$$\frac{m\overline{AD}}{m\overline{BD}} = \frac{m\overline{BD}}{m\overline{CD}} \text{ or}$$

$$(m\overline{BD})^2 = m\overline{AD} \times m\overline{CD}$$

E.g.

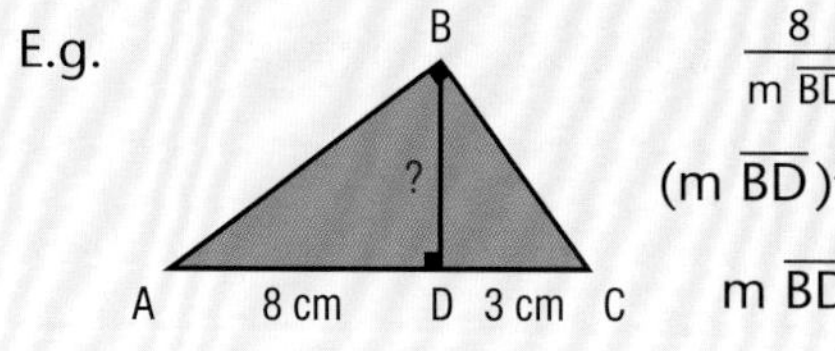

$$\frac{8}{m\overline{BD}} = \frac{m\overline{BD}}{3}$$

$$(m\overline{BD})^2 = 24$$

$$m\overline{BD} = \sqrt{24} \text{ cm or } \approx 4.9 \text{ cm}$$

3. In a right triangle, the product of the length of the hypotenuse and its corresponding altitude is equal to the product of the length of the legs, meaning:

$$m\overline{AC} \times m\overline{BD} = m\overline{AB} \times m\overline{BC}$$

E.g.

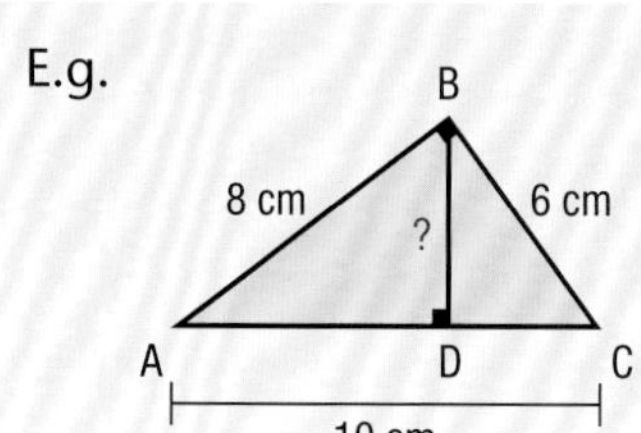

$$10 \times m\overline{BD} = 8 \times 6$$

$$m\overline{BD} = 4.8 \text{ cm}$$

practice 2.5

1 The adjacent figure contains three similar triangles.

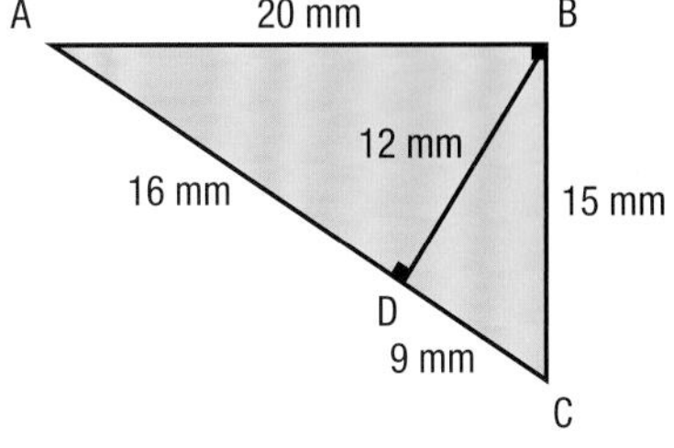

a) Identify all three right triangles using their vertices.

b) Establish the proportions for each of the corresponding sides.

c) Calculate the ratio of similarity for each pair of similar triangles.

d) For each pair of similar triangles, determine the geometric statement that concludes that these triangles are in fact similar.

2 Triangle ABC shown at right has a right angle located at vertex C. Determine the following measurements and indicate the geometric statement on which you based your calculations.

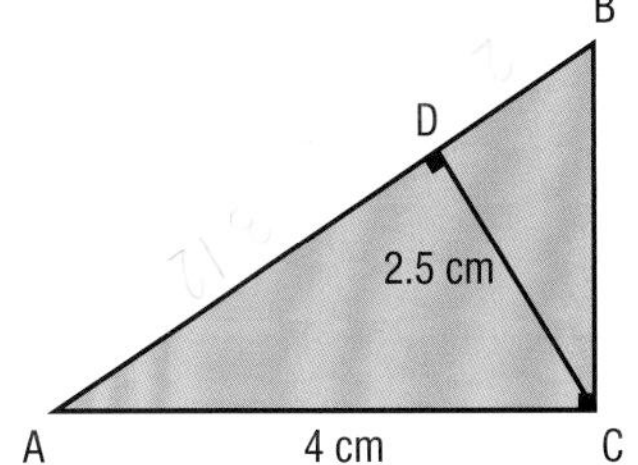

a) m $\overline{BD}$

b) m $\overline{CB}$

3 In the right triangle below, what is the length of segment AD?

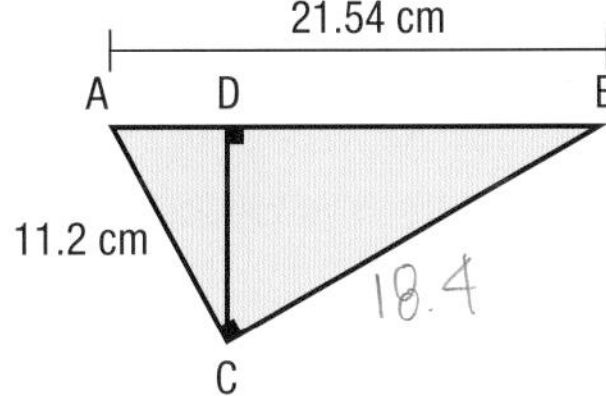

4 In the adjacent equilateral triangle, determine the length of segment CE.

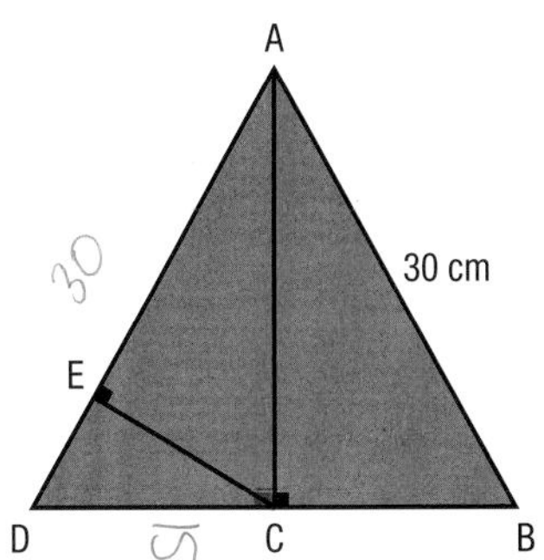

5 Determine the area of the adjacent isosceles trapezoid ABCD.

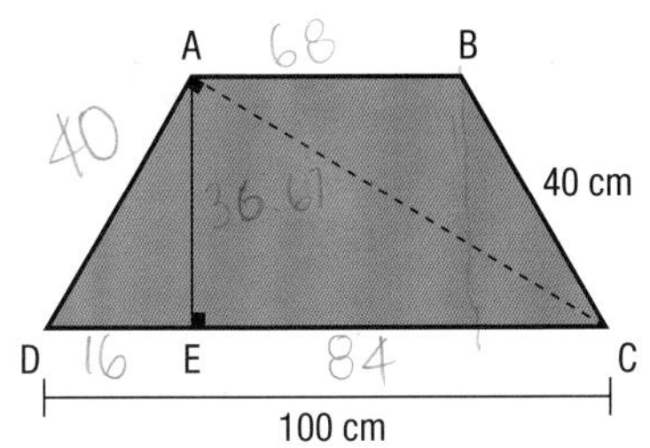

6 A computer-controlled saw must cut triangle GBE from the rectangular steel sheet as shown on the right. At what distance from point D should the cut be started?

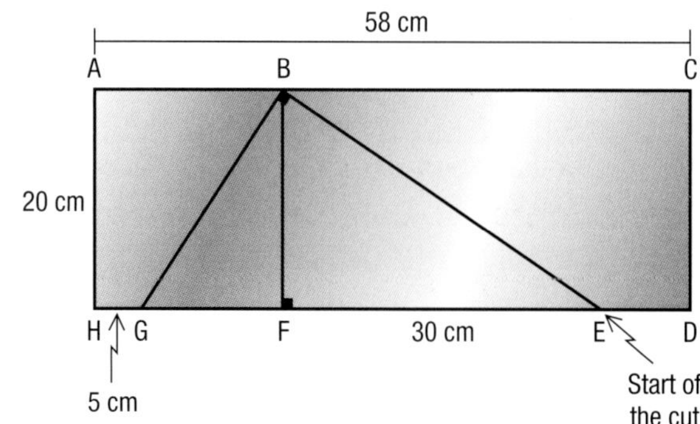

7 By referring to the right triangle illustrated below, complete the following table.

Length of segments

	a	*b*	*c*	*m*	*n*	*h*
a)	9	12				
b)				4		8
c)	10					6

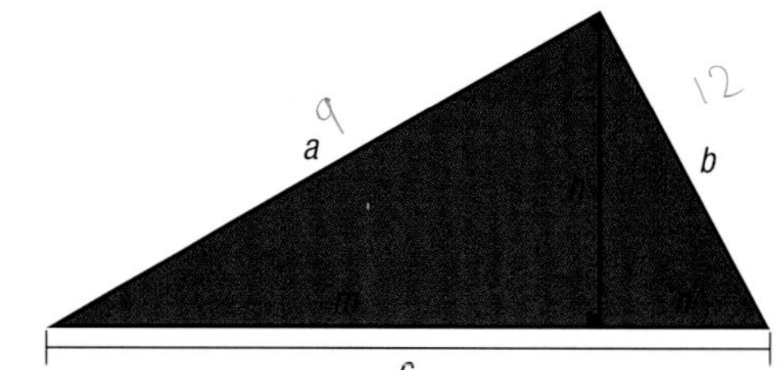

8 In the adjacent right triangle ABC, altitude BE and median BD have been drawn. What is the perimeter of triangle BDE?

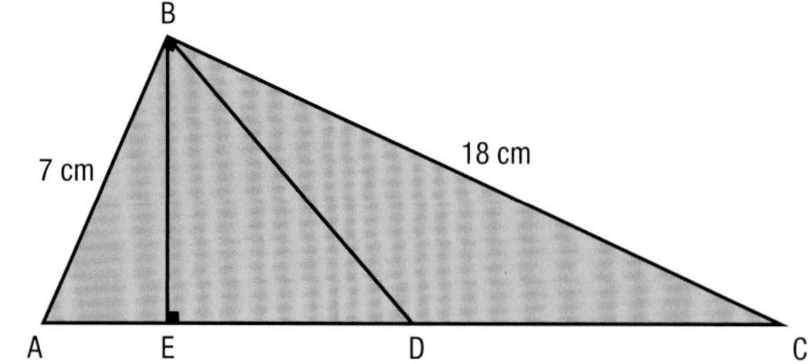

9 For each of the right triangles below, do the following:

1) Determine the length associated with *x*.

2) Identify the geometric statement that allows for your calculations.

a)

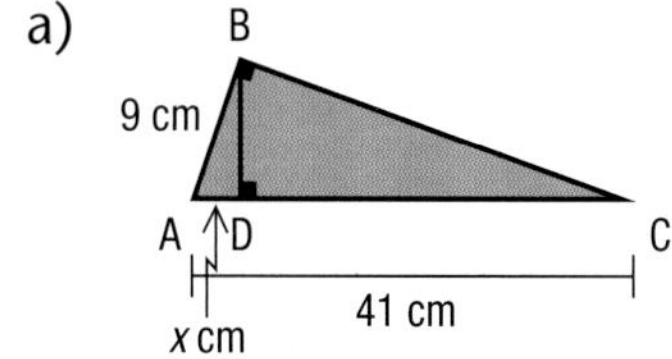

b)

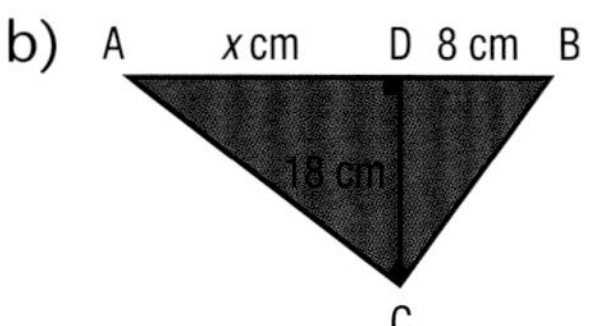

c)

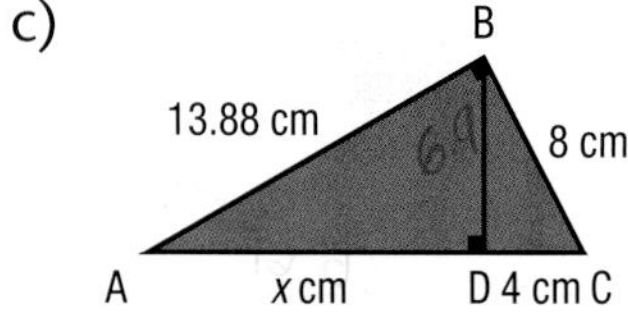

d)

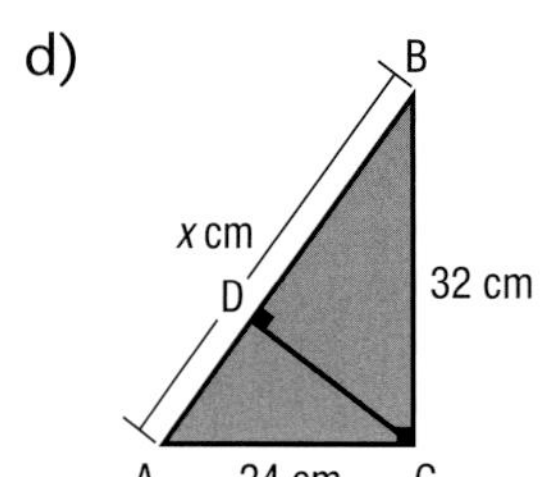

10 **SOUNDING BALLOON** A sounding balloon is an instrument used in meteorology to collect data at high altitudes. This type of balloon is filled with helium and can reach altitudes of up to 30 km. The diagram below represents a sounding balloon that is transmitting information to three weather stations. Without using the Pythagorean theorem, determine:

a) the distance between the balloon and Station **A**

b) the distance between the balloon and Station **C**

c) the height of the balloon

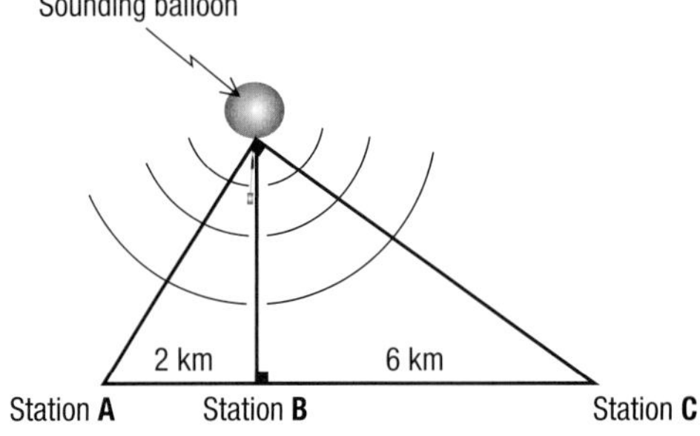

11 Referring to the circle with centre O shown at right, what is the area of the blue section?

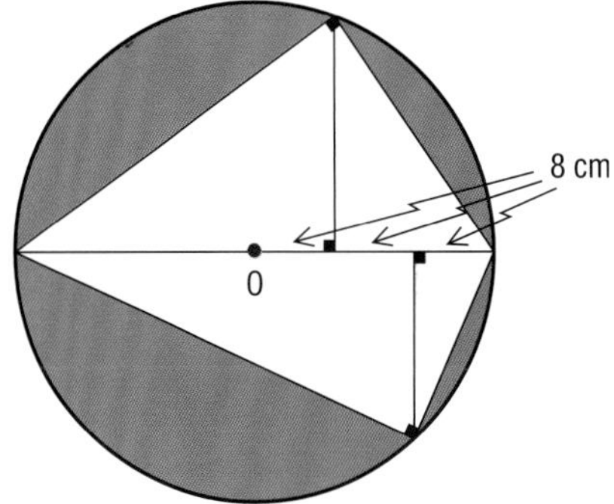

12 The diagram below depicts a crosscut of a hydroelectric dam. Point C indicates the minimum water level that must be maintained in the basin.

a) Determine the height of the dam.

b) What is the minimum water level to be maintained in the basin?

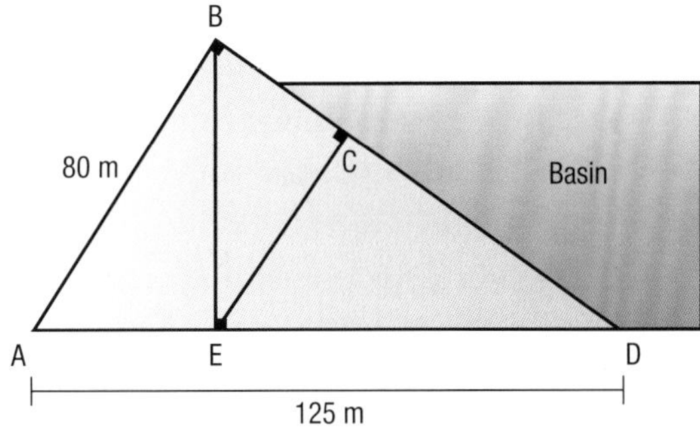

13 A DJ's console overlooks the dance floor as shown in the adjacent diagram. Two stairways allow people to go up and make their requests. How high is the console?

14 A landscape contractor is attaching ropes to saplings he has just planted.

Situation ①

Ropes **A** and **B** are equal in length.

a) At what distance from the foot of the tree must he anchor the two ropes if the attachment point on the tree is located 5 m above the ground?

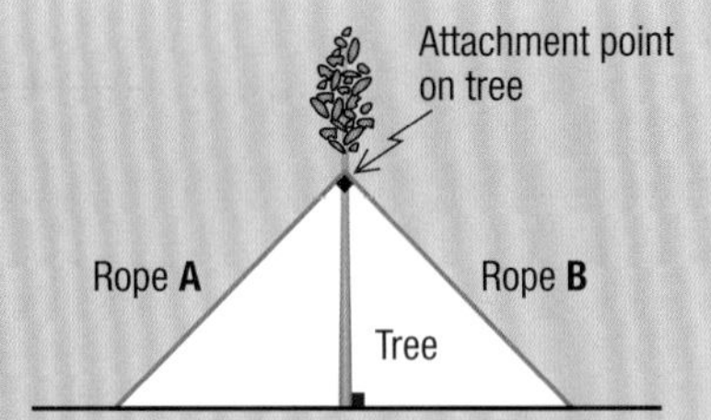

Situation ②

Ropes **A** and **B** are not equal in length.

b) If the attachment point on the tree is located 5 m above the ground, and the distance between the foot of the tree and the anchor point of Rope **A** is 3 m, what is the distance between the foot of the tree and the anchor point of Rope **B**?

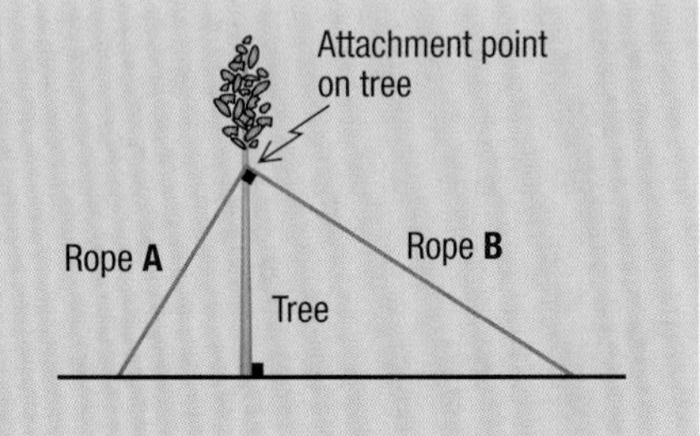

Situation ③

Ropes **A** and **B** are not equal in length.
Rope **A** measures 13 m and the distance between its anchor point and the foot of the tree is 12 m.

c) How high above the ground is the attachment point for these ropes located?

d) What is the distance between the foot of the tree and the anchor point of Rope **B**?

e) What is the length of Rope **B**?

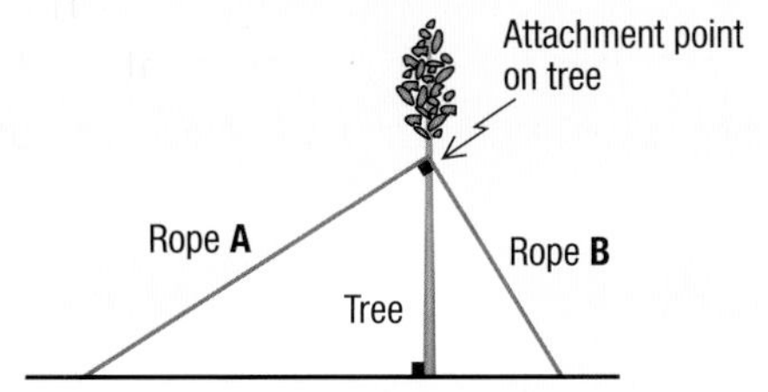

15 CAMPER VAN A company specializes in manufacturing modified camper vans similar to the one shown in the adjacent picture. Considering that when the roof is closed, the interior headroom in the camper van is 1.37 m, determine the maximum headroom when the roof is open.

16 The left-side view of a recycling container is shown in the adjacent diagram. Determine the volume of this container, considering that its base is a square with sides of 4 m each.

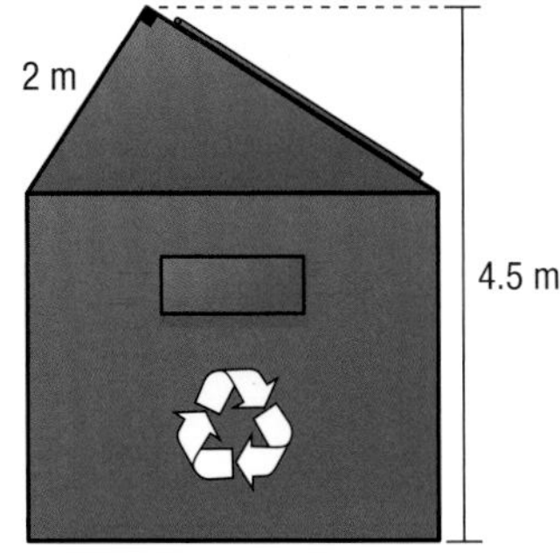

17 **COLLECTOR'S ITEM** The nickel shown below dates back to 1951 and is in the shape of a regular dodecagon. Determine:

a) the length of one side of this coin

b) the length of segment CD

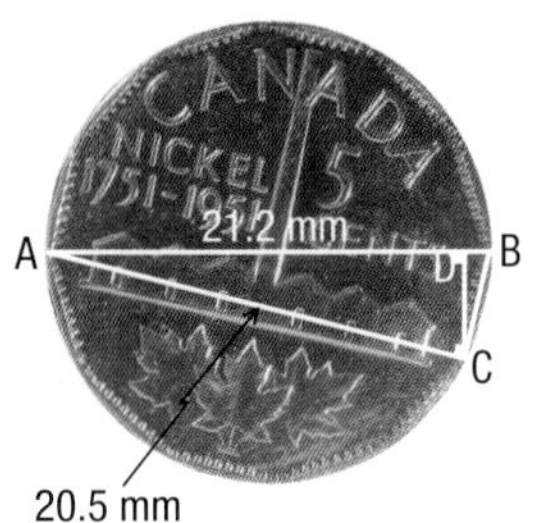

18 At a ski resort, there is a beginner's slope such as the one depicted in the adjacent illustration. If side AE forms a right angle with side AD of parallelogram ABCD, what is the distance covered when skiing down the slope?

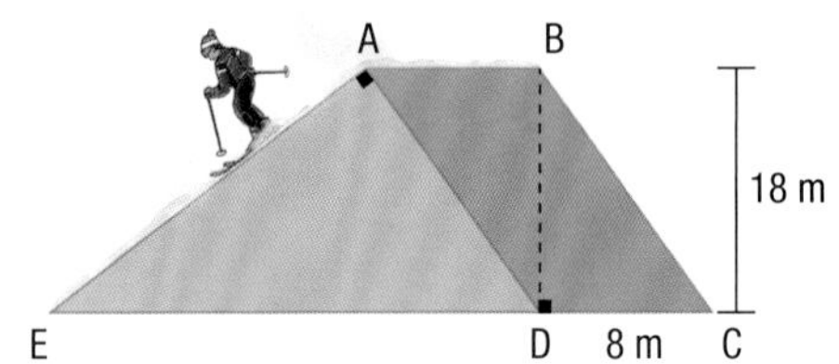

19 The adjacent diagram shows the explosion of a pyrotechnical device during a fireworks display. Using the measurements indicated, determine the height at which this device explodes.

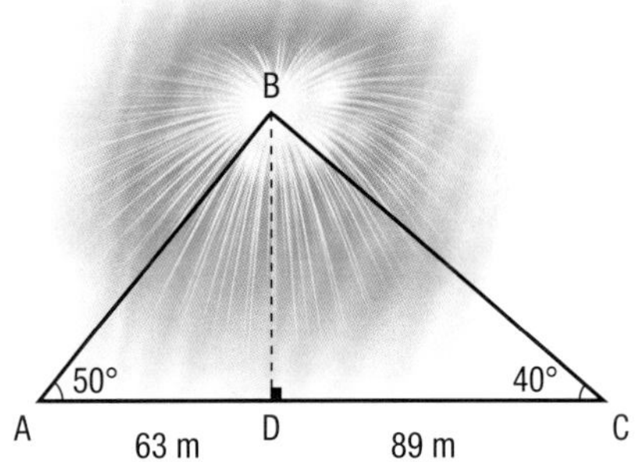

20 A grandstand is set up to accommodate spectators for the final of a soccer tournament. Loudspeakers are located at point F. How high are they from the ground?

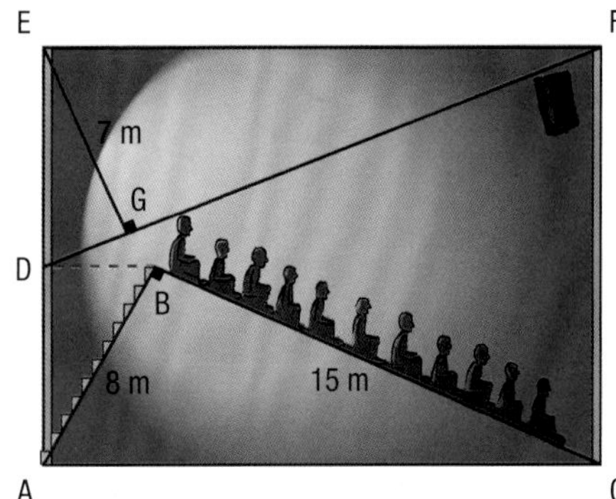

21 **KITE** The origins of kite flying go back nearly 3000 years. At that time, kites were made of bamboo and silk and were used to draw the attention of spirits. As well, the military used kites to scare the enemy.

On the kite pictured in the adjacent diagram, what is the length of the wooden spreader BD?

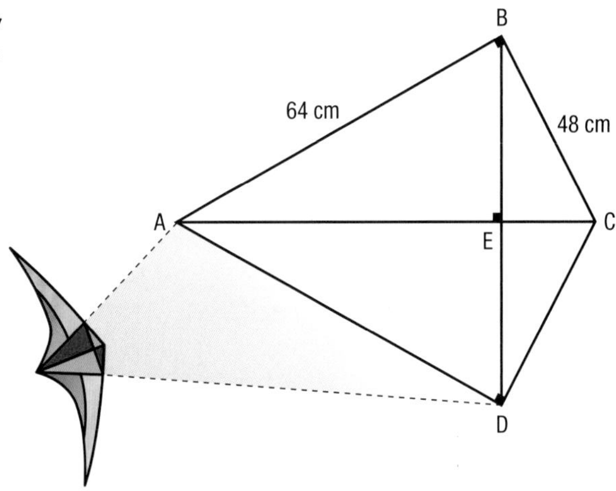

Chronicle of the

past

The Ionian School

Since the beginning of time, mankind has sought answers to questions regarding the origins, the causes and the fundamental basis of all things. To answer these questions, humans turned towards religion and philosophy. It is often said that philosophy is the mother of all sciences. In fact a number of philosophers were also mathematicians. In the 6th century BCE, three philosophers from the Greek Ionian town of Miletus founded the Ionian School.

Significant contributions to Greek culture came from Ionia where the foundations of philosophical and scientific thought were laid.

It seems that Thales was a merchant during the first part of his life. The fortune he acquired allowed him to study and travel. He lived in Egypt for a while where he studied mathematics and Egyptian astronomy.

Thales of Miletus (ca. 625 - ca. 546 BCE)

Thales is known, among other things, for the theorem that bears his name. The different configurations of this theorem enable us to establish ratios of length and to find certain missing measurements using proportions. Below are two possible configurations of this theorem:

Configuration ①

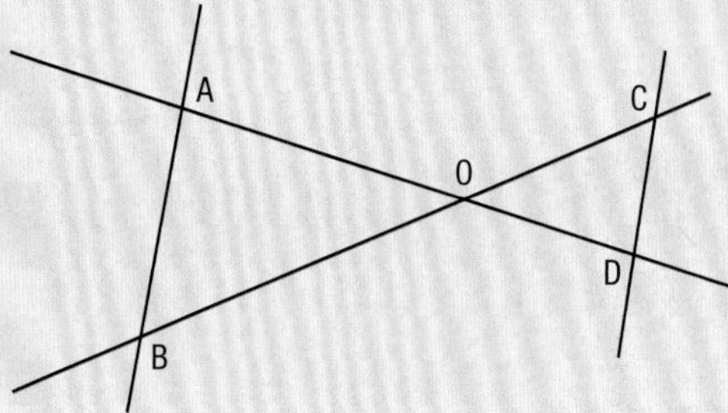

Configuration ②

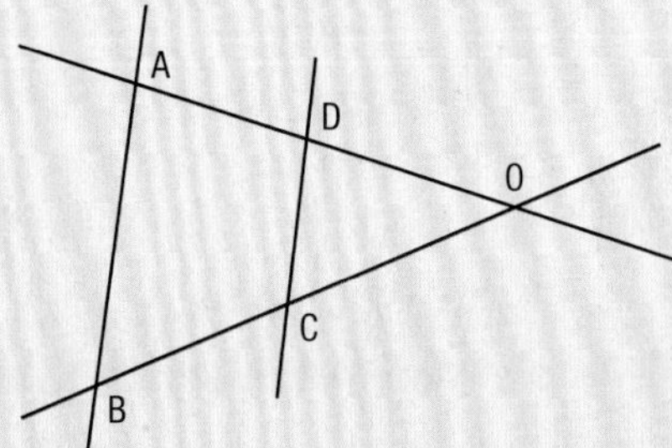

If lines AB and CD are parallel, then $\frac{m\ \overline{AO}}{m\ \overline{DO}} = \frac{m\ \overline{BO}}{m\ \overline{CO}}$.

Thales of Miletus was also a physicist and astronomer. He is said to be the first to assert that the Moon reflected the Sun's light. He became famous by predicting a solar eclipse which took place in 585 BCE.

Anaximander of Miletus (ca. 610 - ca. 547 BCE)

Anaximander, a student of Thales, later taught Pythagoras. Anaximander's interest in cosmology made him one of the fathers of astronomy, and this same interest led him to introduce the gnomon in Greece. Anaximander had gnomons built in Lacedemone (Sparta) to indicate solstices and equinoxes.

The shadow reflected in segment AD is generated at noon, solar time, during the summer solstice, normally on June 21. The shadow reflected in segment AC is generated by the gnomon at noon, solar time, during the winter solstice, normally on December 21.

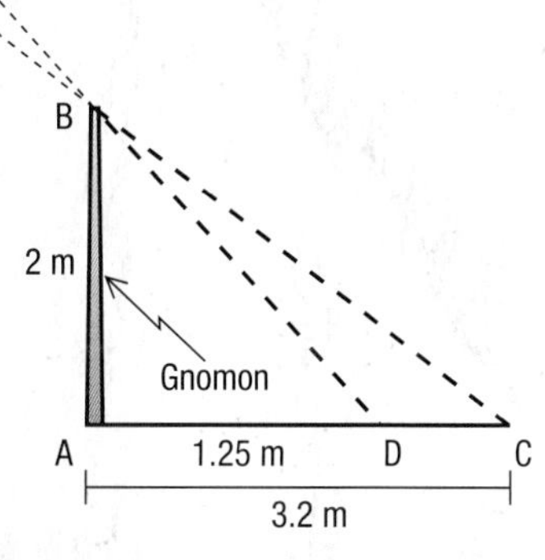

A lunar crater was named in honour of Anaximander.

Anaximene (ca. 585 - ca. 525 BCE)

Anaximene was one of the last disciples of the Ionian School.
Like Anaximander, there are almost no written references of his work, and his contribution to mathematics is uncertain. Anaximene limited himself to vulgarizing the theories of its predecessors. He perfected the gnomon, explained rainbows as light hitting on condensed air (water) and was an inspirationto the Pythagoreans.

1. According to Thales' Configuration ① theorem, if m $\overline{AO}$ = 3.9 cm, m $\overline{DO}$ = 2.4 cm and m $\overline{CO}$ = 2.9 cm, determine the length of $\overline{BO}$.

2. By using Thales' Configuration ② theorem:

a) Show that $\triangle ABO \sim \triangle DCO$.

b) Can you state that $\frac{m\overline{AO}}{m\overline{DO}} = \frac{m\overline{AB}}{m\overline{DC}}$? Explain your answer.

3. By using Anaximander's gnomon diagram, show that triangles ABC and ADB are similar.

In the workplace Civil engineering technicians

The profession

A civil engineering technician's work is related to soils, materials, structures and infrastructures. For instance, his or her job might involve producing technical statements and collecting measurements in anticipation of building new roads or new structures, and then generating computer representations of this information. On some projects, the civil engineering technician can be found supervising the work on the construction site and monitoring compliance with the requirements set at the design stage.

Structural design

When designing construction plans, a technician takes care to choose appropriate materials for the project. For instance, steel trusses and wooden beams are often used in the construction of certain types of buildings or in bridge design.

Figure ①: Wooden beam

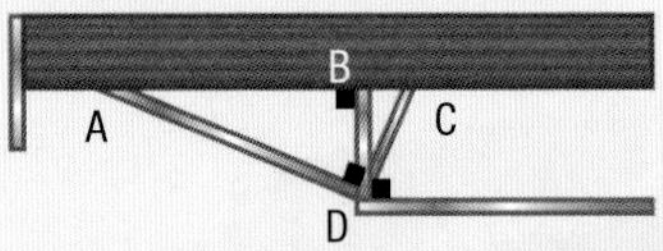

Figure ②: Steel truss

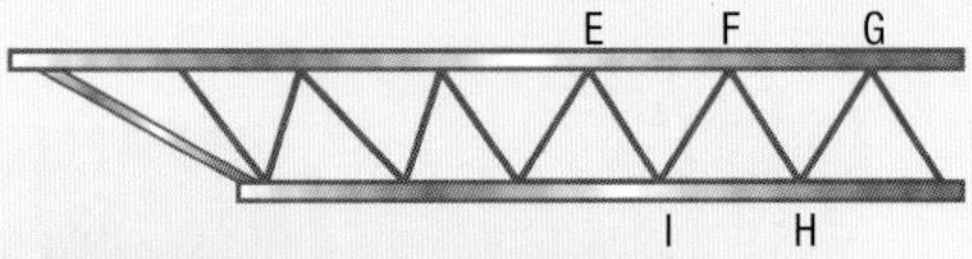

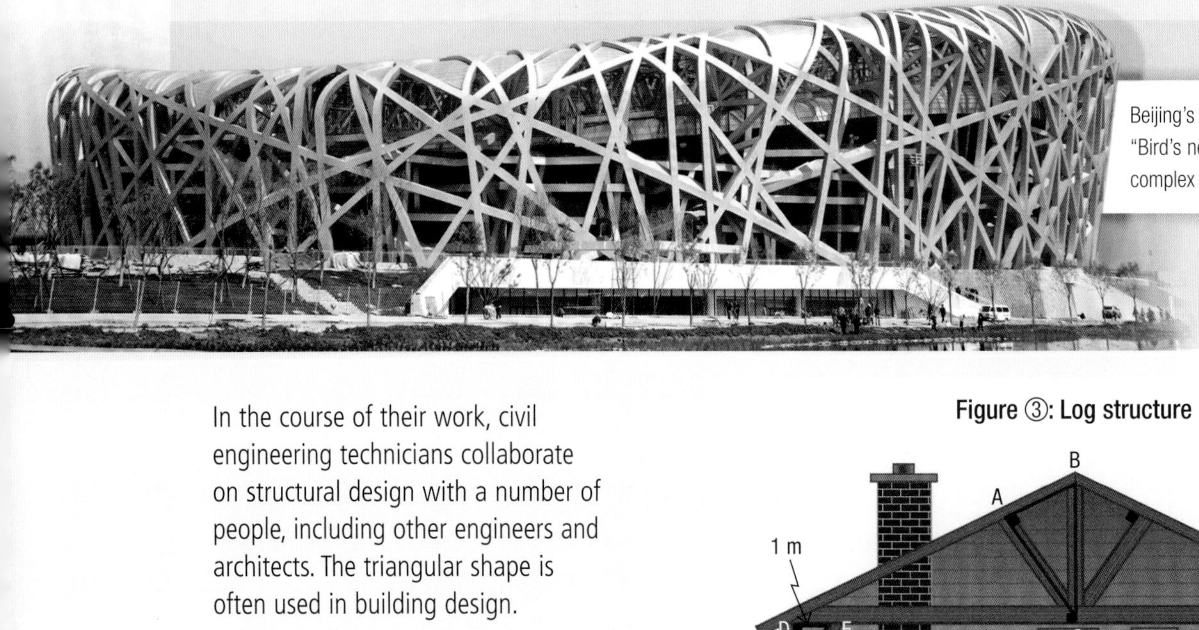

Beijing's Olympic Stadium, called the "Bird's nest," is an example of a highly complex structure.

In the course of their work, civil engineering technicians collaborate on structural design with a number of people, including other engineers and architects. The triangular shape is often used in building design.

The adjacent diagram illustrates the roof structure of a log cabin.

Figure ③: Log structure

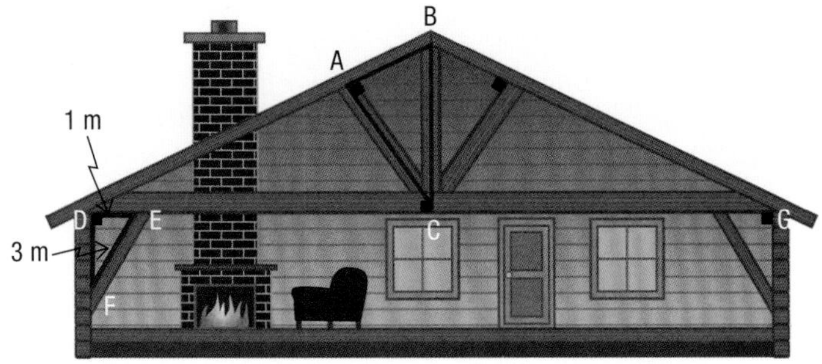

Infrastructure

Civil engineering technicians participate in the preparation of blueprints and specifications for improvements or construction projects on waterwork and sewage infrastructures. For instance, they might determine the position of underground pipes so as to ensure the adequate flow of waste water. In this type of project, the technician sometimes use a laser and target so that the pipe is placed with the desired inclination.

Figure ④: Infrastructure of a street

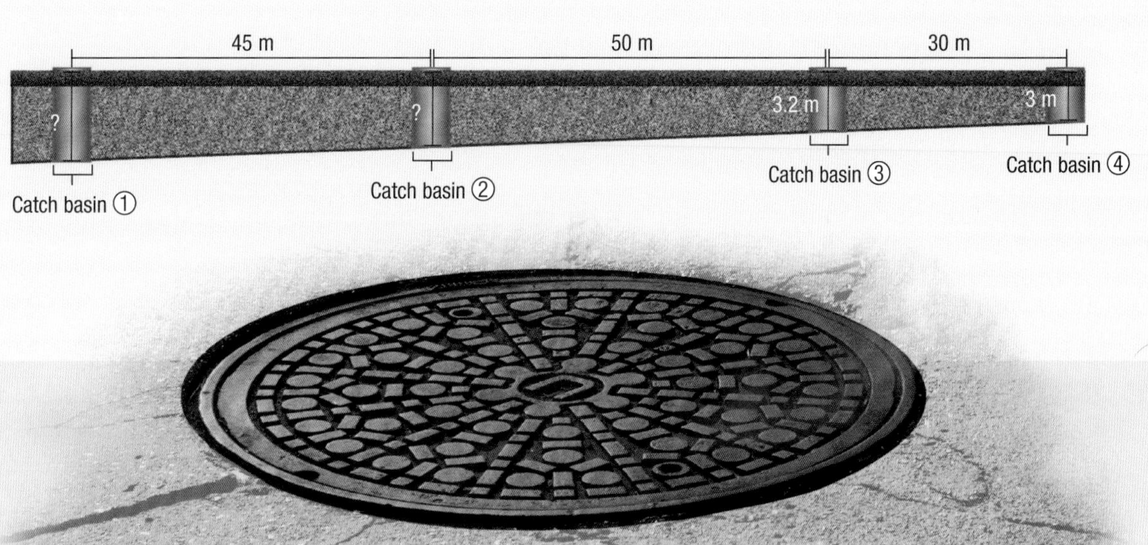

1. In Figure ①, m $\overline{BD}$ = 30 cm and m $\overline{BC}$ = 12 cm. Determine the length of the steel rod AD.

2. In Figure ②, considering that $\overline{EI}$ and $\overline{FH}$ are parallel, what can you say about triangles EFI and FGH?

3. In Figure ③, triangles ABC and DEF are similar. Determine the length of the beam corresponding to segment AC if the beam represented by segment DG measures 10 m.

4. In Figure ④, determine the depth of catch basins ① and ②.

overview

1 Are the polygons identified below necessarily congruent? For each case, explain your answer with the help of a drawing.

a) two squares

b) two squares having one common side

c) two rectangles whose corresponding diagonals are congruent

d) two rectangles with equal perimeters

e) two parallelograms where all the corresponding angles are congruent

2 Indicate whether the following statements are true or false. In the case of a false statement, provide a counter-example.

a) Two rhombuses with the same perimeter are congruent.

b) Two congruent rhombuses have the same area.

c) If the corresponding diagonals of two rhombuses are congruent, then the rhombuses are congruent.

d) Rhombuses are congruent if their corresponding angles are congruent.

e) Two right triangles are congruent if their hypotenuses are congruent.

3 In the adjacent diagram, AB // CD // EF and GH // IJ // KL.

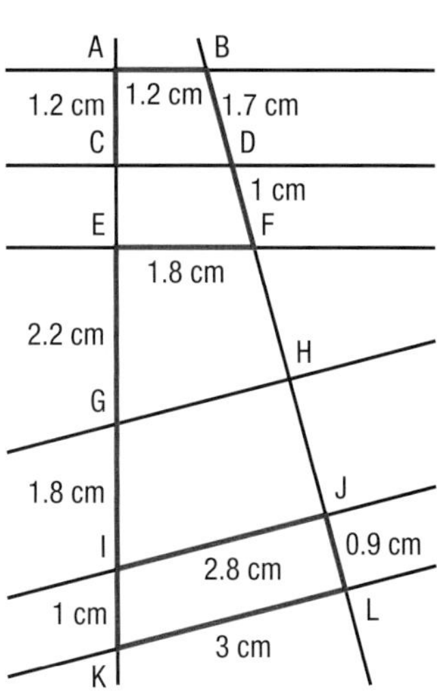

a) Determine:
1) m $\overline{CE}$
2) m $\overline{HJ}$
3) m $\overline{CD}$
4) m $\overline{GH}$

b) Is it possible to calculate m $\overline{FH}$? Explain your answer.

4 In each case, determine the value of m that would result in a perfect square trinomial.

a) $mx^2 + 60x + 225$

b) $\frac{25}{4}a^2 - ma + 64$

c) $400x^2 + 1000x + m$

d) $\frac{4}{9}b^2 - mb + 16$

e) $3x^2 + 6x + m$

f) $(100x)^2 + mx + 100^2$

5 Determine the simplified algebraic expression that represents the ratio between the volume of the sphere and the area of the sphere as shown in the adjacent illustration.

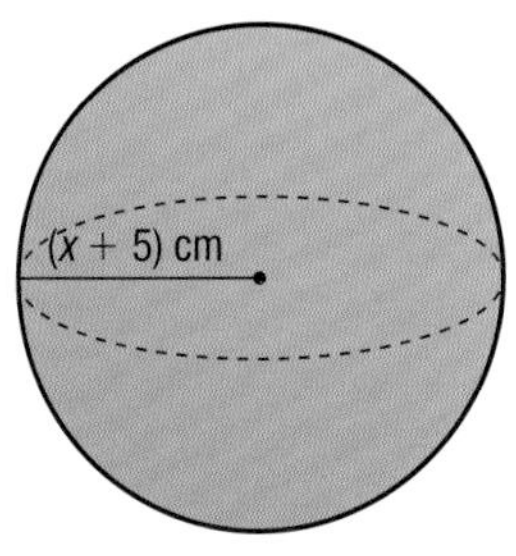

6 In each case, indicate whether the two triangles are congruent, similar or neither. If they are congruent or similar, provide the geometric statement that justifies your claim.

a)

3.24 cm

3.84 cm

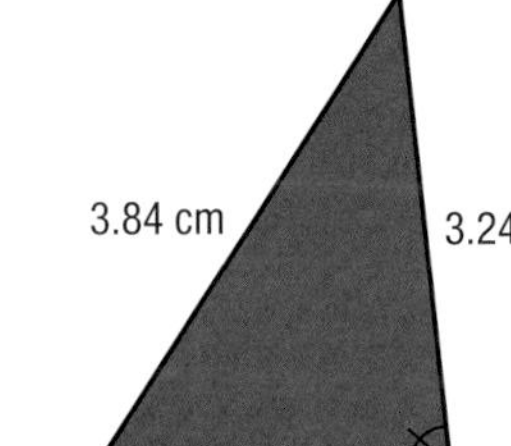

b)

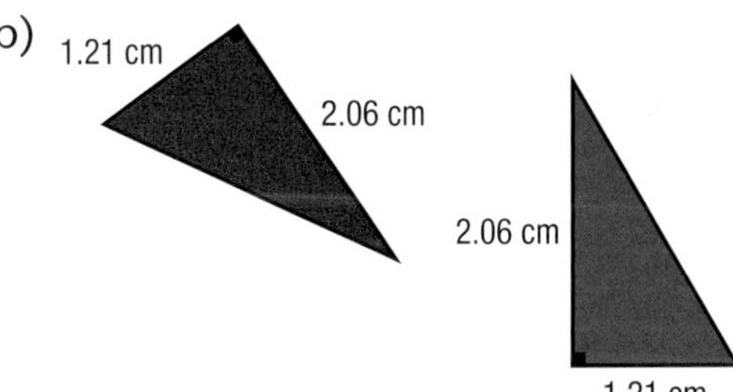

c)

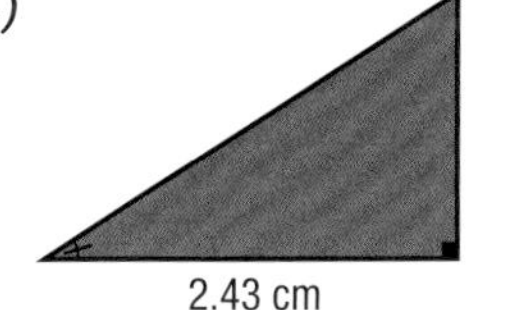

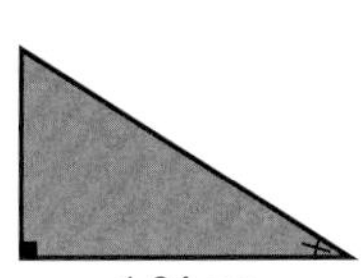

d)

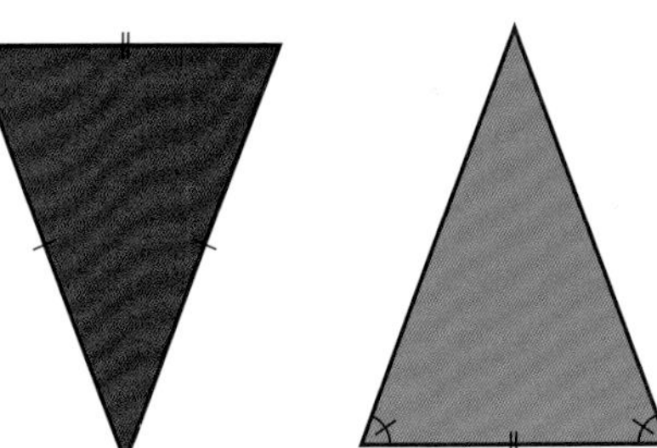

e)

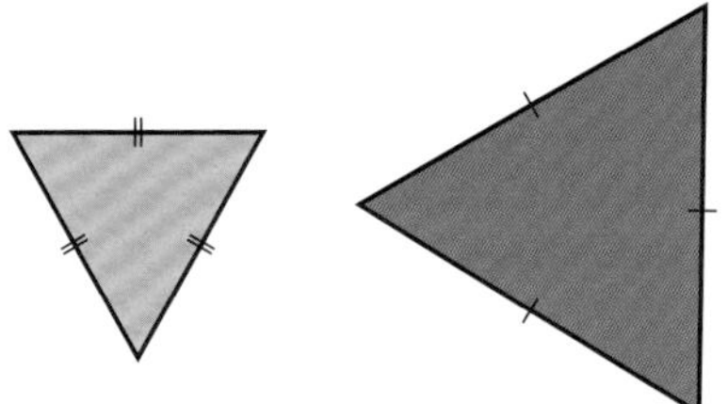

f)

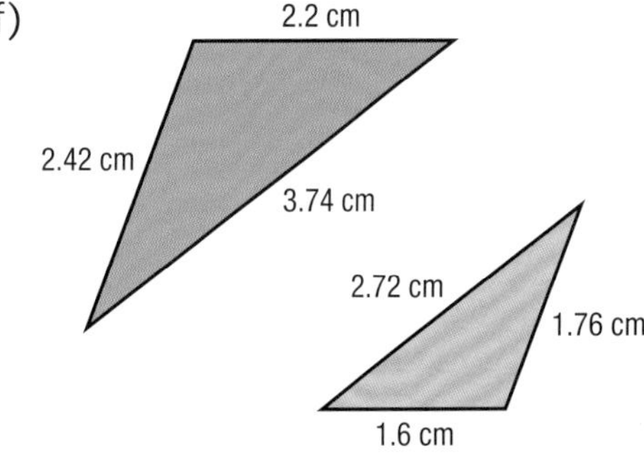

7 For each pair of triangles below, provide the geometric statement that allows you to state that the triangles are congruent.

a)

b)

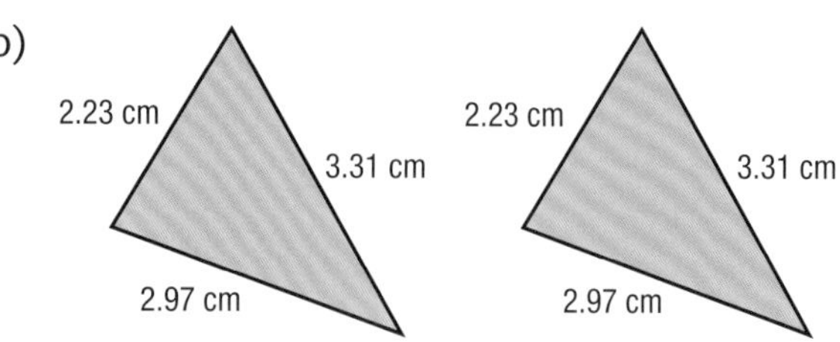

c)

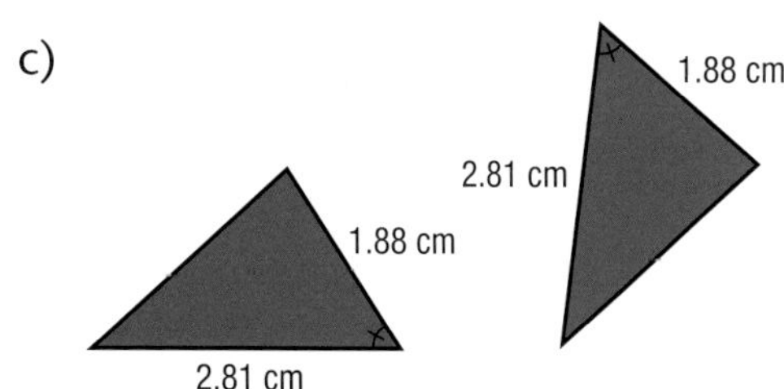

d)

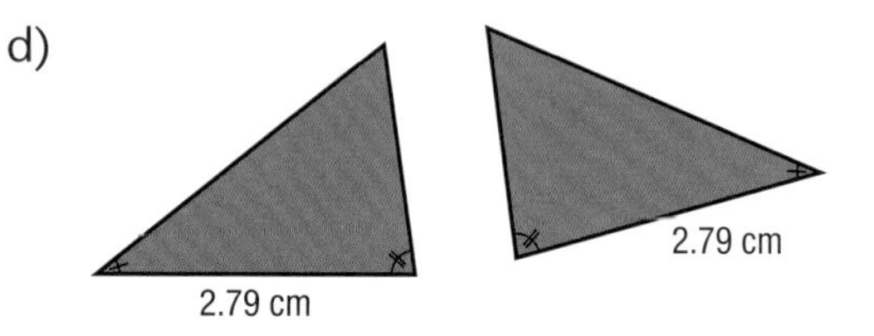

8 Factor the following polynomials.

a) $ax + ay + bx + by$

b) $3x^4 - 6x^3$

c) $4xy + 5x + 12y + 15$

d) $25a^8 - 64b^4$

e) $bc - 2b + 3c - 6$

f) $2z^2 - 8z + 8$

g) $4ax + 2ay - 6x - 3y$

h) $7a^2x^2 + xy - 14a^3xy - 2ay^2$

i) $(x - 1)^2 + 3(x - 1)$

j) $12b^2 - 4ab - 3bx^2 + ax^2$

k) $2m^2 - 4m + 2$

l) $3z^5 - 6z^4 - 5z^3 + 10z^2$

m) $(x - y)^2 - 169$

n) $(m^2 - 2mn + n^2) - 1$

9 The area of a rhombus is $(5xy + 2y - 20x - 8)$ dm^2.

a) Determine an algebraic expression that represents the length of each diagonal in this rhombus.

b) What are the lengths of the diagonals of this rhombus if $x = 1$ and $y = 7$?

c) For what values of variables x and y can this rhombus exist?

10 There are two camp sites on an island. A dock has been built on the shore so that the path connecting Site **A**, the dock and Site **B** is as short as possible. What distance would a person cover walking from Site **A** to Site **B** along this path?

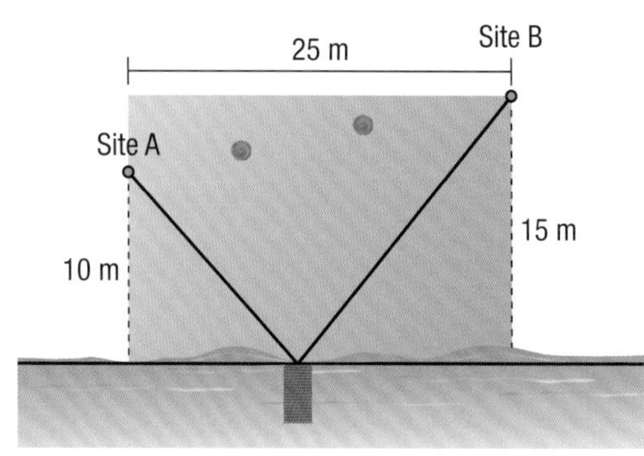

11 A company's new logo has been painted on the semi-trailer shown below and it is in the shape of a right prism with a rectangular base. What is the capacity of this semi-trailer?

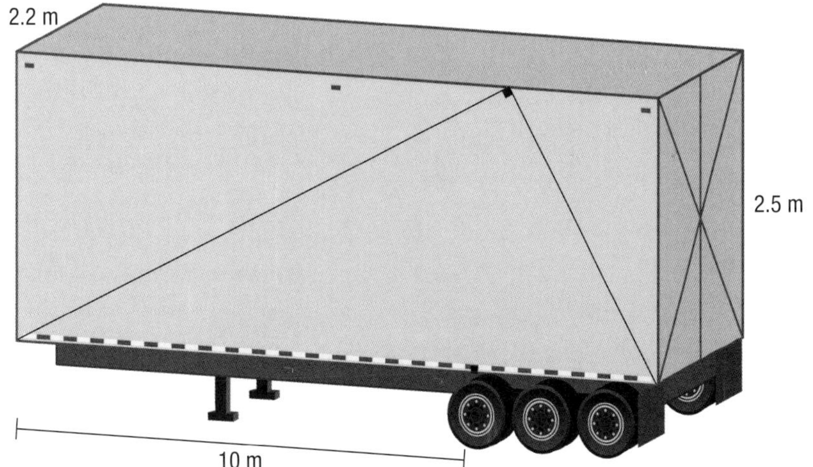

12 Using the two adjacent rectangles, determine a simplified algebraic expression that represents the ratio between:

a) the perimeter of Rectangle **A** and that of Rectangle **B**

b) the area of Rectangle **B** and that of Rectangle **A**

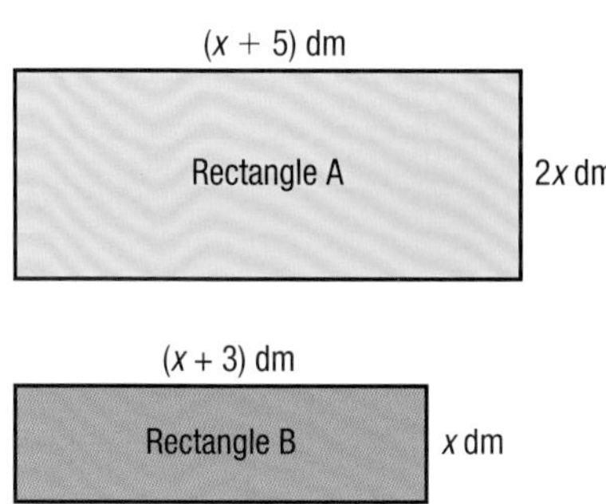

13 For each pair of triangles below, calculate the value of x.

a)

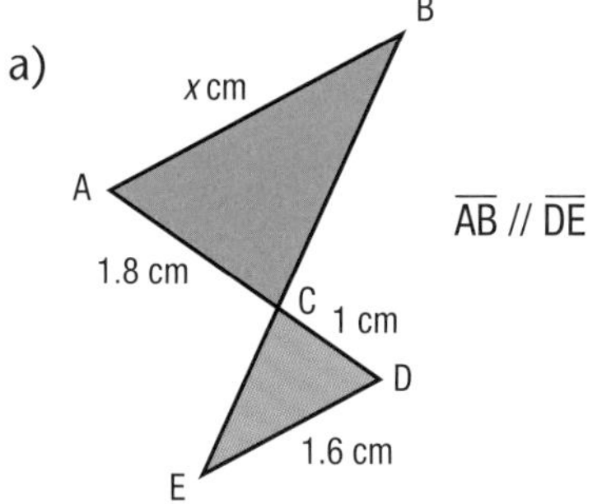

b)

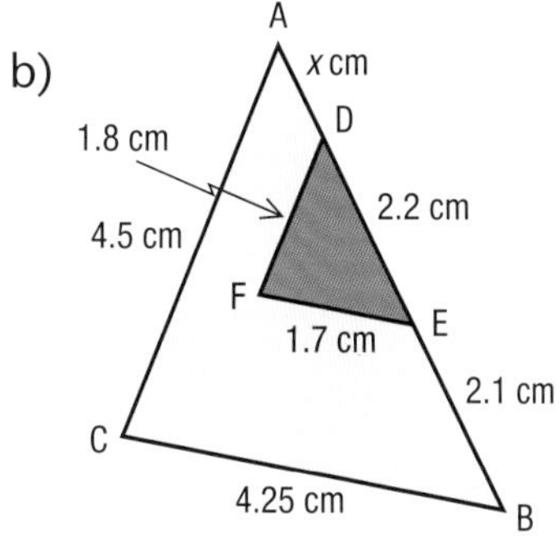

c)

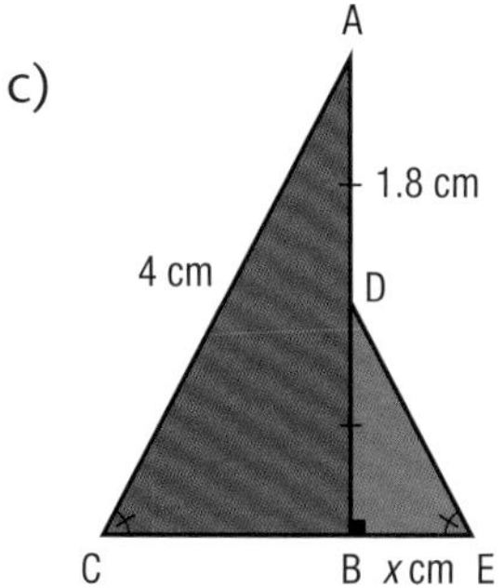

d)

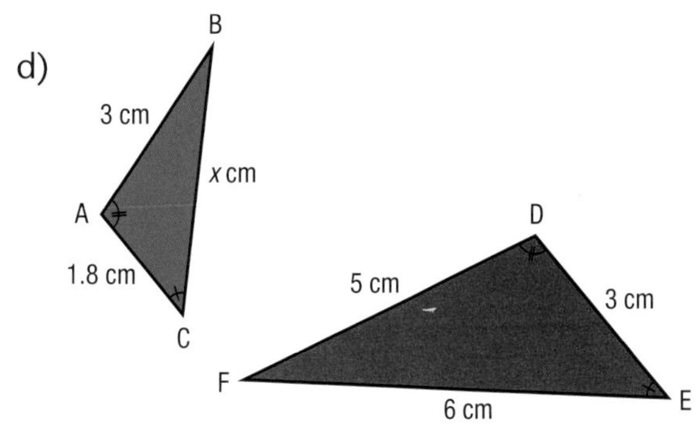

14 The front of the cottage shown in the illustration below is in the shape of an isosceles triangle whose base measures 8 m. The width of the first floor is 6 m. What distance separates the two floors of this cottage?

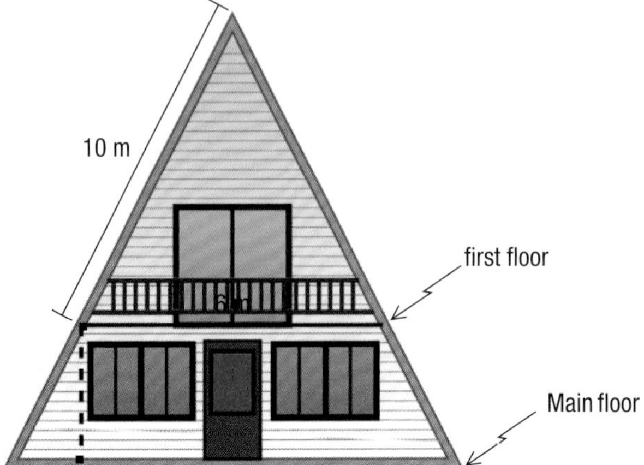

15 The algebraic expressions below correspond to the area of the strip of wallpaper ABEF and the strip of wallpaper BCDE respectively. Suggest an algebraic expression that would represent the perimeter of strip ACDF.

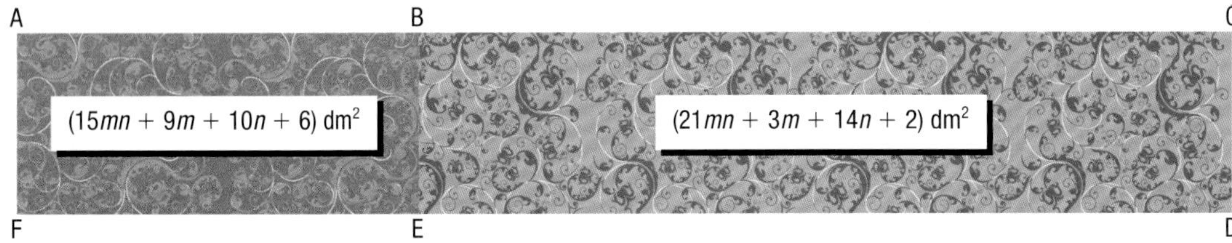

16 A student uses a 30 cm ruler and the shadows produced by the Sun to determine the height of a school. This activity is represented in the adjacent illustration. How tall is this school?

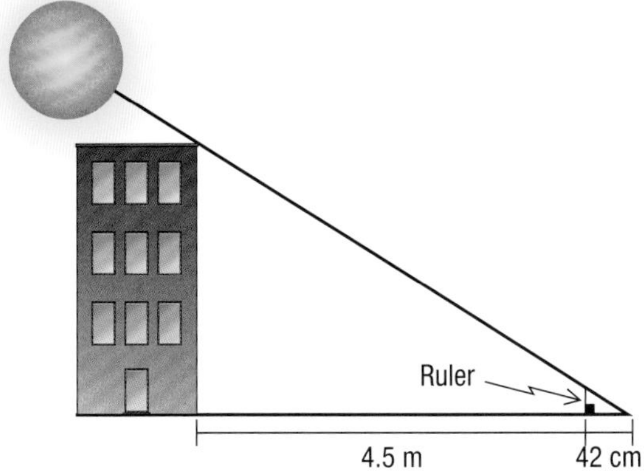

17 The adjacent diagram represents a sports team's flag. The total area of this flag is represented by the expression $(4ab + 8a + 4b + 8)$ dm^2.
What is the area of the green triangle?

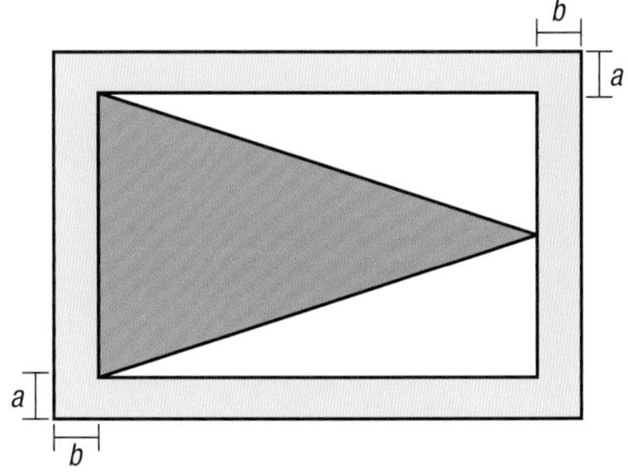

18 In the figure below, $\overline{BE} // \overline{CD}$. Determine the length of segments AB, BE and CD.

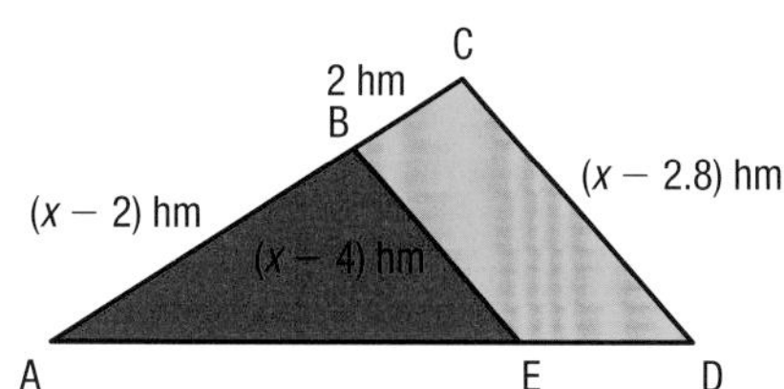

19 **THE PINHOLE CAMERA** The principle of the pinhole camera may have helped Joseph Nicéphore Niépce take the very first photograph in history. The principle is as follows: light enters through a tiny hole pierced in a box and projects an inverted image of the targeted object on a strip of film placed inside the box. What is the height of the image of the person pictured on the film as shown below?

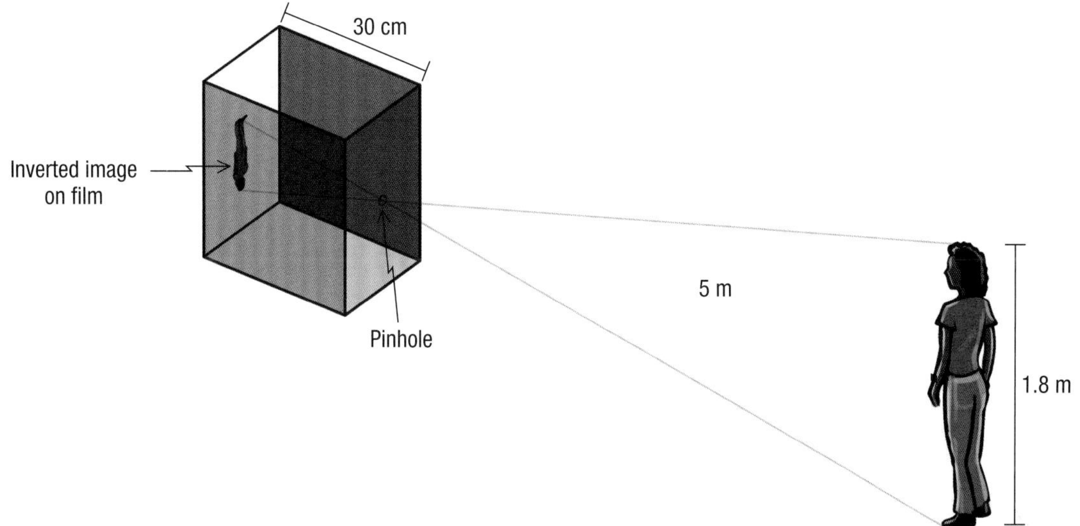

20 In the adjacent figure, m $\overline{AB}$ = m $\overline{OC}$ and m $\overline{BC}$ = 45 cm. Determine the length of segment DO.

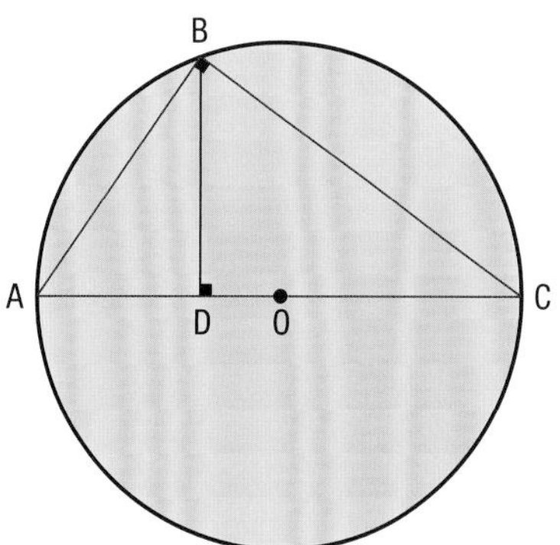

21 Following is some information about the stained glass window represented in the adjacent illustration:

- The total area of the stained glass window corresponds to the expression $(4a^2 + 16a + 16)$ dm^2.
- The blue border is a dm wide.

What is the area of the orange triangle?

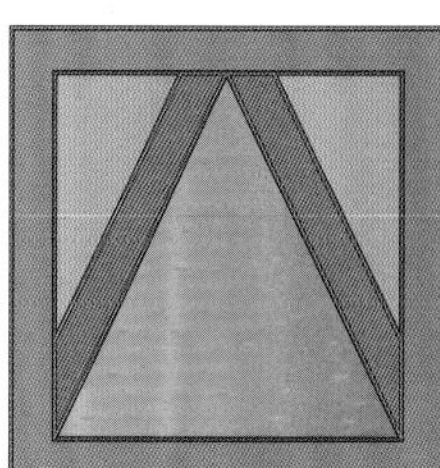

22 Express the area of the recordable area of the compact disc shown in the adjacent illustration as a product of irreducible factors.

23 It was by comparing the ratio of the distances covered by two falling objects with the ratio of their corresponding times that Galileo succeeded in determining whether one object fell faster than the other. If a moving object covers a distance represented by the expression $(x^3 - x^2y + 3x - 3y)$ m, determine an algebraic expression that represents:

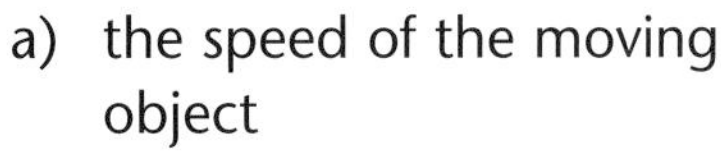

a) the speed of the moving object

b) the time of movement

Galileo (1564-1642) was an Italian physicist and astronomer. Among other things, he is famous for defending Copernicus' idea that the Earth is not at the centre of the universe.

24 In the figure below, segment BE is a median of triangle ABC.

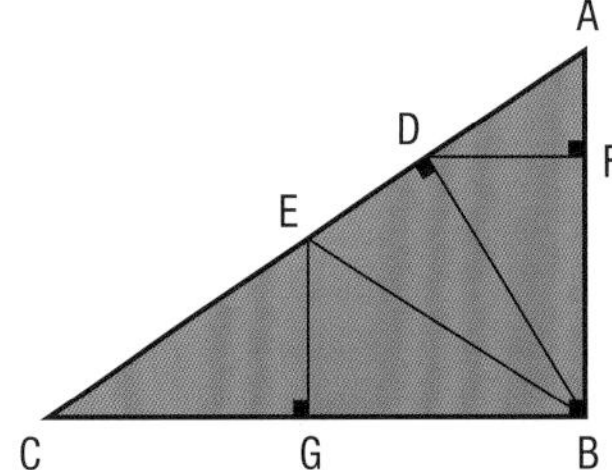

a) If possible, identify a triangle that is similar to triangle:

1) ABC 2) EBG 3) BED

b) If possible, identify a triangle that is congruent to triangle:

1) ECG 2) FDB

25 An individual wants to use outdoor underground electric wires to connect his shed and pool filter to the back of his house. The diagram below illustrates this situation. Determine the shortest length of wire that must be purchased.

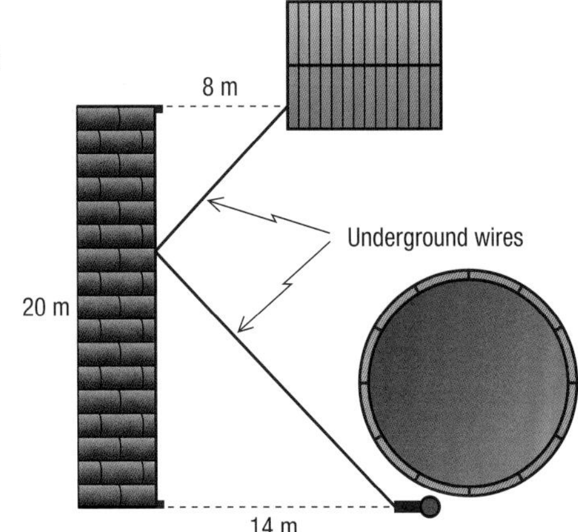

26 **THE ICE STORM** In January 1998, Québec faced an ice storm that was one of its greatest meteorological disasters ever. Dozens of electrical towers like the one shown in the adjacent diagram collapsed under the weight of ice.

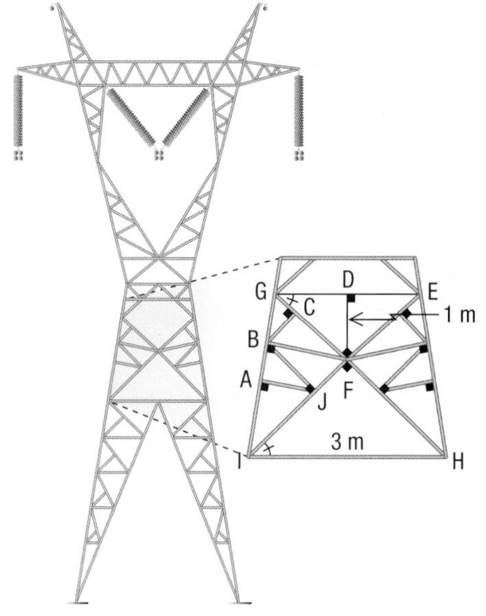

Based on the information presented in the adjacent diagram, and considering that the horizontal segments are parallel to each other, determine the length of segment:

a) BC

b) AJ

27 The perimeter of the adjacent regular octagon is 64 cm.

a) Prove that $\Delta BRA \cong \Delta DOP$.

b) Prove that $\Delta SRQ \sim \Delta GFP$.

c) Determine:

1) m $\overline{BR}$ 2) m $\overline{CS}$

3) m $\overline{GQ}$ 4) m $\overline{EC}$

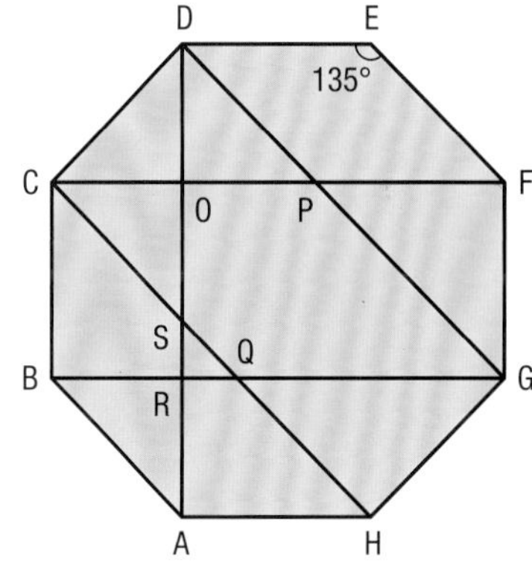

28 Three right triangles have been assembled so as to form the adjacent rectangle. What geometric statement allows you to state that:

a) $\Delta ACD \cong \Delta CAB$?

b) $\Delta ACD \sim \Delta BAE$?

c) Calculate:

1) m $\overline{CE}$ 2) m $\overline{AE}$ 3) m $\overline{BE}$

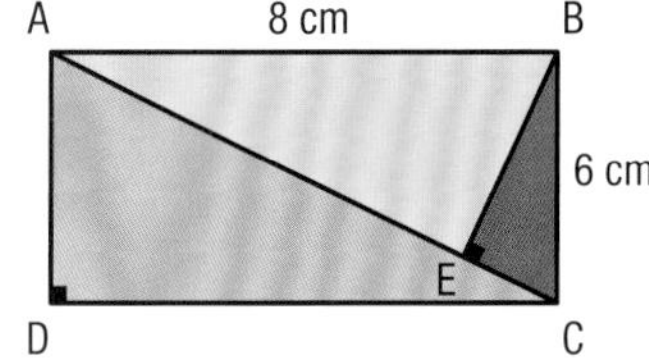

29 As seen on the right, rectangle A'B'C'D' is the image of rectangle ABCD by a dilatation with centre P and a ratio of 2. If the area of the projected image corresponds to the expression $(60xy + 24y + 20x + 8)$ cm, determine an algebraic expression that represents the perimeter of the initial rectangle.

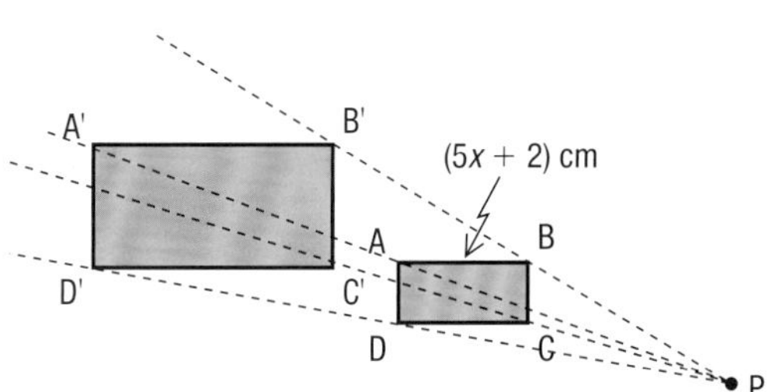

bank of problems

30 As shown in the adjacent diagram, there are two antennas located near a radio station on Simard Street. Determine the distance between the two antennas considering that the technicians used the minimum length of cable, 100 m, to connect the two antennas to the radio station.

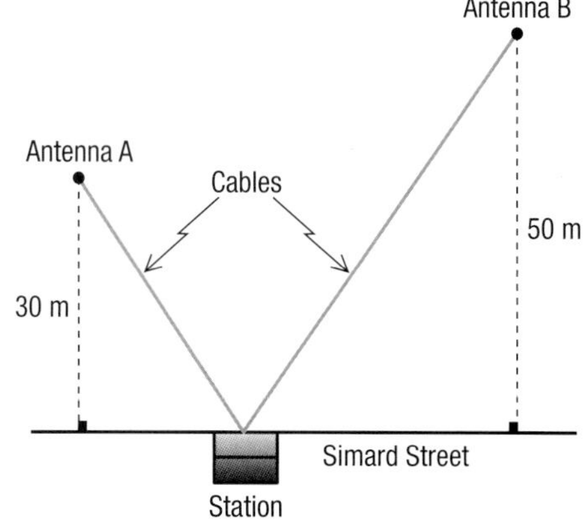

31 The diagram below represents the aluminium structure of a hang-glider. Considering that points B, F and H are respectively the midpoints of segments AD, DG, and AG, prove that the length of rod CI is 1.38 m.

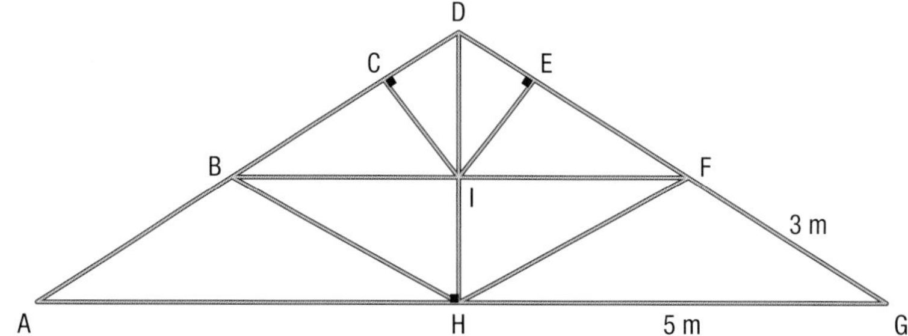

32 During an evening cross-country training session, Thomas runs at a speed of 14.4 km/h. He passes a lamppost, and from that moment on he observes the length of his shadow on the ground. Five seconds later, he notices that the length of his shadow matches his actual height. If Thomas is 1.8 m tall, what is the height of the lamppost?

33 A company specializing in window manufacturing produces a model with decorative strips. The total area of this window, excluding the frame, is $(99xy + 81x + 77y + 63)$ cm^2. Express the area of the blue section using a product of factors.

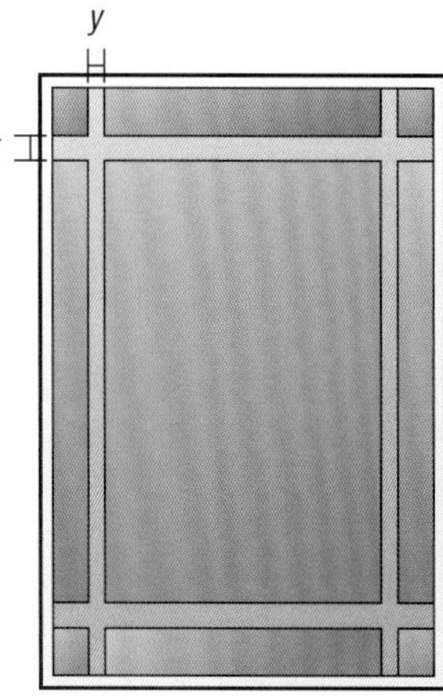

34 The figures below illustrate the model for a wheel rim at its design stage and as the finished product. In the blueprint for the rim, ABCDE is a regular pentagon.

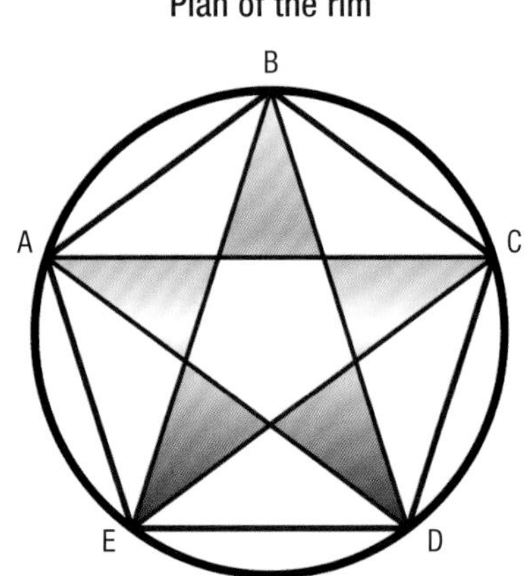

Using the blueprint, prove that diagonals AC and BE are congruent.

35 An antenna is placed in the centre of the roof of both Building **A** and Building **B** shown below. Each building is in the shape of a regular square-based prism. Determine the length of one side of the base of each building.

Some antennas are protected by a waterproof shelter, called a radome.

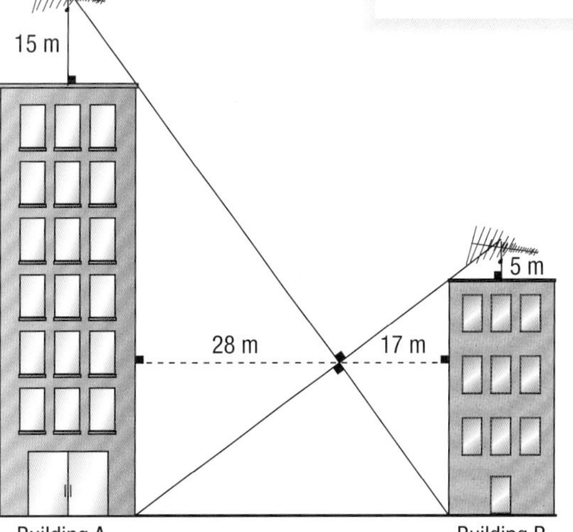

36 As illustrated in the adjacent diagram, a metal blade spins around axis BD. What is the volume of the right circular cone formed by the rotation of this blade? Explain your procedure.

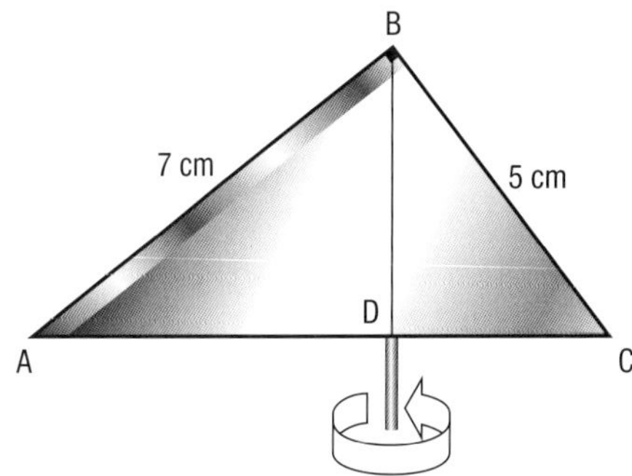

VISION 3

From lines to systems of equations

How do you determine the distance travelled by a hummingbird during its migration? How do you search for the company that provides the best service? How do you calculate the cost of connecting a municipality to a gas pipeline? How do you compare the fuel consumption of two airplanes? In "Vision 3," you will discover various concepts linked to analytic geometry. You will learn to calculate the distance between two points and to find the coordinates of a point of division that divides a segment into a given ratio. You will also explain situations using systems of equations and inequalities.

Arithmetic and algebra	Geometry	Statistics	Probability
• Solving systems of first-degree equations in two variables: substitution and elimination methods • First-degree inequalities in two variables	• Distance between two points • Point of division • Slope of a segment or a line • Equation of a line • Parallel lines, perpendicular lines, and perpendicular bisectors		

LEARNING AND EVALUATION SITUATIONS — GPS [Global Positioning System] 212

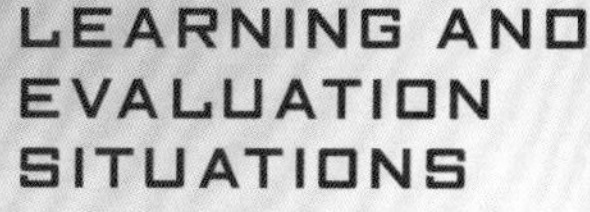

Chronicle of the past — René Descartes 190

In the workplace — Power linepersons 192

REVISION 3

PRIOR LEARNING 1 An eco-friendly choice

To protect the environment, more and more people are opting for hybrid cars as a means of transportation. To meet the growing demand, many car manufacturers have had to increase their production. Below is some information on the finances of one of these companies:

A hybrid vehicle is one that uses several different sources of energy to operate. For example, the vehicle might use electricity and gas.

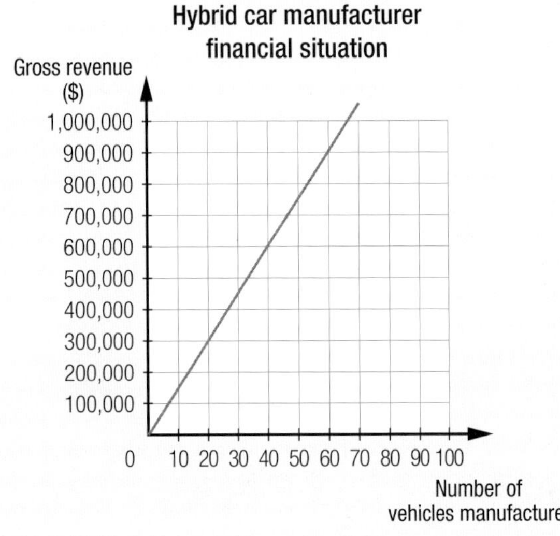

Hybrid car manufacturer production cost

Number of vehicles manufactured	Production cost ($)
0	6,000
30	453,000
50	751,000
70	1,049,000

a. Determine the equation that describes the relationship between:

1) the number of vehicles manufactured and the gross revenue

2) the number of vehicles manufactured and the production cost

b. How many vehicles will this manufacturer have to build to make a profit?

More and more cities are starting to use hybrid buses like the one shown in the adjacent illustration for their public transportation services.

PRIOR LEARNING 2 Performance bonuses

Many companies motivate their employees by paying them a lump sum bonus in addition to their regular salary. Following is some information on the monthly amounts received by an employee who was paid a total of $3,050 in lump sum bonuses from January to June:

- The amount received in February is $550 less than double the amount received in January.
- The amount received in March is $100 more than the amount received in January.
- The amount received in April is the same as that received in February.
- The amount received in May is the same as the amount received in January.
- The amount received in June is $50 more than the amount received in January.

a. Assuming a is the amount received in January, for each of the next five months determine the algebraic expression that represents the amount received using this variable.

b. What are the lump sum bonuses this employee received each month?

SOLVING SYSTEMS OF EQUATIONS

Different strategies can be used to solve a system of first-degree or linear equations in two variables, in other words, to find the values of the variables that simultaneously satisfy both equations.

Graphical representation

In a graphical representation, the coordinates of the intersection point of two lines represent the solution to the system of equations. Graphical representation often only provides an approximation of the solution.

E.g. $y = -3x + 10$
$y = 4x - 4$

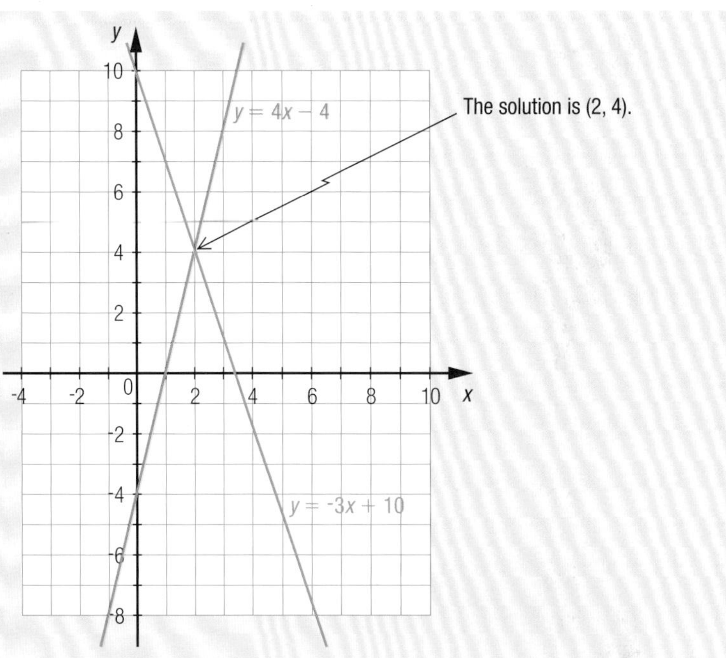

Table of values

It is possible to solve a system of equations by constructing a table of values. Find the value of the independent variable for which the dependent variables are equal.

E.g. $y = 2x + 1$
$y = -4x + 7$

x	-2	-1	0	**1**	2	3	4
y	-3	-1	1	**3**	5	7	9
y	15	11	7	**3**	-1	-5	-9

The solution is (1, 3).

Comparison method

The comparison method allows you to solve a system of equations by comparing algebraic expressions when they are of the form $\begin{array}{l} y = a_1x + b_1 \\ y = a_2x + b_2 \end{array}$.

E.g. To solve the system $\begin{array}{l} y = -110x + 1900 \\ y = -150x + 2400 \end{array}$ using the comparison method, do the following:	
1. Compare the two algebraic expressions containing the variable that is not isolated.	$-110x + 1900 = -150x + 2400$
2. Solve the resulting equation.	$-110x + 1900 = -150x + 2400$ $40x = 500$ $x = 12.5$
3. Substitute the value obtained into one of the original equations to determine the value of the other variable.	$y = -150 \times \mathbf{12.5} + 2400$ $y = 525$ Therefore, the solution is (12.5, 525).
4. Validate the solution by substituting 12.5 for x and 525 for y in each of the original equations. $\mathbf{525} = -110 \times \mathbf{12.5} + 1900$ $\mathbf{525} = -150 \times \mathbf{12.5} + 2400$	

FIRST-DEGREE INEQUALITIES IN ONE VARIABLE

Finding the set of values that satisfies an inequality means that you have **solved** this inequality. Sometimes it is necessary to use inequalities to find the solution to a problem. Follow the procedure below.

1. Identify the unknowns.	E.g. The perimeter of a rectangular plot of land is at least 178 m. The length is 5 m more than triple its width. Calculate the possible dimensions of the plot. The unknowns are: • the width of the plot • the length of the plot
2. Represent each unknown quantity with a variable or an algebraic expression involving variables.	The width of the plot (in m): x The length of the plot (in m): $3x + 5$
3. Construct an inequality that represents the situation.	$2(x + 3x + 5) \geq 178$
4. Solve the inequality following the rules of inequality transformations.	$2(x + 3x + 5) \geq 178$ $2(4x + 5) \geq 178$ $8x + 10 \geq 178$ $8x \geq 168$ $x \geq 21$
5. State the solution taking the context into account.	You can deduce that the width of the plot must be at least 21 m. For example, the plot's dimensions could be 21 m by 68 m.

knowledge in action

1 Solve the following systems of equations.

a) $y = 30 - x$
$y = x + 2$

b) $y = {}^{-}7x + 10$
$y = x - 10$

c) $y = {}^{-}x$
$y = x + 1$

d) $y = 17 - 3x$
$y = {}^{-}x - 3$

e) $y = \frac{x}{2}$
$y = {}^{-}3x + 4$

f) $y = \frac{2x - 17}{25}$
$y = \frac{x - 6}{15}$

g) $y = 3.2x + 4.4$
$y = 6.4x - 2$

h) $y = -12.3x + 3$
$y = -0.3x - 9$

i) $y = \frac{6}{5}x + 8$
$y = \frac{5}{2}x + \frac{1}{5}$

2 In each case, find the solution to the system of equations.

a) System of equations ①

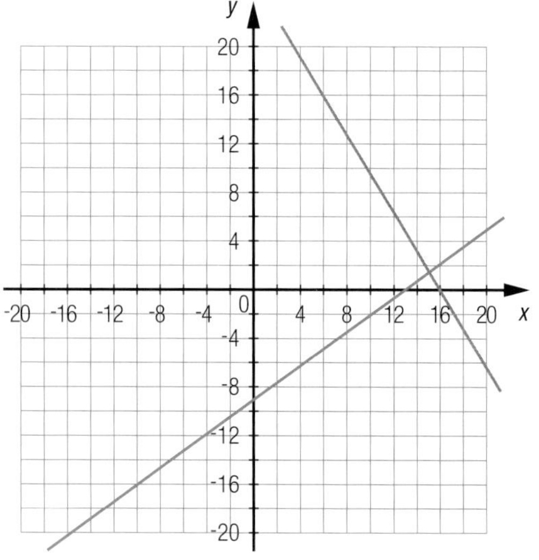

b) System of equations ②

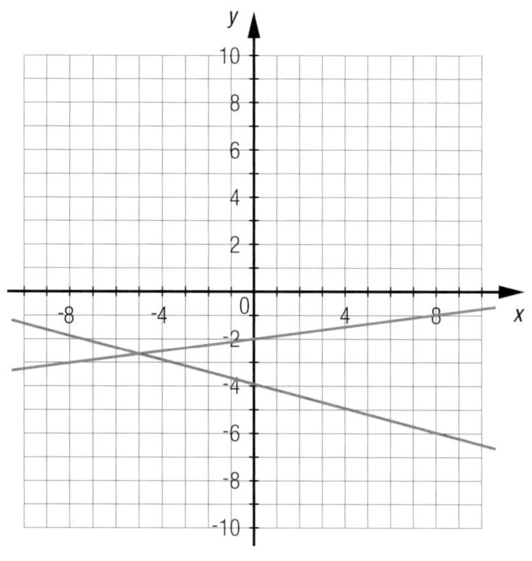

c) System of equations ③

x	-30	-15	0	30	60
y_1	160	130	100	40	-20
y_2	85	130	175	265	355

d) System of equations ④

x	-38	-36	-34	-32	-30
y_1	78	76	74	72	70
y_2	84	80	76	72	68

e) System of equations ⑤

x	0	5	10	15	20
y_1	15	115	215	315	415

x	2	12	22	32	42
y_2	80	530	980	1430	1880

f) System of equations ⑥

x	-20	-18	-16	-14	-12
y_1	512	461	410	359	308

x	-13	-7	10	19	24
y_2	43.5	22.5	-37	-68.5	-86

3 Express each of the following statements as an inequality.

a) The value of d is a minimum of 2.

b) The value of p is less than 19.

c) The value of a is at most 3.

d) The value of g is at least equal to 8.

e) The value of t is smaller than v.

f) The value of r is not greater than b.

g) Twice the value of c is greater than -6.

h) The value of s is at least 5 more than m.

4 Match the number lines in the left column with the inequalities in the right column.

	Number line		Inequality
A	7	**1**	$h \geq 7$
B	-7 7	**2**	$h < -7$
C	7	**3**	$h \geq -7$
D	-7 7	**4**	$h < 7$
E	-7	**5**	$-7 \leq h < 7$
F	-7	**6**	$7 > h > -7$

5 Solve the following inequalities.

a) $24 + 3x > 18$

b) $3a - 12 \leq -2a + 28$

c) $-(t - 1) + 7 \geq -11$

d) $-14 + 0.2b < 8$

e) $-3.4m - 7.2 \leq 7.08$

f) $18 \geq 36 - 3(c + 11)$

g) $-\frac{5n + 1}{2} < 9$

h) $\frac{2x - 4}{3} > -\frac{x - 2}{2}$

i) $\frac{4x - 5}{2} \geq 3$

j) $\frac{-x + 5}{4} \leq \frac{2x + 1}{3}$

k) $\frac{2}{x + 1} > 6$

l) $\frac{7x}{2x + 6} < 8$

6 At the airport, two planes are ready to take off. Plane **A** is carrying 1850 L of fuel and consumes 500 L/h. Plane **B** is carrying 1000 L of fuel and consumes 100 L/h. Plane **A** departs, and Plane **B** leaves 5 minutes later.

a) Identify the unknowns in this situation and represent them using different variables.

b) Represent the situation using a system of equations.

c) When will the two planes be carrying the same amount of fuel?

Reconnaissance aircraft are usually used to keep watch over enemy operations. For example, these aircraft patrol and sometimes photograph the airspace and waterways of restricted areas. Today, satellites have largely replaced these reconnaissance aircraft.

7 For each polygon below, measured in centimetres, find the value(s) of the variable x where:

1) the polygon exists
2) the perimeter of the polygon is greater than 100 m
3) the area of the polygon is less than or equal to 200 cm²

a) **Rectangle**

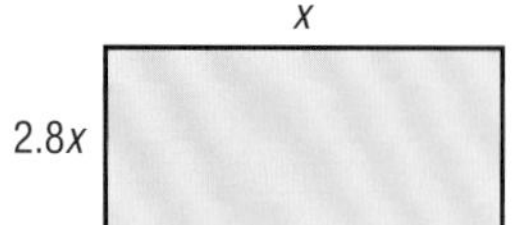

b) **Parallelogram**

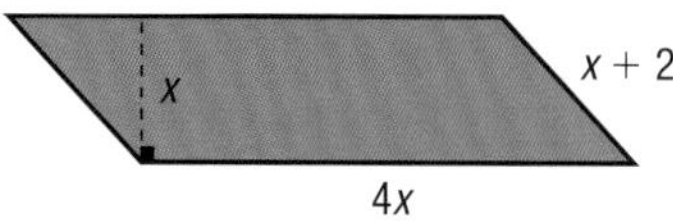

c) **Right triangle**

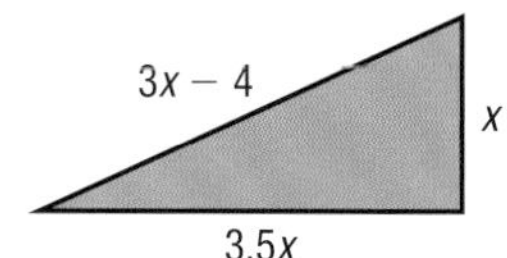

d) **Equilateral triangle**

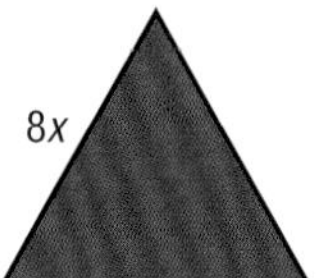

8 Malika and Ian take horseback riding lessons together. When their lesson is over, they decide to race their horses over a distance of 500 m. However, since Malika's horse is faster than Ian's, Ian will start 75 m ahead of Malika. Malika's horse gallops at a speed of 6.8 m/s while Ian's horse gallops at 5.3 m/s. Assume that these speeds are constant throughout the race.

a) Use a graphical representation to show the distance covered by:
   1) Ian and his horse as a function of time
   2) Malika and her horse as a function of time

b) Determine the distance from the endpoint at which the two riders will be side by side.

c) Calculate the time taken by:
   1) Ian and his horse to finish the race
   2) Malika and her horse to finish the race

At the 2008 Peking Olympic Games, Eric Lamaze won a Gold medal for individual show-jumping and a Silver in team show-jumping.

SECTION 3.1 Points and segments in the Cartesian plane

This section is related to LES 5.

PROBLEM Electricity

In July 1884, journalist Pierre Giffard reported that the first French city to use electricity as its light source was Bellegarde-sur-Valserine. In September 1885, the second city to use electricity as its light source was La Roche-sur-Foron. In 1886, Bourganeuf was added to the list.

A hydroelectric dam was built at cascade des Jarrauds. During that time, the authorities had to use a high-tension cable to power all the cities shown in the Cartesian plane below. The scale is in kilometres.

This advertising poster was published in 1897 to promote electric lighting in homes.

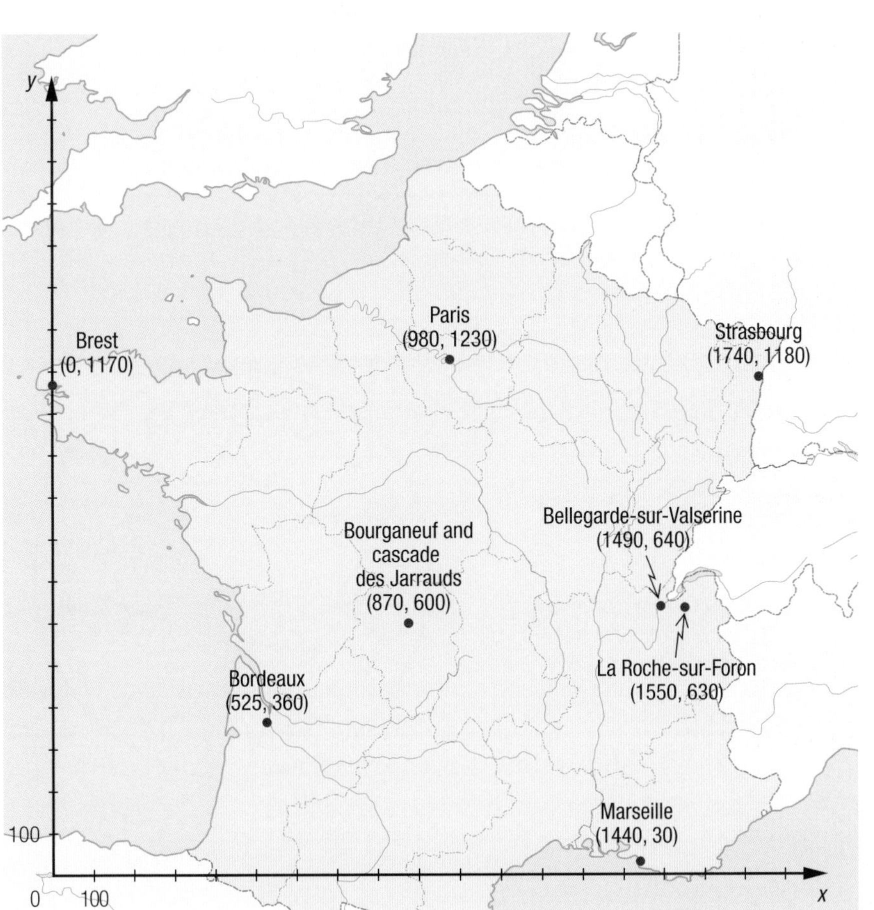

What is the shortest possible length of the cable for this project?

ACTIVITY 1 Imported cars

Imported cars make up a large part of Canadian car sales. Vehicles imported from the United States are usually transported by truck while those from overseas are usually shipped by boat. A gangway is used to load and unload the cars that are shipped by boat.

The graph below depicts a car travelling from point A to point B on a gangway. The scale is in metres.

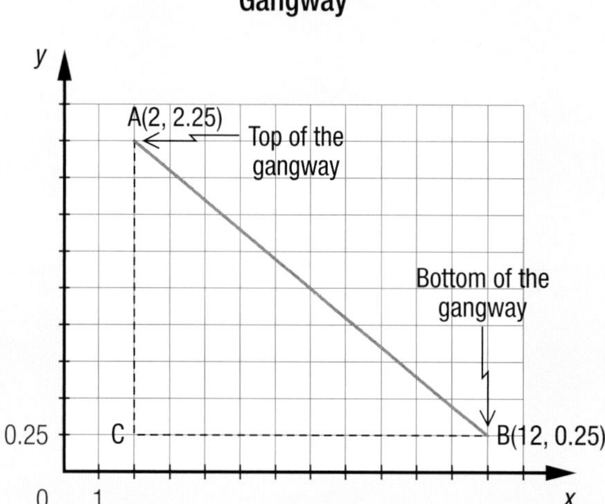

a. What is the vertical change, meaning what is the number associated with the vertical displacement of the car?

b. What is the horizontal change, meaning what is the number associated with the horizontal displacement of the car?

c. What are the coordinates of point C?

d. What type of triangle is triangle ABC?

e. Calculate:

1) the distance between point A and point C
2) the distance between point B and point C
3) the distance the car travelled over the gangway

Cars imported into Canada must comply with the Canadian automobile safety standards, meaning standards regarding seat belts or a vehicle's braking system.

ACTIVITY 2 Land surveying

In 1626, Samuel de Champlain, considered Canada's first land surveyor, surveyed the three first seigneuries (long, narrow plots of land along the St-Lawrence River). Today, land surveying is required to delimit or describe a region or any piece of land used for a specific purpose such as animal reserves or energy transport corridors.

Samuel de Champlain (1570-1635)
Cartographer, explorer and Governor of New France

A surveyor must place Survey Marker **M,** the midpoint of $\overline{AB}$, on the land shown in the adjacent graph.

The preliminary work done by the surveyor is shown in red. The scale is in metres.

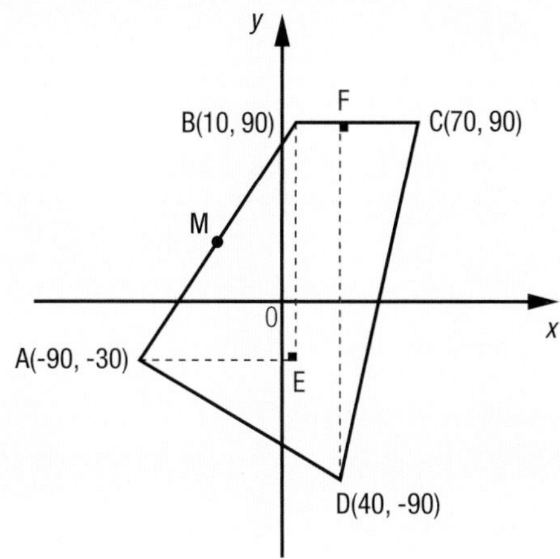

a. What are the coordinates of point E?

b. Find the coordinates:

1) of the midpoint of $\overline{AE}$
2) of the midpoint of $\overline{BE}$
3) of Survey Marker **M**

The seigneurial system was introduced in Québec in 1623. The most important duty of the seigneur (or lord) was to "build his house and home in the seigneurie," in other words, to build a manor and to live there, or to name a representative to act in his place.

On the same lot, a Survey Marker **P** must be placed at a distance $\frac{2}{3}$ the length of $\overline{DC}$, that is $\frac{2}{3}$ of the length of the segment starting at point D.

c. Will Survey Marker **P** be closer to point C or point D? Explain your answer.

d. At what ratio will Survey Marker **P** divide $\overline{DC}$?

e. Find the coordinates:

1) of the point situated $\frac{2}{3}$ along $\overline{DF}$
2) of the point situated $\frac{2}{3}$ along $\overline{FC}$
3) of Survey Marker **P**

ACTIVITY 3 Algebra or geometry?

Analytic geometry allows you to solve geometry problems using algebraic calculations.

French mathematicians René Descartes and Pierre de Fermat made many discoveries related to analytic geometry. However, it is independently that they developed certain concepts related to the field of mathematics.

René Descartes (1596-1650)
French mathematician, physicist and philosopher and one of the founders of modern philosophy.

Pierre de Fermat (1601-1665)
French jurist and mathematician. He was known as the "Prince of Amateurs."

Descartes combined algebra and geometry to prove that the midpoint of the hypotenuse of a right triangle is equidistant from the three vertices of the triangle.

To prove this, draw a right triangle ABC on a Cartesian plane as shown below.

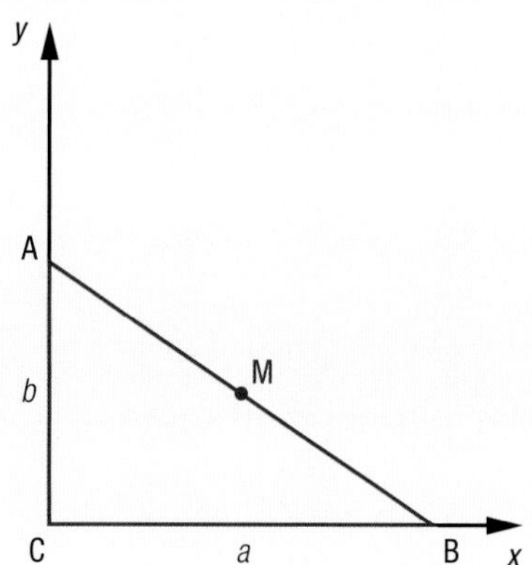

a. What has to be proven?

b. Find the coordinates of each of the vertices of the triangle ABC.

c. Find the coordinates of the midpoint M.

d. Find the algebraic expression that corresponds to the length of:

1) $\overline{AM}$ 2) $\overline{BM}$ 3) $\overline{CM}$

e. What can you conclude by comparing the three lengths found in **d.**?

f. Why is it preferable to align two sides of triangle ABC with the axes of the Cartesian plane?

Techno math

Dynamic geometry software allows you to draw figures in a Cartesian plane. By using the tools SHOW AXES, SEGMENT, MIDPOINT, COORDINATES, PERPENDICULAR LINE, TRIANGLE, DISTANCE AND SLOPE, you can draw a segment in a Cartesian plane and display its slope.

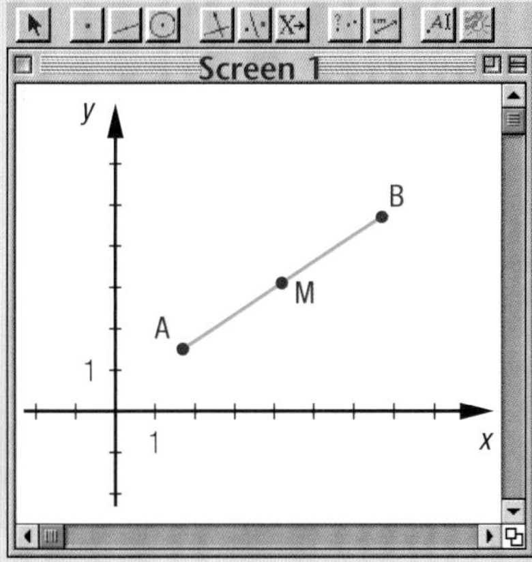

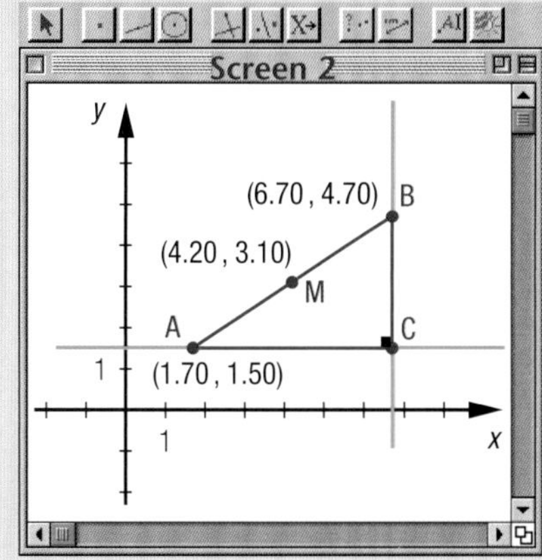

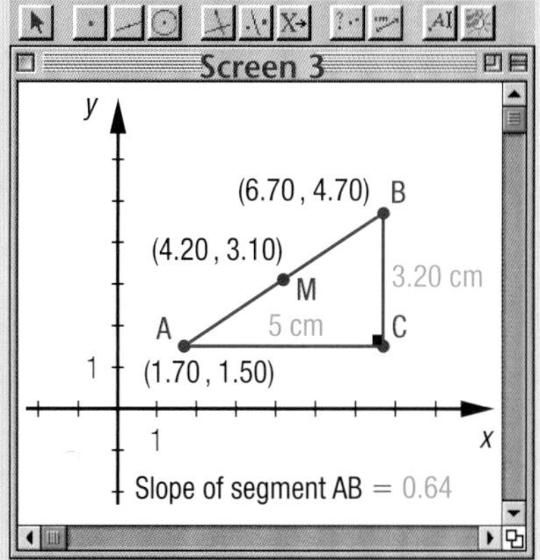

By modifying the position or the inclination of segment AB, changes related to the coordinates of the points and the slope can be observed.

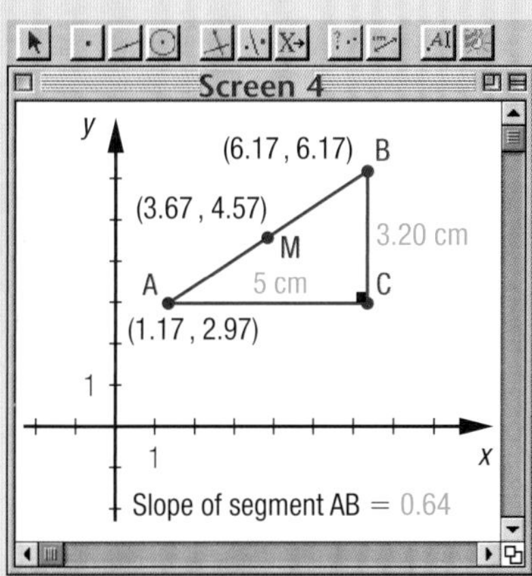

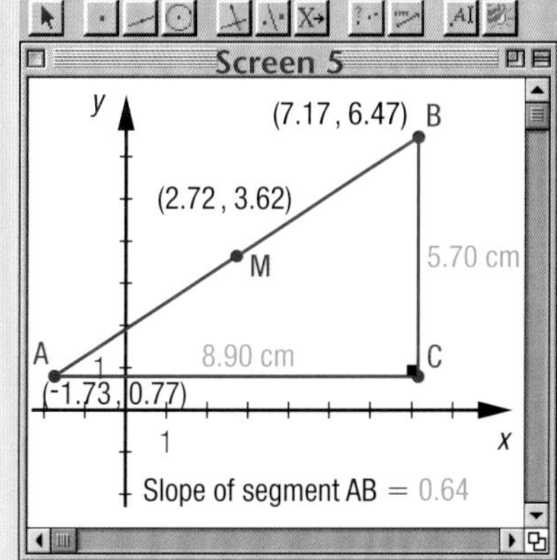

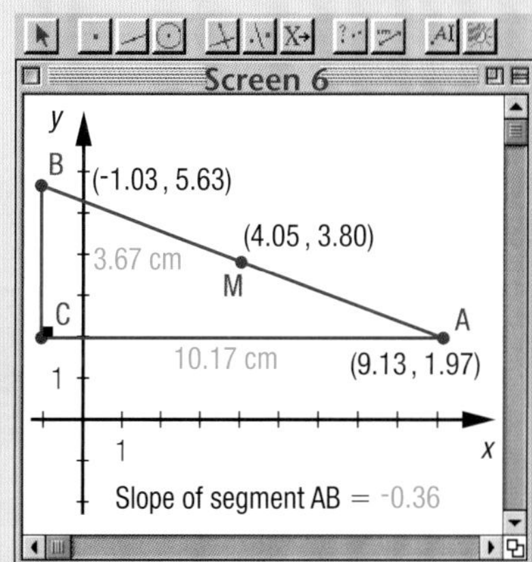

a. Compared to Screen **3**, what changes have been made to:

1) Screen **4**?
2) Screen **5**?
3) Screen **6**?

b. For Screens **3** through **6**, verify that:

1) the *x*-coordinate of M corresponds to the mean of the *x*-coordinates of A and B
2) the *y*-coordinate of M corresponds to the mean of the *y*-coordinates of A and B

c. For each of Screens **4** through **6**, do the following:

1) compare the ratio $\frac{\text{change in } y\text{-coordinates from A to B}}{\text{change in } x\text{-coordinates from A to B}}$ and the slope of segment AB
2) find the length of segment AB

d. Using dynamic geometry software, determine:

1) if there is a relationship between the slope of a segment and its length
2) what happens to the slope of a segment when it is parallel to the *x*-axis
3) what happens to the slope of a segment when it is parallel to the *y*-axis

knowledge 3.1

CHANGES TO THE *X*- AND *Y*- COORDINATES

For a point A(x_1, y_1) and a point B(x_2, y_2), note the following:

- The **change in the *x*-values** from A to B is: $\Delta x = x_2 - x_1$.
- The **change in the *y*-values** from A to B is: $\Delta y = y_2 - y_1$.

SLOPE OF A SEGMENT

The slope of a segment whose endpoints are A(x_1, y_1) and B(x_2, y_2) is a number that describes its slope, rate of change, gradient, inclination or steepness. The slope equals the ratio of the change in the *y*-value to the change in the *x*-value. The slope of a segment joining two points can be calculated by using the following formula.

$$\text{Slope of } \overline{AB} = \frac{\Delta y}{\Delta x} = \frac{y_2 - y_1}{x_2 - x_1}$$

E.g. The slope of segment AB whose endpoints are A(1, 6) and B(-7, 12) is calculated as follows.

$$\text{Slope of } \overline{AB} = \frac{y_2 - y_1}{x_2 - x_1} = \frac{12 - 6}{-7 - 1} = -\frac{3}{4}$$

DISTANCE BETWEEN TWO POINTS

The distance between point A and point B equals the length of the segment joining these two points. This length is expressed as a positive number.

The distance *d* between point A(x_1, y_1) and point B(x_2, y_2) is calculated using the following formula.

The absolute value of a real number allows you to consider this number without taking its sign into account. The absolute value of a number is expressed by placing it between two vertical lines. For example: $|3| = 3$ and $|-3| = 3$.

$$d(A, B) = \sqrt{(x_2 - x_1)^2 + (y_2 - y_1)^2}$$

E.g. The distance between point A(3, 4) and point B(-2, 6) is calculated as follows.

$$d(A, B) = \sqrt{(x_2 - x_1)^2 + (y_2 - y_1)^2} = \sqrt{(-2 - 3)^2 + (6 - 4)^2} = \sqrt{29} \text{ or } \approx 5.39 \text{ u}$$

POINT OF DIVISION

The position of any point of division on a segment can be found using a fraction or a ratio.

E.g.

In the adjacent graphical representation, note the following:

- Point P is situated $\frac{3}{5}$ along segment AB.
- Point P divides segment AB into a ratio of 3:2.
- Point P is situated $\frac{2}{5}$ along segment BA.
- Point P divides segment BA into a ratio of 2:3.

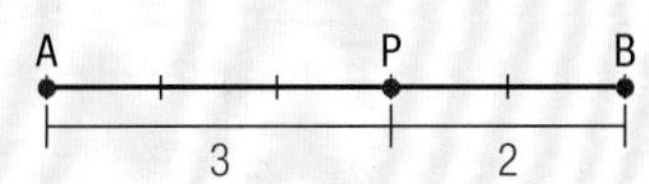

Point of division P is on segment AB whose endpoints are $A(x_1, y_1)$ and $B(x_2, y_2)$. If point P is located at a fraction $\left(\frac{a}{b}\right)$ of the distance between points A and B, its coordinates are:

$$\left(x_1 + \frac{a}{b} \times \Delta x, y_1 + \frac{a}{b} \times \Delta y\right)$$

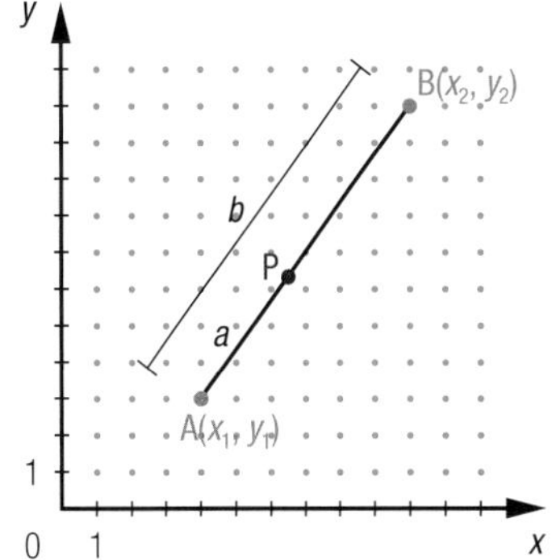

E.g.

1. The coordinates of midpoint M of segment AB, whose endpoints are A(-2, -4) and B(4, 6), can be found by doing the following calculations.

   $\left(x_1 + \frac{a}{b} \times \Delta x, y_1 + \frac{a}{b} \times \Delta y\right)$

   $\left(-2 + \frac{1}{2}(4 - -2), -4 + \frac{1}{2}(6 - -4)\right)$

   (1, 1)

   The coordinates of midpoint M are (1, 1).

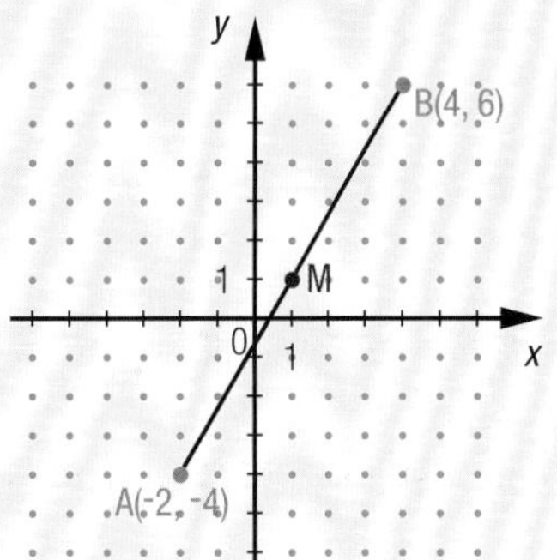

2. Point P divides segment AB in a ratio of 3:1. Its endpoints are A(3, 7) and B(-4, -10). The coordinates of point P can be found by doing the following calculations.

   The ratio 3:1 corresponds to the fraction $\frac{3}{4}$.

   $\left(x_1 + \frac{a}{b} \times \Delta x, y_1 + \frac{a}{b} \times \Delta y\right)$

   $\left(3 + \frac{3}{4}(-4 - 3), 7 + \frac{3}{4}(-10 - 7)\right)$

   (-2.25, -5.75)

   The coordinates of point P are (-2.25, -5.75).

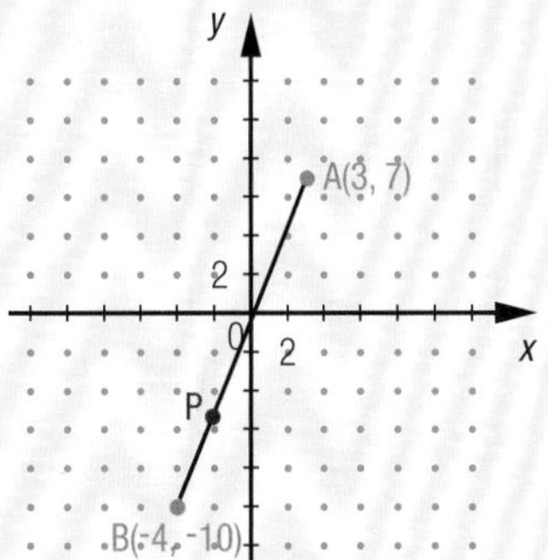

practice 3.1

1 To find the length of a segment whose endpoints are A(3, 4) and B(6, 8), Joanne solves $\sqrt{(6-3)^2+(8-4)^2}$ and Tess solves $\sqrt{(3-6)^2+(4-8)^2}$. Explain why they obtain the same result.

2 Using points whose coordinates are indicated in the adjacent box, determine:

A(5, 6)	B(-2, 9)
C(-4, 7)	D(-9, -11)
E(-50, -50)	F(30, 40)

a) m $\overline{AB}$

b) m $\overline{CD}$

c) m $\overline{EF}$

d) m $\overline{BE}$

e) the distance between the midpoint of $\overline{AB}$ and the midpoint of $\overline{CD}$

3 In each case, determine the slope of the line whose:

a) *x*-intercept is: 5
y-intercept is: -3

b) *x*-intercept is: -7
y-intercept is: -1

c) *x*-intercept is: $\frac{1}{3}$
y-intercept is: 2

d) *x*-intercept is: π
y-intercept is: $-\pi$

4 Determine the type of triangle or quadrilateral whose vertices are:

a) A(10, 20), B(10, 60), C(40, 20)

b) D(0, 0), E$(1, \sqrt{3})$, F(2, 0)

c) G(0, -3), H(-1, 1), I(3, 2)

d) J(2, 1), K(1, -1), L(-2, -1)

e) A(3, -2), B(4, -1), C(7, 0), D(6, -1)

f) E(6, 1), F(3, -4), G(-2, -7), H(1, -2)

g) I(6, 10), J(5, 13), K(11, 15), L(12, 12)

h) M(-2, -1), N(-1, 1), O(3, 1), P(4, -1)

i) U(6, 8), V(4, 12), W(8, 14), X(10, 10)

j) A(0, 14), B(-10, 24), C(-12, 38), D(-2, 28)

5 Find the coordinates of the point:

a) situated at the midpoint of segment AB whose endpoints are A(5, 6) and B(13, 8)

b) situated $\frac{2}{3}$ along segment CD whose endpoints are C(6, 8) and D(12, 17)

c) situated $\frac{3}{5}$ along segment FE whose endpoints are F(-3, -1) and E(2, 14)

d) that divides segment HG whose endpoints are H(3, -2) and G(-2, 3) in a ratio of 2 : 3

6 In the Cartesian plane, point P(5, 8) is situated $\frac{3}{4}$ along segment EH. Determine the coordinates of endpoint H if those of endpoint E are:

a) (-1, 7) b) (2, 2) c) (-1, 4.25)

d) (17, 44) e) (2, -4) f) $\left(\frac{1}{4}, \frac{2}{5}\right)$

7 Prove that the midpoints of the sides of the adjacent quadrilateral ABCD are the vertices of a parallelogram.

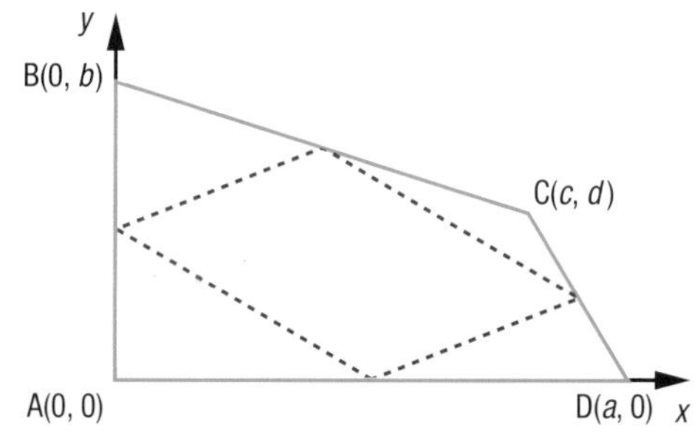

8 In the adjacent Cartesian plane, the roads connecting Cities **A**, **B** and **C** are shown. As a strategy to relieve traffic congestion at peak hours, a secondary road was built from City **C** to the midpoint of segment AB. The scale is in kilometres.

a) What are the coordinates of the point where the new road and the road connecting Cities **A** and **B** will meet?

b) Calculate the length of the new road.

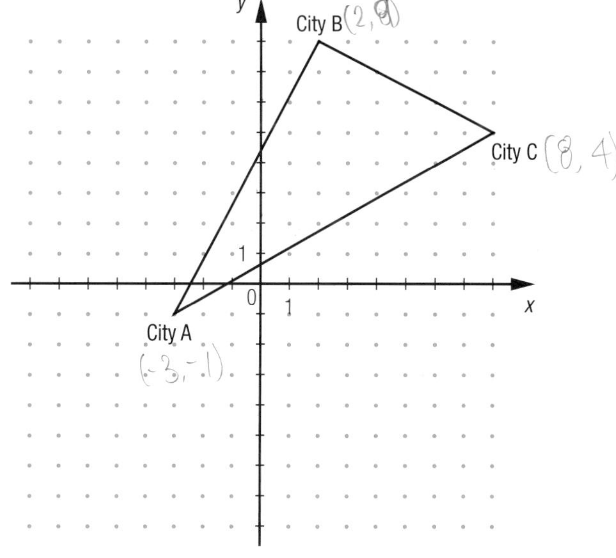

9 The slope of a segment is a number that defines its inclination.

a) What is the angle formed by segment AB and the horizontal axis?

b) What is the horizontal change from A to B?

c) What is the vertical change from A to B?

d) What do you notice when you establish the ratio of the vertical change to the horizontal change?

e) What conclusion can you reach about the slope of a segment that is parallel to the *y*-axis?

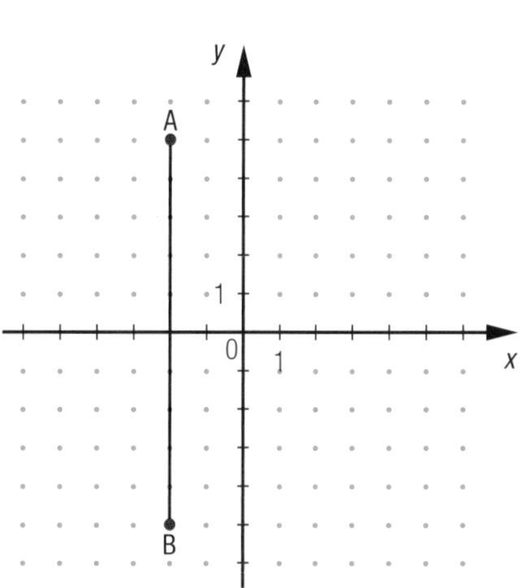

10 In the Cartesian plane shown below, right trapezoid ABCD has been drawn, and the coordinates of its vertices are provided.

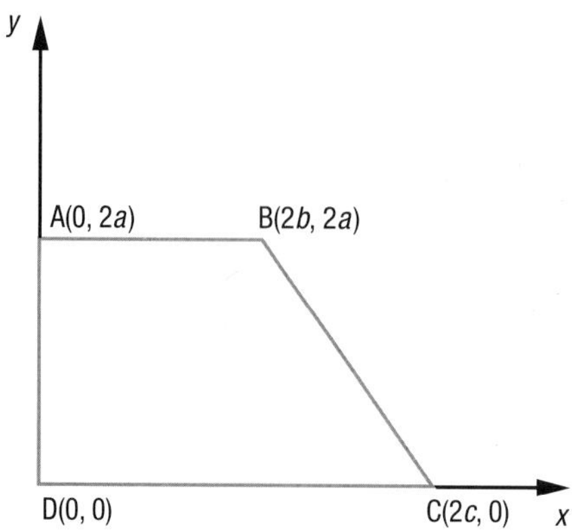

a) Find the midpoints of $\overline{AD}$ and $\overline{BC}$, and label these points M_1 and M_2 respectively.

b) Calculate the slope of segment:

1) AB
2) M_1M_2
3) DC

c) Using an algebraic expression, define the length of:

1) AB
2) M_1M_2
3) DC

d) Is it correct to state that the segment connecting the midpoints of the non-parallel sides of a right trapezoid is parallel to the bases and that the length of this segment is equal to half the sum of the length of its bases?

11 An electrician installs electrical outlet boxes on a wall. The outlets are one above the other. Below are the distances between each of the outlet boxes and the ceiling:

1st outlet	2nd outlet	3rd outlet	4th outlet	5th outlet
5 cm	15 cm	25 cm	35 cm	45 cm

If the 1st outlet corresponds to point A and the 5th outlet to point B:

a) In what ratio does the 2nd outlet divide segment AB?

b) In what ratio does the 4th outlet divide segment BA?

c) Find the location of the 3rd outlet using a fraction based on the length of $\overline{AB}$.

Electricians carry out electrical projects such as designing electrical systems and preparing plans using computer software.

12 An animal conservation agent has used a Cartesian plane to repesent the territory he is in charge of. From his location at point A(-1, -9), he uses a tranquilizer gun on a caribou situated at point B(-5, 5). If the distance between point A and point B corresponds to the maximum range of the tranquilizer gun, what is the area the agent can cover without moving?

The caribou is part of the Cervidae family, which includes four other species that are indigenous to Canada: moose, North American elk, white-tailed deer, and mule deer.

13 As represented in the adjacent graph, the following information relates to the path of objects in an assembly line:

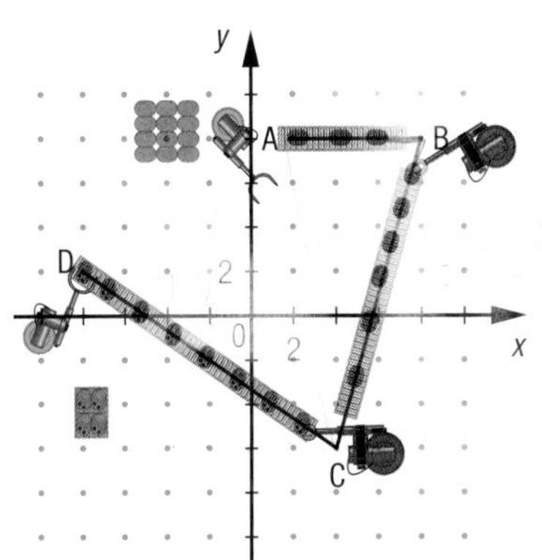

- An object starts at point A, stops at point B for 8 seconds, at point C for 9 seconds, then is removed from the assembly line at point D.
- A new object starts at point A before the object above reaches point D.
- The travel time needed to go from point A to point B is 10 seconds, from point B to point C is 12 seconds, and from point C to point D is 20 seconds.

An object starts at point A on the assembly line. Considering that the scale is in metres, answer the following:

a) What distance must the object cover in order to reach point D?

b) At what speed is the object travelling between points:

1) A and B?

2) B and C?

3) C and D?

c) How many objects can be processed through this assembly line in 2 h?

14 **COLLEGE STUDIES** Interest in post-secondary studies varies according to a variety of factors including economic activity. Considering that the situation from 2005 to 2009 can be represented graphically by a linear relation, determine the number of students who were enrolled in college in 2007.

College (CEGEP) enrollment

Year	Number of students
2000	159 617
2004	154 026
2005	153 290
2009	176 473

15 The adjacent diagram represents a safety strip along the side of a road. The slope of segment AB is -4. Considering that the scale is in metres, determine:

a) the coordinates of point B

b) the distance between points B and C

c) the distance between points A and B

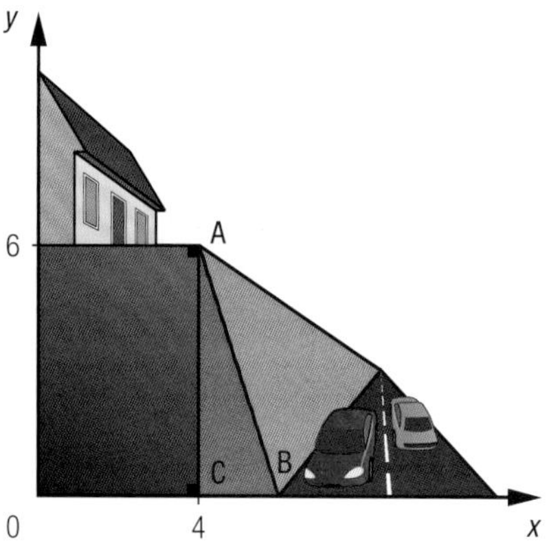

16 The adjacent graph shows a section of a railway track between City **A** and City **B**. The scale is in kilometres.

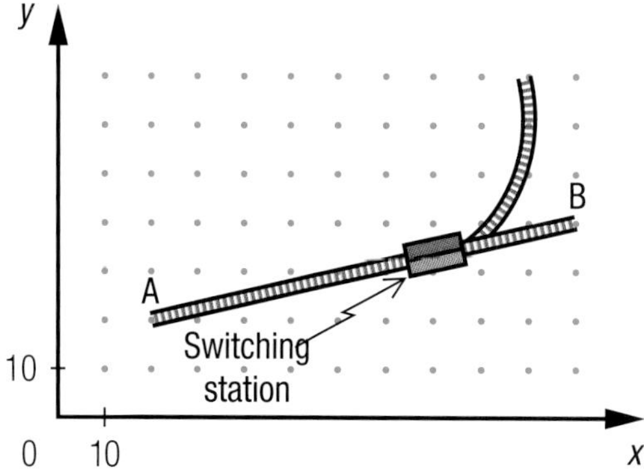

a) What is the distance between these two cities?

b) What is the distance from the switching station to City **B** if its position divides segment AB in a ratio of 5:2?

17 **CHIROPRACTICS** The traction device shown below is a chiropractic apparatus that helps improve the curvature of a patient's spine.

A patient is lying on a table to which a rope is attached 1.2 m above the table; the slope of the rope is $-\frac{7}{6}$. When traction is applied, the slope of the rope is then $-\frac{8}{9}$.

Position of the patient before traction

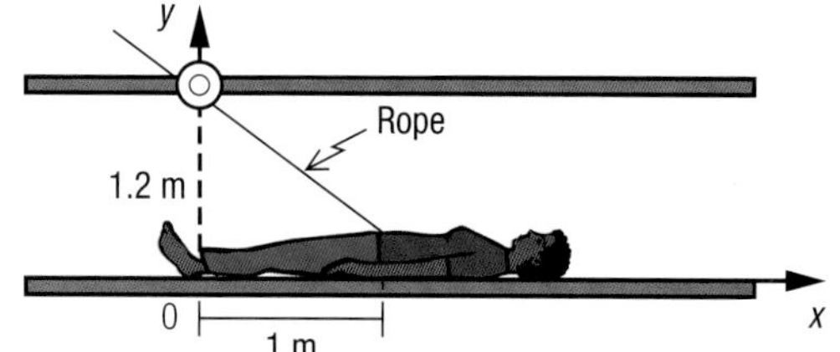

Position of the patient during traction

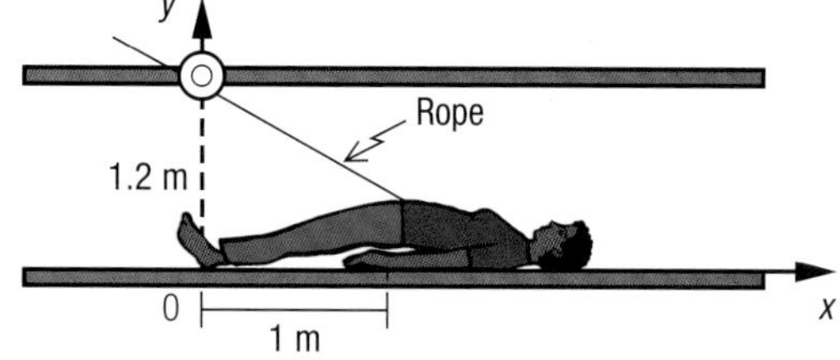

a) At what height above the table is the rope attached to the patient:

1) before traction is applied?

2) while traction is being applied?

b) Find the length of the rope:

1) before traction is applied

2) while traction is being applied

18 In general, a person's height can be determined by measuring certain bones in their body. The adjacent graph shows Andrea's height in relation to the length of her humerus when she was 14, 16, 18 and 20 years old.

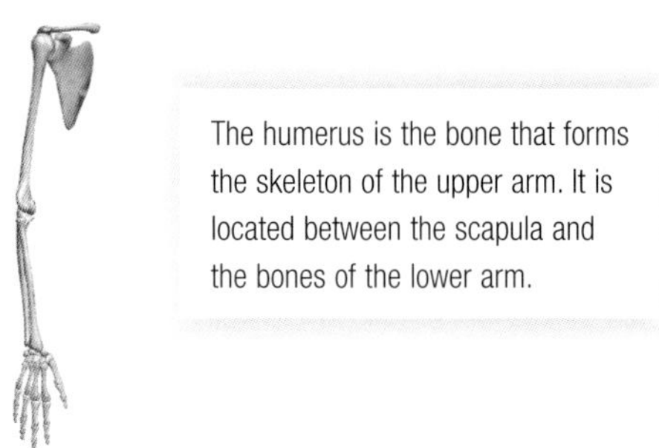

The humerus is the bone that forms the skeleton of the upper arm. It is located between the scapula and the bones of the lower arm.

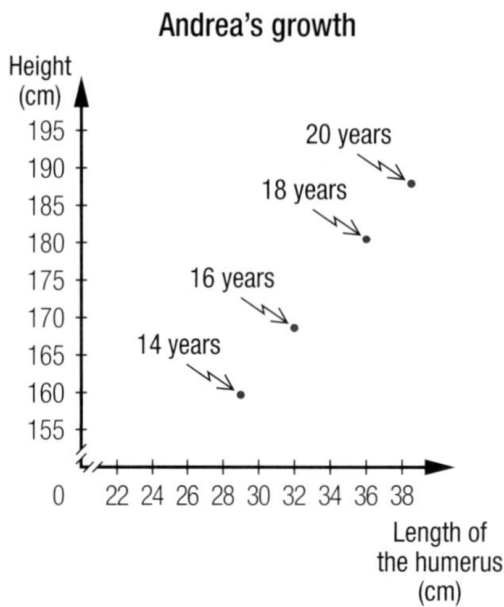

a) How long was Andrea's humerus at age 19?

b) What was Andrea's height at age 17?

c) What was the difference between the length of Andrea's humerus when she was 18 and when she was 15?

19 **MIGRATION** The hummingbird and the monarch butterfly are two migratory species found in Québec. Both of these species migrate to Mexico in the fall. Below are some of the characteristics of these two species:

	Hummingbird	Monarch butterfly
Mean speed (km/h)	60	32
Duration of flight per day (h)	12	3.75

The adjacent map shows the distance, in kilometres, that the two species must travel from Sherbooke to Reynosa, a city in Mexico.

Sherbrooke 0
Reynosa (2407, 2401)

On September 1, the monarch begins its migration. When it has travelled $\frac{3}{5}$ of the distance between the two cities, the hummingbird leaves Sherbrooke and flies in the direction of Reynosa.

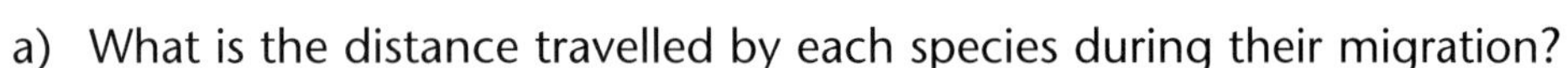

a) What is the distance travelled by each species during their migration?

b) When the hummingbird leaves for Reynosa, what are the coordinates of the monarch's location?

c) When will the hummingbird catch up to the monarch?

The ruby-throated hummingbird is the only species of bird capable of flying in place and backwards. It beats its wings at an average rate of 65 beats/second which makes them seem almost invisible.

SECTION 3.2 Lines in the Cartesian plane

This section is related to LES 5 and 6.

PROBLEM Irrigating agricultural land

An agriculture technician analyzes the chemical composition of soils, the daily precipitation, the temperature, the humidity rate, and the number of plants in the ground in order to suggest ways that farmers can improve the quality and quantity of their harvests.

An agricultural technician tells a strawberry producer that she should be irrigating her fields using an overhead sprinkling method. This method imitates rain. In planning the installation of the irrigation system, the field was represented in a Cartesian plane.

An overhead sprinkling system does not make use of gravity. Instead, it uses pipes through which water circulates under great pressure. The pipes feed the sprinkler system and the water is sprinkled on the plants as a fine rain. This irrigation method helps limit water consumption.

The three lines defined below represent the boundaries of the field. The scale of the Cartesian plane used for this representation is in hectometres.

Line A

Passes through points P(0, 10) and R(4, 2).

Line B

Passes through point R(4, 2) and its slope is -0.5.

Line C

Passes through point P(0, 10) and its slope is 0.5.

The technician recommends watering the field during the night following a sunny day. According to the data collected, 12 274 L of water are needed for her to water an area of 10 hm^2.

How much water will the irrigation system use to water this field after a sunny day?

In Québec, some 700 farmers grow more than 2300 hectares of strawberries. It is the third largest fruit industry after apples and blueberries.

ACTIVITY 1 Laser cutting

The automobile and aeronautic industries use programmable machines to cut parts. These machines are equipped with a lens-focussed laser that is powerful and precise. Lasers can also cut steel, plastic, and glass parts without burrs and do so with a precision of nearly one tenth of a millimetre.

A sheet of stainless steel is put into a programmable machine. The cutting instructions are programmed using coordinates and equations. The diagram below represents a piece that has to be cut from a steel sheet. The scale is in centimetres.

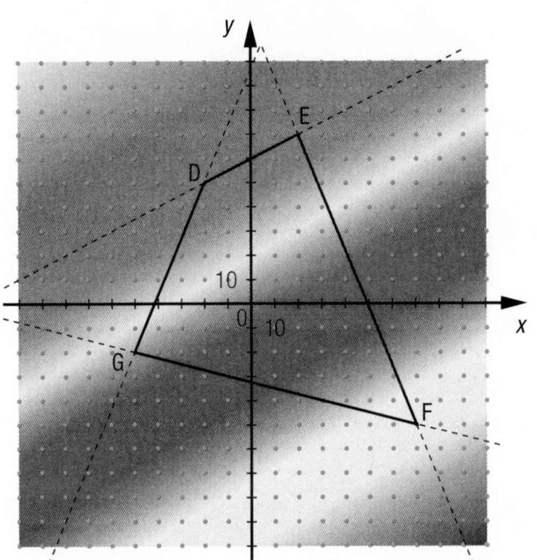

The laser, which is at the base of a light amplifier, can be used in many ways such as a micro-machining tool, a bar code reader, a remote sensor, for printing, for medicine and for sound and light shows.

a. Define the equation of line EF in function form $y = ax + b$.

b. What are the slope and the y-intercept of this line?

c. What algebraic manipulation must be made to the equation found in **a.**:

1) to find the x-intercept of the line?

2) to express the equation in general form $Ax + By + C = 0$?

Written in general form $Ax + By + C = 0$, the equation of line DE is $x - 2y + 120 = 0$.

d. In this equation, what is the value of:

1) **A?** 2) **B?** 3) **C?**

e. What algebraic manipulations must be made to the equation $x - 2y + 120 = 0$ to express it in function form $y = ax + b$?

f. What relationship can be established between the values found in **d.** and:

1) the slope of line DE?

2) the y-intercept of line DE?

3) the x-intercept of line DE?

ACTIVITY 2 Pierre de Fermat and lines

In his writings, Pierre de Fermat stated that two lines in the same plane are parallel if they never intersect. Based on this idea, Fermat discovered a relationship between the slopes of two parallel lines and the slopes of two perpendicular lines.

As a magistrate in Toulouse, France, Pierre de Fermat was a famous lawyer who was interested in science. Along with Descartes, he invented analytic geometry and, at the same time as Pascal, he developed a probability theory. Pierre de Fermat was also interested in arithmetic, and he is the man to whom we owe Fermat's optical principle.

Pierre de Fermat (1601-1655)

Triangle ABC is defined by lines l_1, l_2 and l_3.

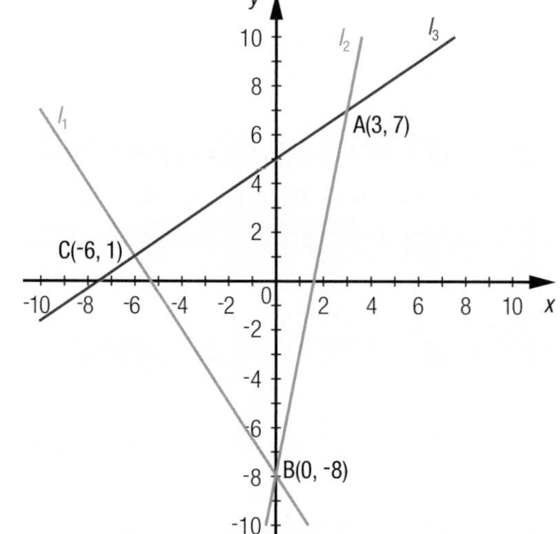

a. Using the Pythagorean theorem, prove that this is a right triangle at vertex C.

b. Find the slope of lines:

1) l_1 2) l_2 3) l_3

c. Verify that the product of the slopes of lines l_1 and l_3 is -1.

d. 1) What are the coordinates of the midpoint of segment AB?

2) Considering that the slope of the perpendicular bisector of segment AB is -0.2, determine the equation of this perpendicular bisector.

3) Verify that the product of the slopes of line l_2 and the prependicular bisector is -1.

e. What conclusion can you draw about the slopes of two perpendicular lines?

The adjacent quadrilateral DEFG consists of lines l_4, l_5, l_6 and l_7.

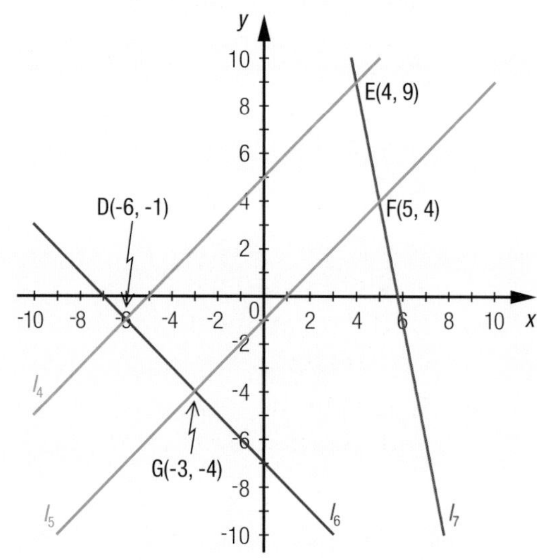

f. 1) Find the slopes of lines l_4 and l_5. What do you notice?

2) What can be said about the position of lines l_4 and l_5 in relation to each other?

g. Verify that the product of the slopes of lines l_4 and l_6 is -1.

h. What type of quadrilateral is DEFG?

ACTIVITY 3 Connecting to a gas pipeline

A gas pipeline is a pipe that is used to transport gas over long distances. Gas pipelines connect gas deposits, distribution centres and industrial or urban areas.

The total length of all the pipelines in the world is estimated at 1 000 000 km, that is, approximately 25 times the circumference of the Earth.

A municipality wants to connect itself to a gas pipeline so it can use natural gas. To evaluate the shortest length and the lowest possible cost for the connection, the municipal engineer superimposes a Cartesian plane onto a map of the area. The scale is in kilometres.

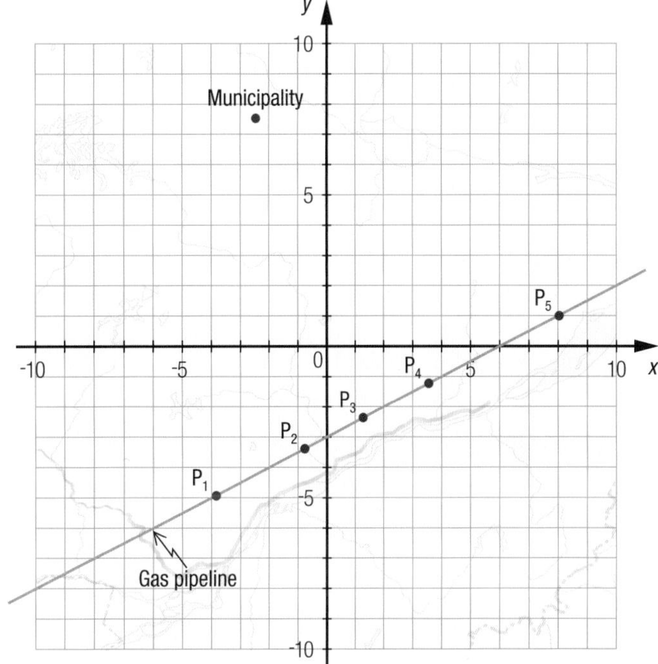

a. Referring to points P_1 through P_5:

1) which is farthest from the municipality?
2) which is closest to the municipality?
3) between which two points should the municipality connect itself to the pipeline?

b. 1) Using a copy of the diagram, draw a line representing the shortest distance between the municipality and the gas pipeline.
2) Describe the characteristics of this line.
3) Compare the line and its characteristics with those of your peers.

c. To calculate the shortest length of the connection, the engineer suggests the following procedure.

1. Determine the slope of the line that is perpendicular to the pipeline.
2. Determine the equation of the line that is perpendicular to the pipeline and passes through the municipality.
3. Find the coordinates of the intersection point between the gas pipeline and this perpendicular line.
4. Calculate the length of the segment that connects the intersection point and the municipality.

It costs the municipality \$31,000/km to connect to the gas pipeline. Using the procedure described above, determine the lowest cost for the connection.

Techno math

Dynamic geometry software allows you to draw lines on a Cartesian plane and manipulate them. By using the tools SHOW AXES, POINT, PARALLEL LINES, PERPENDICULAR LINES and EQUATION, you can draw parallel and perpendicular lines and present their equations in various forms.

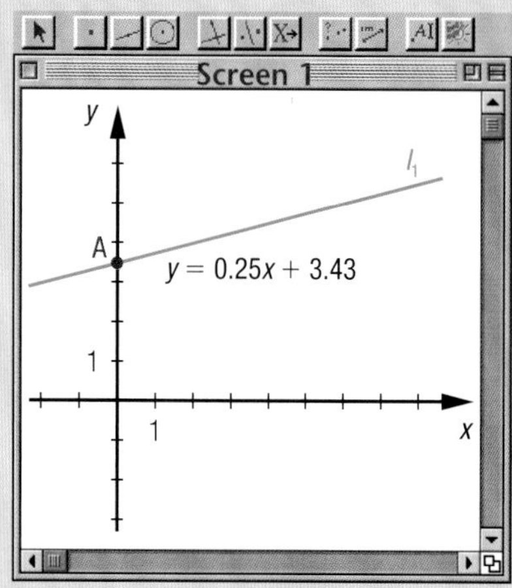

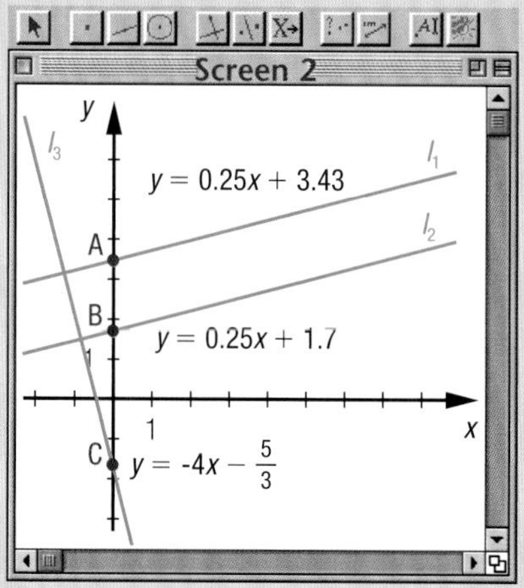

By modifying the inclination of line l_1 or the position of points A, B and C, you can observe various changes that occur to the equations of the lines.

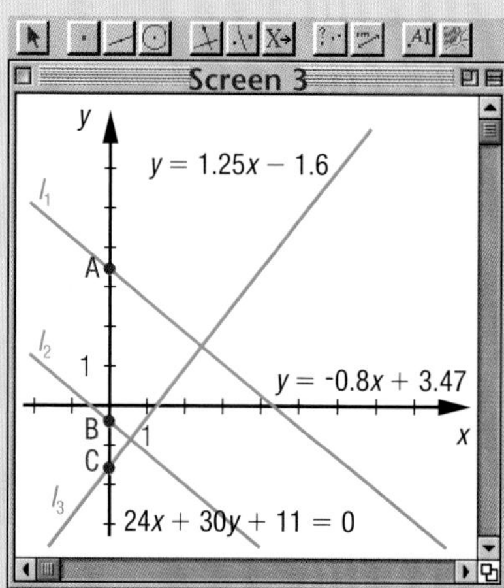

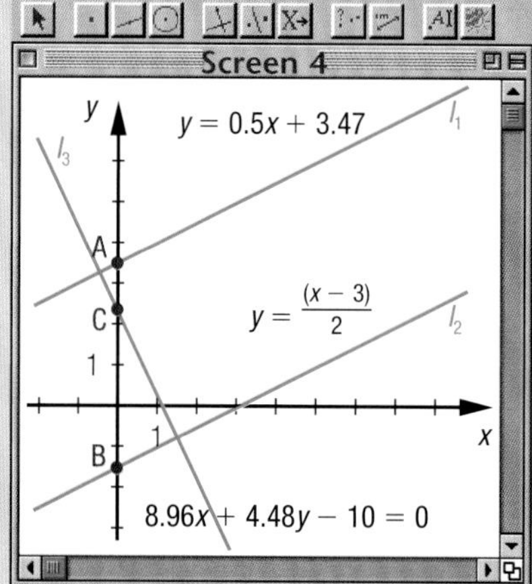

a. Find the coordinates of four points on the line displayed in Screen **1**.

b. On Screen **2**, how do the equations of lines l_1 and l_2 prove that $l_1 // l_2$?

c. On Screen **2**, verify that the product of the slopes of lines l_1 and l_3 is -1.

d. Using function form, $y = ax + b$, write an equation for line l_2 displayed in Screen **3**.

e. Determine the *y*-intercept of each of the lines displayed in Screen **4**.

f. On Screen **4**, given that $l_1 \perp l_3$, what can be said about the position of line l_2 in relation to the other lines?

g. Using dynamic geometry software, determine what happens to the equation of a line that is:

1) parallel to the *x*-axis
2) parallel to the *y*-axis

EQUATION OF A LINE

There are various ways of writing the equation of a line. Below are two:

Equation type	Equation	Relationship between parameters	Characteristics
Function or standard form	$y = ax + b$	Slope: a y-intercept: b x-intercept: $-\frac{b}{a}$	This equation type can be used to describe any non-vertical line.
General form	$Ax + By + C = 0$	Slope: $-\frac{A}{B}$ y-intercept: $-\frac{C}{B}$ x-intercept: $-\frac{C}{A}$	This equation type can be used to describe any line.

Using algebraic manipulations, it is possible to convert the equation of a line from general form to standard form and vice versa.

E.g.

1) Using algebraic manipulations, the equation $y = 62.7x - 41$ can be expressed in general form.

$$y = 62.7x - 41$$
$$y - y = 62.7x - 41 - y$$
$$0 = 62.7x - y - 41$$

Therefore, the equation is $62.7x - y - 41 = 0$.

2) Using algebraic manipulations, the equation $3x + 4y - 4 = 0$ can be expressed in standard form.

$$3x + 4y - 4 = 0$$
$$3x + 4y - 4 - 3x = 0 - 3x$$
$$4y - 4 = -3x$$
$$4y - 4 + 4 = -3x + 4$$
$$4y = -3x + 4$$
$$\frac{4y}{4} = \frac{-3x + 4}{4}$$

Therefore, the equation is $y = -\frac{3}{4}x + 1$.

3) The equation $3x + 4y - 4 = 0$ can also be expressed in standard form using the parameters A = 3, B = 4 and C = -4.

Slope: $-\frac{A}{B} = -\frac{3}{4}$

y-intercept: $-\frac{C}{B} = -\frac{-4}{4} = 1$

Therefore, the equation is $y = -\frac{3}{4}x + 1$.

PARALLEL LINES, PERPENDICULAR LINES AND THE PERPENDICULAR BISECTOR OF A SEGMENT

In a Cartesian plane, note the following:

- Two lines with the same slope are parallel.
- Two lines with slopes that are the negative reciprocal of each other are perpendicular. The product of the slopes of these two lines is -1.
- A segment and its perpendicular bisector are perpendicular and their slopes are the negative reciprocal of each other; the product of their slopes is -1.

E.g.

1) Line l_1 with equation $y = -2x - 3$ and line l_2 with equation $y = -2x + 5$ are parallel because their slopes are the same; each line has a slope of -2.
2) Line l_1 with equation $y = -2x - 3$ and line l_4 with equation $y = \frac{1}{2}x - 7$ are perpendicular because the slope of one is the negative reciprocal of the other: $-2 \times \frac{1}{2} = -1$.
3) Line l_3 with equation $y = -3x + 25$ is the perpendicular bisector of segment AB because it passes through its midpoint (6, 7) and their slopes are the negative reciprocal of each other: $-3 \times \frac{1}{3} = -1$.

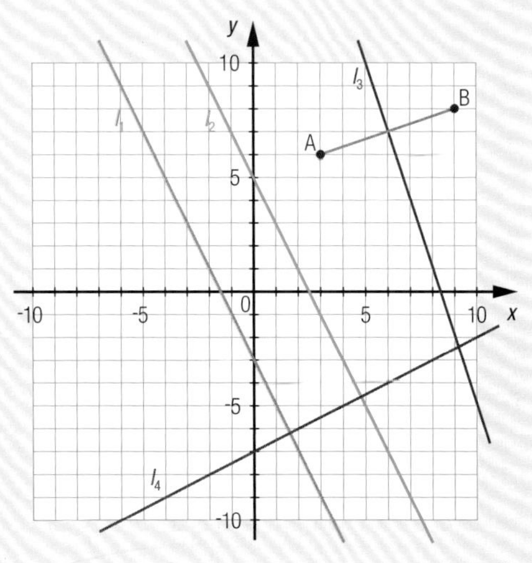

DISTANCE FROM A POINT TO A LINE

The distance from a point to a line corresponds to the shortest distance separating them. Following is a method of determining the shortest distance between a point P and a line l_1:

1. Determine the equation of the line that passes through point P and is perpendicular to line l_1.	E.g. • the slope of a line that is perpendicular to l_1: -4 • the equation of a line perpendicular to l_1 and that passes through point P(-7, 5): $y = -4x - 23$	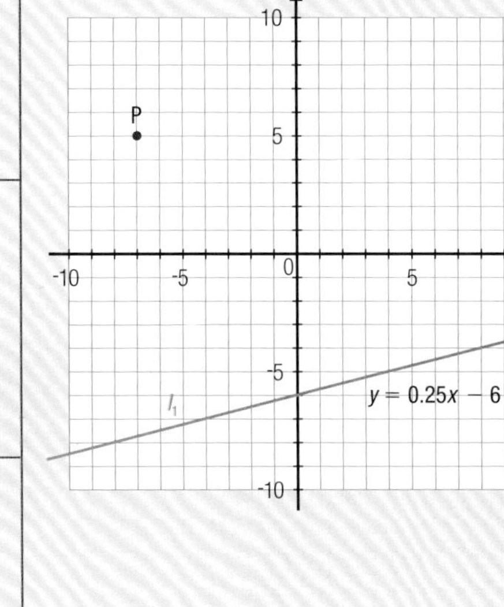
2. Find the coordinates of the intersection point of the original line and the line that is perpendicular to it passing though point P.	• the coordinates of the intersection point: (-4, -7)	
3. Calculate the distance between the intersection point and point P.	• $\sqrt{(-7 - -4)^2 + (5 - -7)^2}$ ≈ 12.37 u • The distance between point P and line l_1 is ≈ 12.37 u.	

practice 3.2

1 Determine the equation of the line in each of the graphical representations below.

a) Graph ①

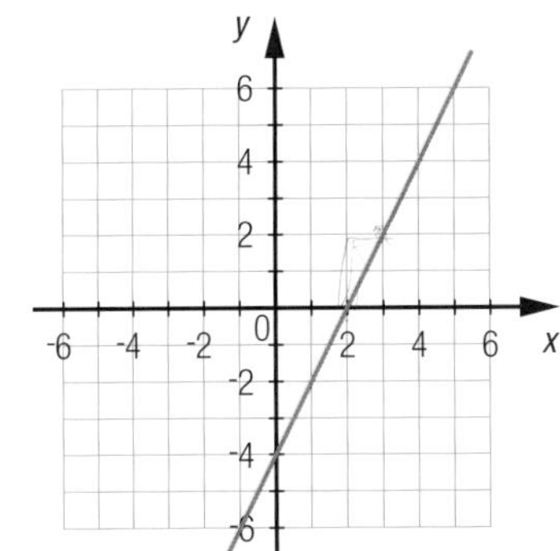

b) Graph ②

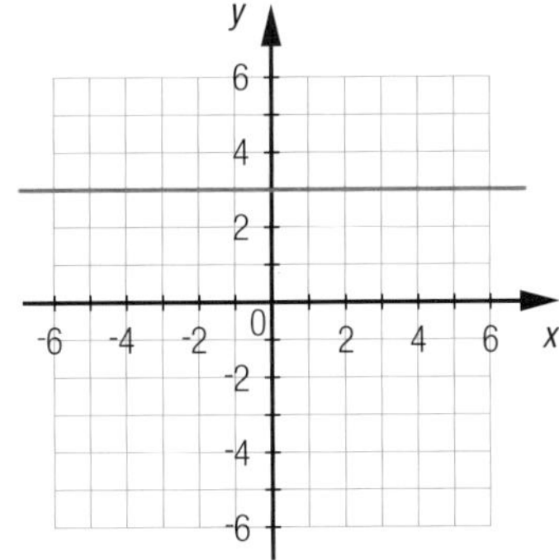

c) Graph ③

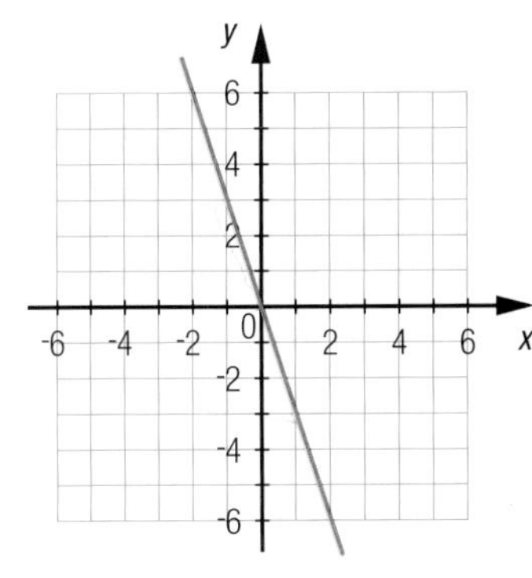

d) Graph ④

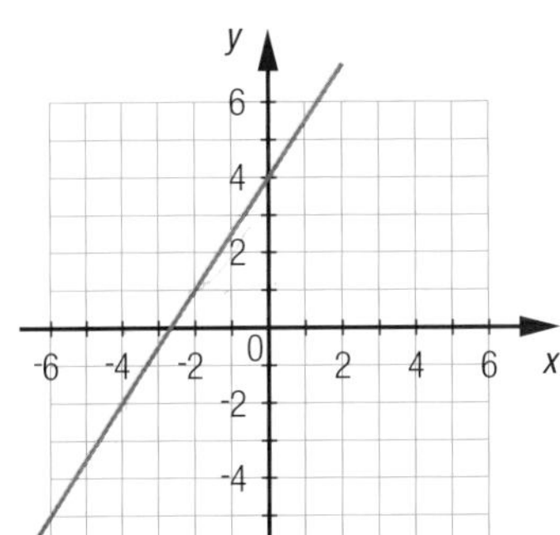

e) Graph ⑤

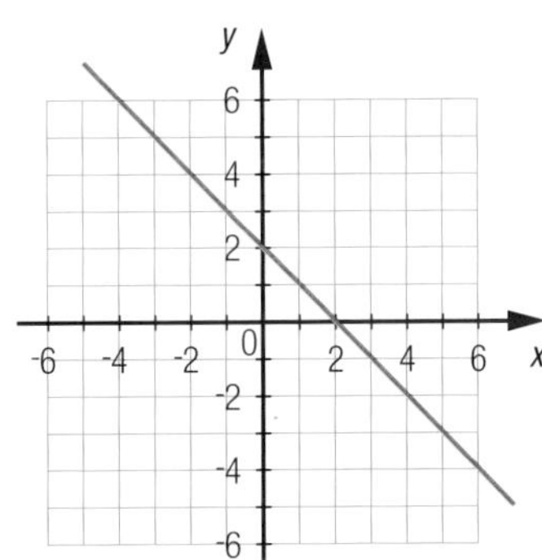

f) Graph ⑥

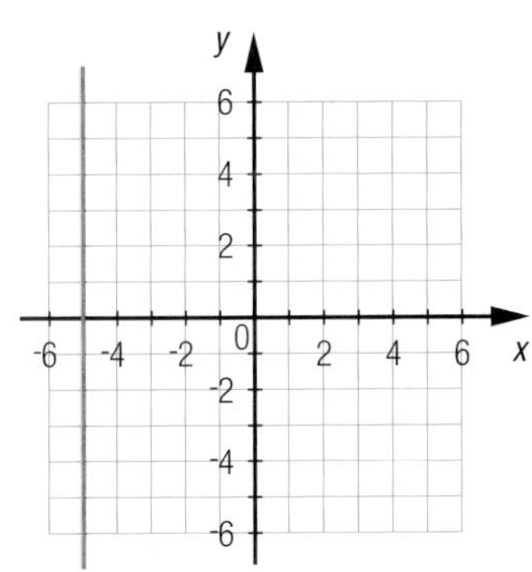

2 Determine the slope, y-intercept and x-intercept of the line whose equation is:

a) $y = -x + 23$

b) $y = -12x + 5$

c) $x + y = 15$

d) $6x - 2y = 19$

e) $x + y - 7 = 0$

f) $3x - 4y + 5 = 0$

g) $y = x$

h) $y = 5.7$

i) $-21x + \frac{2}{7}y + 48 = 0$

3 For each of the following, determine the equation of each line:

1) in standard form 2) in general form

a) the line whose slope is 8 and whose y-intercept is -3

b) the line whose slope is -2 and passes through point P(12, 5)

c) the line whose slope is $\frac{4}{5}$ and passes through point R(6, 10)

d) the line that passes through points S(-1, 25) and T(3, -7)

e) the line whose x-intercept is -10 and whose y-intercept is 20

f) the line that passes through the origin and whose slope is -1.3

4 For each of the following, draw the line that has the following characteristics.

a) • x-intercept: 7
 • y-intercept: 5

b) • parallel to the y-axis
 • passes through point P(6, 4)

c) • slope: 0.5
 • x-intercept: 3

d) • slope: $\frac{2}{3}$
 • y-intercept: $^-2$

e) • parallel to x-axis
 • passes though point R($^-6$, $^-7$)

f) • perpendicular to x-axis
 • passes through point S(8, $^-1$)

5 Determine the equation of the line that:

a) passes through point A(0, 3) and is parallel to the line with equation $y = 2x + 5$

b) passes through point B(3, 4) and is parallel to the line with equation $y = 5$

c) passes through point C($^-2$, 8) and is parallel to the line with equation $x = 2$

d) passes through point D(2, 3) and is perpendicular to the line with equation $y = 2x + 5$

e) passes through point E(4, 12) and is perpendicular to the line with equation $4x - 5y + 3 = 0$

f) corresponds to the perpendicular bisector of the segment whose endpoints are F(2, 1) and G(6, $^-9$)

6 In each case, calculate the distance between point P and line l.

a) 
Graph ①

b) 
Graph ②

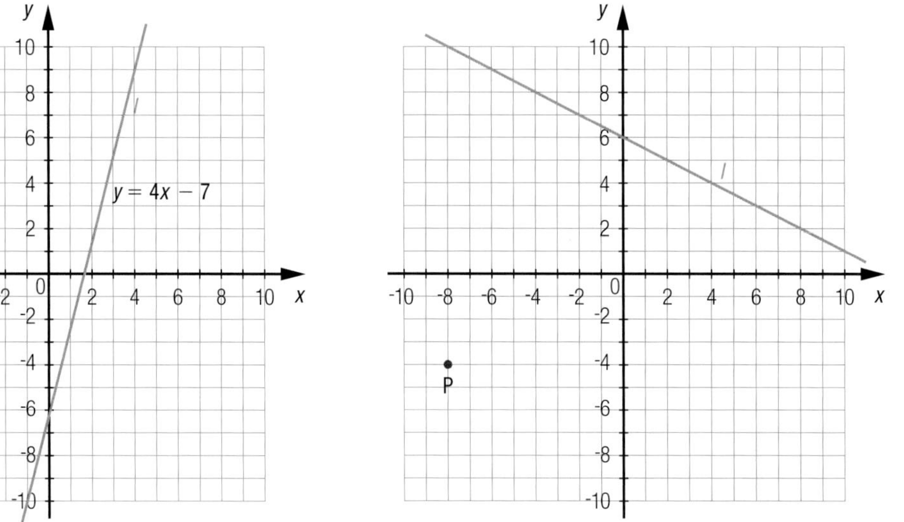

7 Two lines drawn in the Cartesian plane are perpendicular. Find the slope of one of these lines if the slope of the other corresponds to the algebraic expression:

a) $2a$

b) $\frac{^-ab}{2}$

c) $7a^2$

d) $0.4(a + b)$

8 The x-intercept of a line is -12 and its slope is 2.5.

a) What is the y-intercept of this line?

b) Express the equation of this line in general form.

c) Determine the equation of a line that is perpendicular to it.

9 Calculate the area of the figures below.

a)

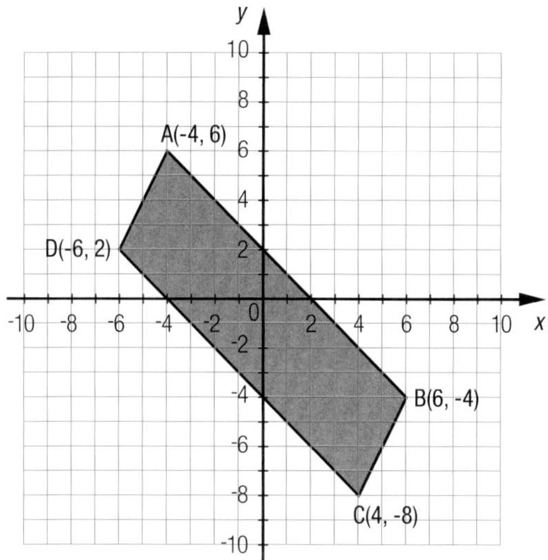

b)

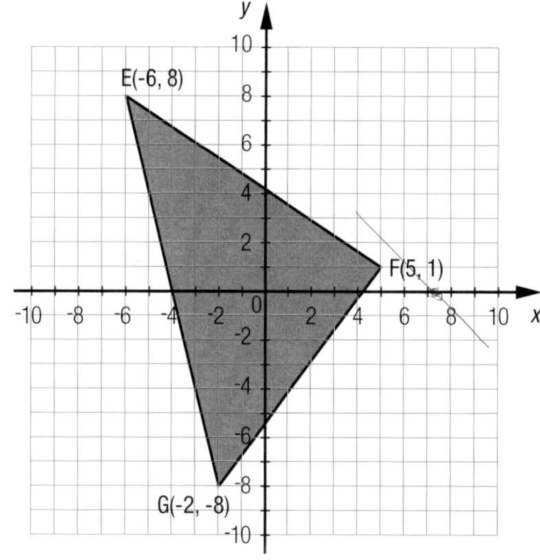

10 Express each equation in general form.

a) $y = -2x - 1$

b) $y = \frac{x}{7} + 2$

c) $y = -\frac{4}{5}x - 33$

11 Express each equation in standard form.

a) $0 = x - y + 15$

b) $x + y = 112$

c) $10x - 2y + 5 = 0$

12 Of the tables of values below, indicate which ones would represent a line that is parallel to the line with equation $3x - 4y - 68 = 0$.

1

X	Y1
2	-15.5
3	-14.75
4	-14
5	-13.25
6	-12.5
7	-11.75
8	-11

X=2

2

X	Y2
-18	10.5
-17	11.25
-16	12
-15	12.75
-14	13.5
-13	14.25
-12	15

X= -18

3

X	Y3
21	30
22	31.333
23	32.667
24	34
25	35.333
26	36.667
27	38

X=21

13 Write the equation in general form of a line that is superimposed on:

a) the x-axis

b) the y-axis

14 In the adjacent Cartesian plane, three lines intersect and define triangle ABC. Determine:

a) the equation of line BC in general form

b) the equation of the perpendicular bisector of BC in standard form

c) if triangle ABC is a right triangle

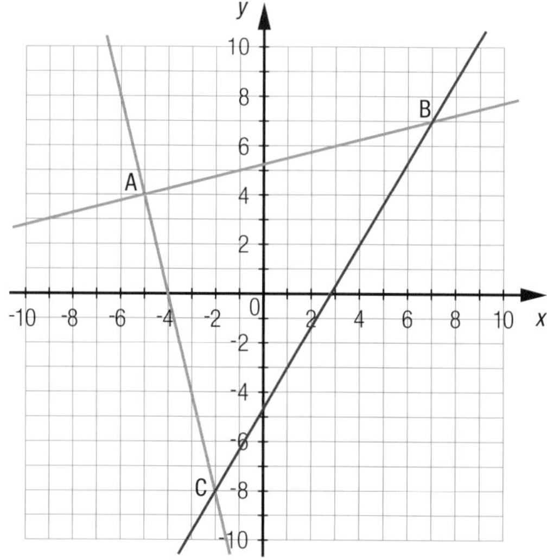

15 In a Cartesian plane where the scale is in centimetres, a line is located 6 cm from point P(18, 15).

a) 1) How many lines, parallel to the *x*-axis, fit the description above?
   2) Write the equations of these lines.

b) 1) How many lines, parallel to the *y*-axis, fit the description above?
   2) Write the equations of these lines.

c) How many lines can fit this description?

d) What geometric figure is formed by the set of points located 6 cm from point P?

16 **AUTOMOBILE INDUSTRY** The automobile industry is using more and more plastic; this makes vehicles lighter and more shock absorbent. The following information shows that plastic makes up an increasingly higher percentage of the mass of a medium-sized car.

Quantity of plastics in cars

Year	1970	1980	1990	2000	2008
Mass of plastic used to build a medium-sized car (%)	6	8	11	14	17

a) Represent this situation using a scatter plot.

b) Draw the line of best fit.

c) Determine the equation of this line.

d) According to this equation, if the mass of a medium-sized car is 8000 kg, what will be the mass of the plastic needed to build it in 2030?

17 The lateral sides of a stainless steel range hood are in the shape of isosceles trapezoids. One of the sides has been drawn in the adjacent Cartesian plane. Considering that the slopes of each of its non-parallel sides are -7.5 and 7.5, respectively, find the area of the side shown.

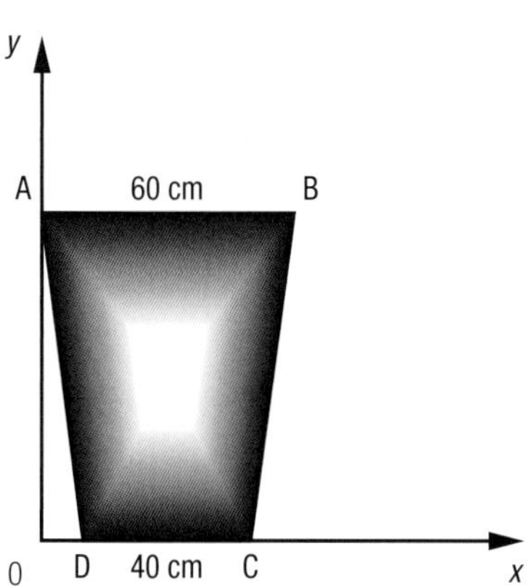

18 For what value of k is the line that passes through points A(-1, 20) and B(k, 5) parallel to a line whose slope is -3?

19 Among the equations in the table below, find those that represent lines that are:

a) parallel

b) perpendicular

1	$y = -7x - 2$
2	$21x + 3y + 6 = 0$
3	$y = -\frac{2}{5}x + 8$
4	$-\frac{7}{2}x - \frac{y}{2} - 1 = 0$
5	$2x + 5y - 40 = 0$

20 a) Describe a procedure that can be used to calculate the distance between two parallel lines.

b) Calculate the distance between the lines whose equations are $y = 8.5x + 20$ and $17x - 2y - 2 = 0$.

21 The adjacent graph provides information concerning the manoeuvres made by a plane approaching a runway.

a) What slope does the pilot choose for his descent?

b) What is the equation of the line, in general form, corresponding to the trajectory of the plane as it descends?

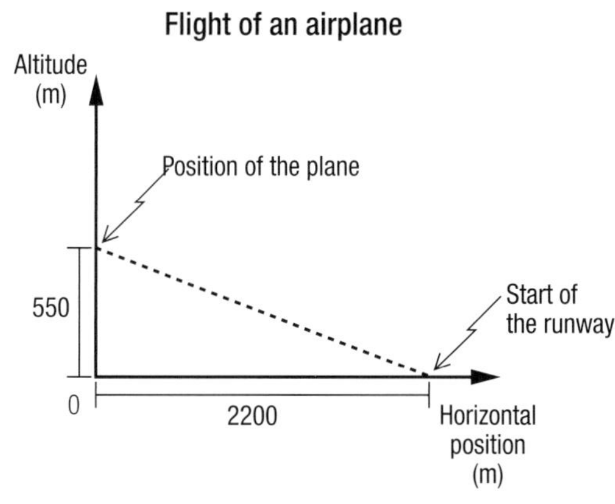

Runways are used for both take-offs and landings. They are generally aligned with the dominant winds to help lift planes during take-off and slow them down as they land using air resistance as an additional brake.

22 When a fire alarm sounds, the firefighters must find the exact location of the fire on a map and determine the shortest route possible to get to the scene. On the graph below, the three main roads of a city and the fire station, represented by point B, are provided. The scale is in kilometres.

A fire breaks out at the intersection of the median from vertex B and the altitude from vertex A.

a) Find the coordinates that define the location of the fire.

b) Calculate the time needed for the firefighters to arrive at the scene if the mean speed of a fire truck is 65 km/h.

c) What is the shortest distance between the fire station and the road AC?

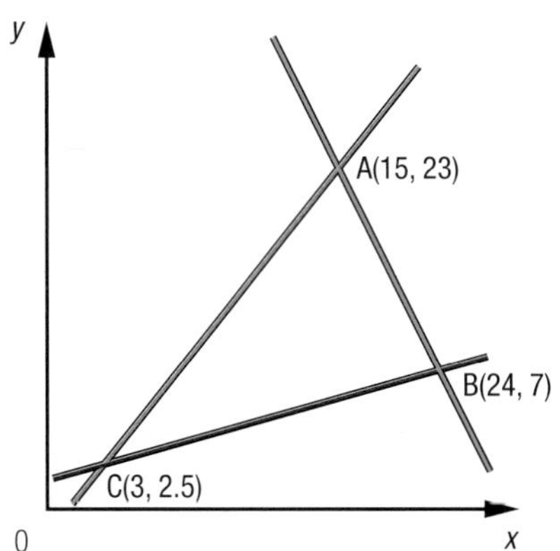

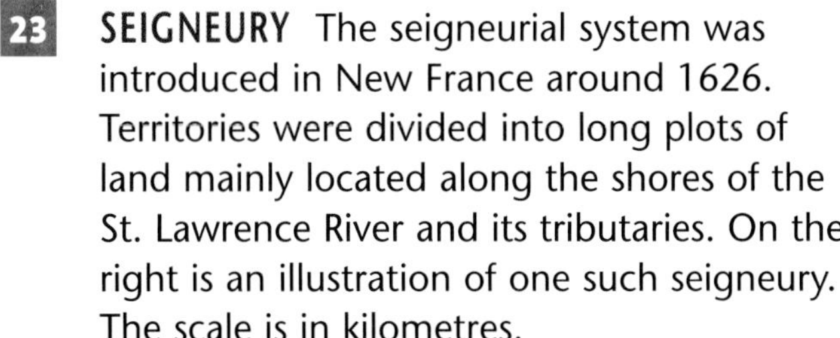

23 **SEIGNEURY** The seigneurial system was introduced in New France around 1626. Territories were divided into long plots of land mainly located along the shores of the St. Lawrence River and its tributaries. On the right is an illustration of one such seigneury. The scale is in kilometres.

a) Determine the equation of the line that passes through points:

1) A and B 2) B and C

3) C and D 4) A and D

b) Find the coordinates of point C.

c) Calculate:

1) the perimeter of the seigneury

2) the area of the seigneury

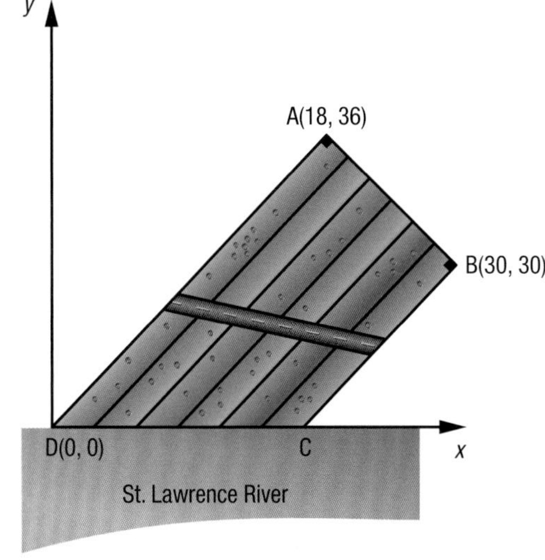

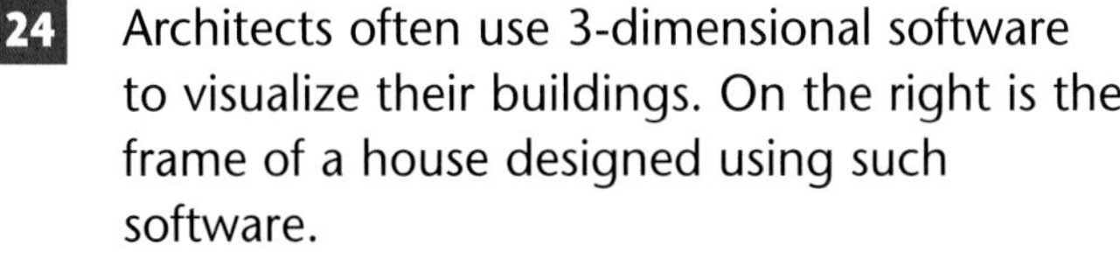

24 Architects often use 3-dimensional software to visualize their buildings. On the right is the frame of a house designed using such software.

a) What is the equation of the line that passes through points A and B?

b) If a person who is 1.77 m tall stands in the centre of the first floor of the house, what is the distance between the top of his head and the highest point of the roof?

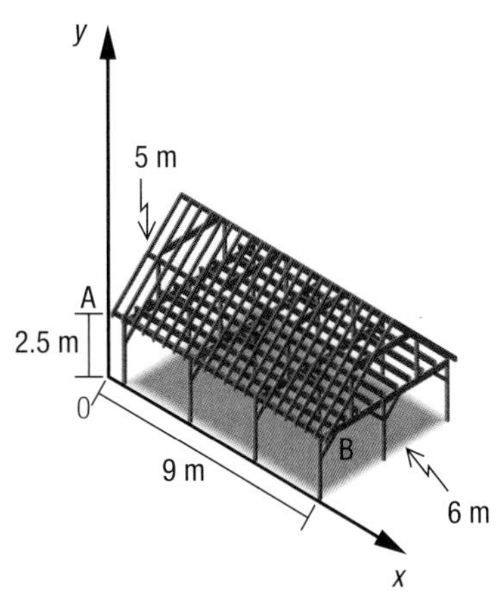

SECTION 3.3 System of equations

This section is related to LES 5 and 6.

PROBLEM Airbus 380

In 2007, the Airbus 380 was the largest commercial aircraft. It was designed to carry up to 850 passengers and its tanks hold nearly 310 000 L of fuel. This plane can fly routes covering distances of up to 15 200 km.

The following representations offer some information on the fuel consumption of two planes during a flight:

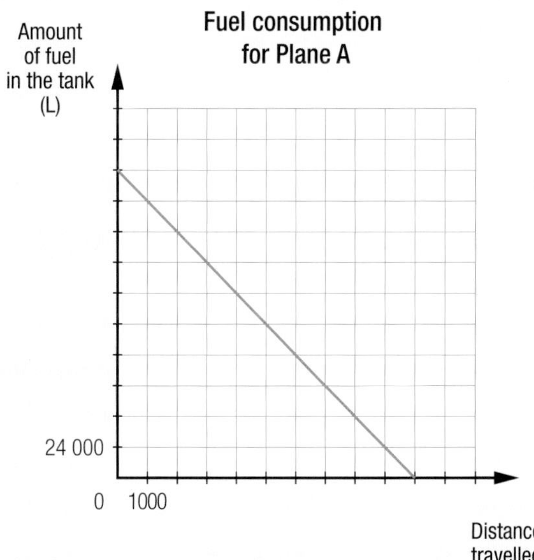

Fuel consumption for Plane B

Distance travelled (km)	Amount of fuel in the tank (L)
0	200 000
300	194 000
750	185 000
900	182 000
1275	174 500

What distance must be travelled so that the tanks of the two planes contain the same amount of fuel?

The Airbus 380 is equipped with an ultramodern cockpit that includes, among other things, LCD and interactive screens and very sophisticated radar systems.

ACTIVITY 1 Crash tests

Before any model of a car is put on the market, it undergoes a large number of performance tests. Experts rate fuel consumption, check braking distance and perform various crash tests in order to provide the best possible product.

Head-on crash test

Specialists use two different speeds for a head-on crash test. The following are some of the requirements that make it possible to establish each of the two impact speeds:

- The mean speed of the vehicle for the two tests must be 80 km/h.
- During the first test, the speed of the vehicle must be 116 less than twice the speed of the vehicle during the second test.

a. Assuming a is the speed of the vehicle during the first test and b the speed of the vehicle during the second test, what system of equations can you establish using the requirements mentioned above?

b. What do you notice about the position of the variables in relation to the equal sign in the equation corresponding to:

1) the first requirement?
2) the second requirement?

c. What equation is obtained by replacing the variable a with its equivalent expression $2b - 116$ in the equation corresponding to the first requirement?

d. Solve the equation obtained in **c.** and interpret the solution in relation to the context.

e. At what speed is the vehicle travelling during the first test?

f. On a Cartesian plane, represent the system of equations that corresponds to this situation and find the intersection point of the two lines. What do you notice?

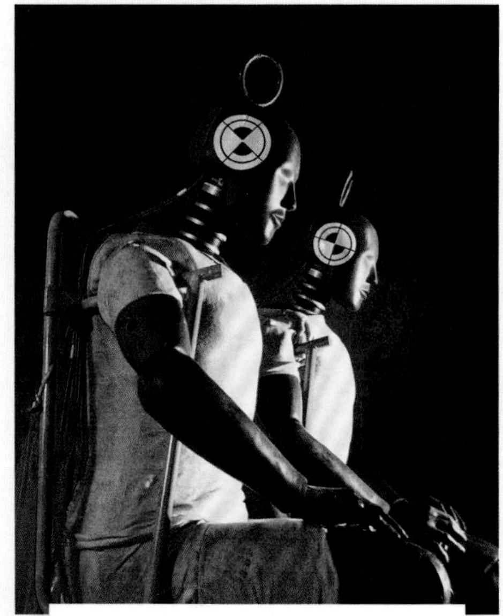

Crash test dummies are used to imitate the behaviour of the human body during crash tests. They represent men, women, and children and may assume the role of driver, passenger or even pedestrian.

ACTIVITY 2 Spring water

Québec has many lakes and rivers. This fresh water resource represents approximately 3% of all the fresh water in the world. In Québec, many companies specialize in the bottling of spring water.

With an area of approximately 2335 km², Lake Mistassini, located near James Bay, is the largest natural lake in Québec.

The following offers some information on a company that produces 500 mL and 1 L bottles:

- In the morning, 30 min are devoted to the production of 500 mL bottles and 60 min to the production of 1 L bottles, for a total production of 1275 bottles.
- In the afternoon, 45 min are devoted to the production of 500 mL bottles and 15 min to the production of 1 L bottles for a total production of 975 bottles.

a. Using the information provided above, what system of equations can you establish considering that x corresponds to the number of 500 mL bottles produced per minute and y to the number of 1 L bottles produced per minute?

b. What do you notice about the position of the variables in relation to the equal sign in the equation corresponding to:

1) the morning production?

2) the afternoon production?

The company's managers want to triple the time devoted to morning production of both bottle formats and double the time devoted to the afternoon production of both bottle formats. This can be represented by the following system of equations.

① $90x + 180y = 3825$

② $90x + 30y = 1950$

c. Will this affect the number of bottles of each format produced per minute? Explain your answer.

d. What do you notice about the coefficient of the x-variable in each of the equations?

e. What equation will you obtain if you subtract each term in Equation ② from its like term in Equation ①?

f. How many bottles of each format will be produced per minute?

ACTIVITY 3 Navigation aids

A lift lock is a hydraulic system that transports boats or ships through waters that are at different levels.

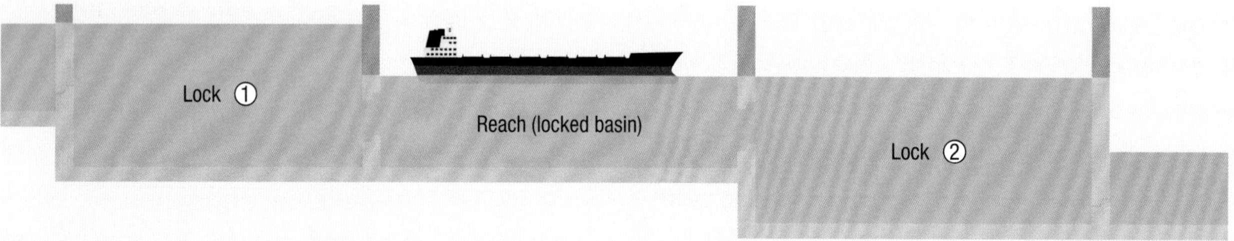

As illustrated above, below is some information about the water level in Lock ① and in the Reach in relation to the time elapsed since the floodgate was opened:

Lock ①

- Water level before the floodgate was opened: 25 m
- Mean change in the water level since the floodgate was opened: 2 m/min

Reach

- Water level before the floodgate was opened: 15 m
- Mean change in the water level since the floodgate was opened: 2 m/min

a. Create a system of equations expressing the water levels l in Lock ① and in the Reach according to the time t that elapsed since the floodgate was opened.

b. Represent this system of equations on a Cartesian plane.

c. Describe the position of the two lines in relation to each other.

d. When will the water in Lock ① and the water in the Reach be at the same level? Explain your answer.

At some point, the operators of the locks notice that the level of the water in Lock ② varies according to the rule $l = 2t + 15$, where l corresponds to the water level in Lock ② and t, to the time that has elapsed since the opening of the floodgate.

e. Create a system of equations expressing the water levels l in Lock ② and in the Reach according to the time t that elapses once the floodgate is opened.

f. Represent this system of equations on a Cartesian plane.

g. Describe the position of the two lines obtained in relation to each other.

h. When will the water in Lock ② and the water in the Reach be at the same level? Explain your answer.

Techno math

A graphing calculator allows you to find the solution to a system of equations using a graphical representation or a table of values.

This screen allows you to edit the equations within a system where *x* is the variable associated with the *x*-axis and *y* is the variable associated with the *y*-axis.

Screen 1

```
X= Plot1  Plot2  Plot3
\Y1=-0.25X+2
\Y2=X+8
\Y3=
\Y4=
\Y5=
\Y6=
\Y7=
```

Screen 2

This screen allows you to define the desired portion of the Cartesian plane.

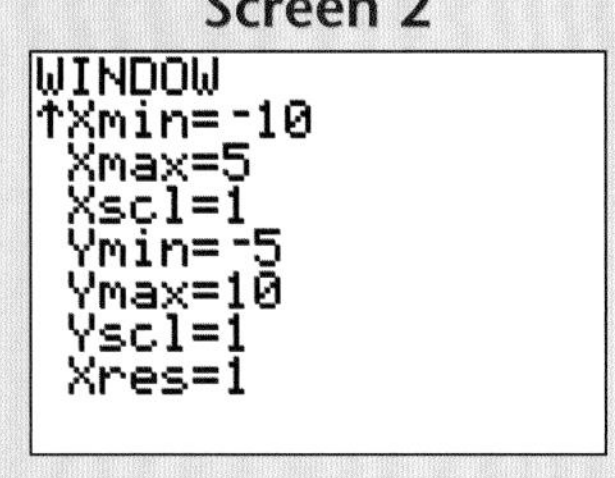

Screen 3

This screen shows the graphical representation of a system of equations. By moving the cursor along the lines, it is possible to approximate the coordinates of the intersection point.

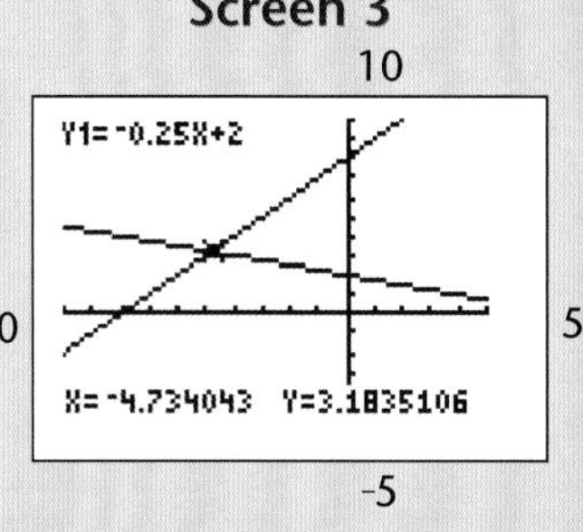

Screen 4

This screen shows the various calculations available on a graphing calculator.

Screen 5

By selecting the two lines and positioning the cursor near the intersection point, the coordinates of this point will be automatically calculated.

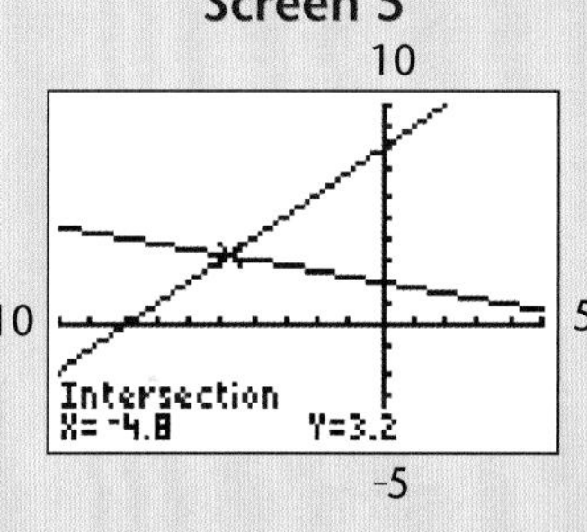

Screen 6

This screen allows you to define the table of values by choosing the starting value and the scale of the *x*-values.

Screen 7

This screen displays the table of values and finds the coordinates of the intersection point.

X	Y1	Y2
-5	3.25	3
-4.9	3.225	3.1
-4.8	3.2	3.2
-4.7	3.175	3.3
-4.6	3.15	3.4
-4.5	3.125	3.5
-4.4	3.1	3.6

X=-4.8

a. According to Screens **5** and **7**, what is the solution to the system of equations displayed in Screen **1**?

b. What do the expressions `Xscl=1` and `Yscl=1` in Screen **2** allow you to determine in Screen **3**?

c. The adjacent screen displays the equations for lines l_1, l_2 and l_3. Using a graphing calculator, find the coordinates of the vertices of the triangle corresponding, respectively, to the intersection points of lines l_1 and l_2, l_1 and l_3, l_2 and l_3.

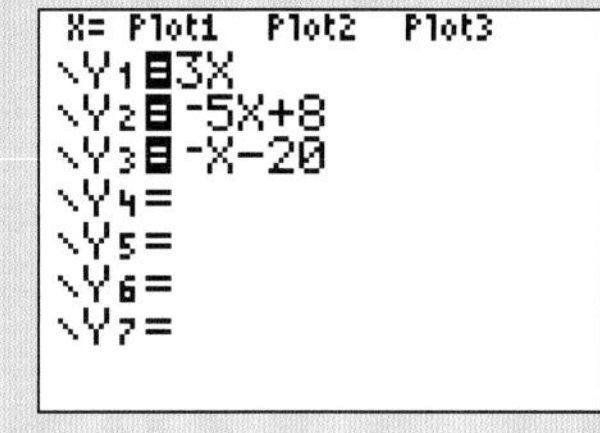

knowledge 3·3

SOLVING SYSTEMS OF EQUATIONS

There are different strategies used to solve a system of two first-degree equations in two variables, in other words, to find a set of values (x, y) that satisfy both equations.

Substitution method

The substitution method uses algebraic manipulations to solve a system of equations

of the form $\begin{array}{l} a_1x + b_1y = c_1 \\ y = a_2x + b_2 \end{array}$ where one of the variables is isolated in one of the equations.

E.g. To solve the system $\begin{array}{l} 4x - 2y = 16 \\ y = x - 5 \end{array}$ using the substitution method, proceed as follows:	
1. If necessary, isolate one of the variables in one of the equations.	$4x - 2y = 16$ $y = x - 5$
2. Substitute this variable into the other equation with the equivalent expression, creating one equation in one variable.	$4x - 2(x - 5) = 16$
3. Solve the resulting equation.	$4x - 2x + 10 = 16$ $x = 3$
4. Substitute the variable obtained into the initial equation in order to solve for the second variable.	$y = 3 - 5$ $y = -2$ The solution is (3, -2).
5. Verify the solution by substituting 3 for x and -2 for y for each of the equations: $4 \times \mathbf{3} - 2 \times \mathbf{-2} = 16$ $\mathbf{-2} = \mathbf{3} - 5$	

Elimination method

The elimination method uses algebraic manipulations to solve a system of equations of the

form $\begin{array}{l} a_1x + b_1y = c_1 \\ a_2x + b_2y = c_2 \end{array}$ where both variables are located on the same side of the equal sign.

E.g.

To solve the system of equations $\begin{matrix} 3x + 5y = 25 \\ x + y = -5 \end{matrix}$ using the elimination method, proceed as follows:

1. If necessary, create a system of equations where the coefficients of one variable are equal or opposite.	$\begin{matrix} 3x + 5y = 25 \\ x + y = -5 \end{matrix} \Rightarrow \begin{matrix} 3x + 5y = 25 \\ 3x + 3y = -15 \end{matrix}$ ($\times 3$)
2. Create an equation in one variable by adding or subtracting the equations.	$\begin{array}{r} 3x + 5y = 25 \\ -\ 3x + 3y = -15 \\ \hline 2y = 40 \end{array}$
3. Solve the resulting equation.	$2y = 40$ $y = 20$
4. Substitute the value obtained into either of the initial equations in order to solve for the second variable.	$3x + 5 \times 20 = 25$ $x = -25$ The solution is (-25, 20).
5. Verify the solution by substituting -25 for x and 20 for y for each of the equations: $3 \times \mathbf{-25} + 5 \times \mathbf{20} = 25$; $\mathbf{-25} + \mathbf{20} = -5$	

SPECIAL SYSTEMS OF EQUATIONS

Non-coinciding parallel lines

Two lines are parallel and non-coinciding when their equations have the same slope but have different y-intercepts. Solving this type of system algebraically leads to an impossibility and therefore has no solution.

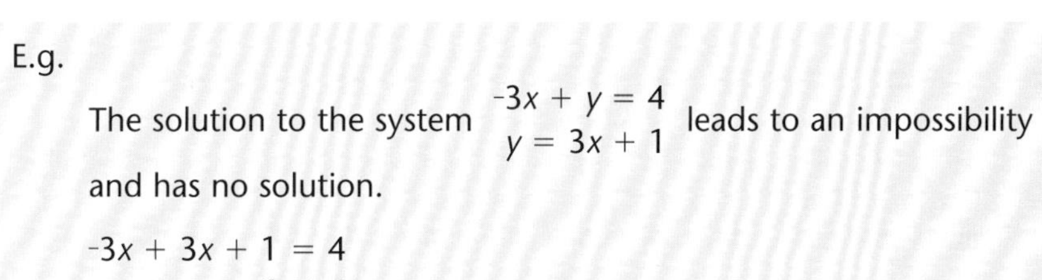

E.g.

The solution to the system $\begin{matrix} -3x + y = 4 \\ y = 3x + 1 \end{matrix}$ leads to an impossibility and has no solution.

$-3x + 3x + 1 = 4$

$\mathbf{1 \neq 4}$

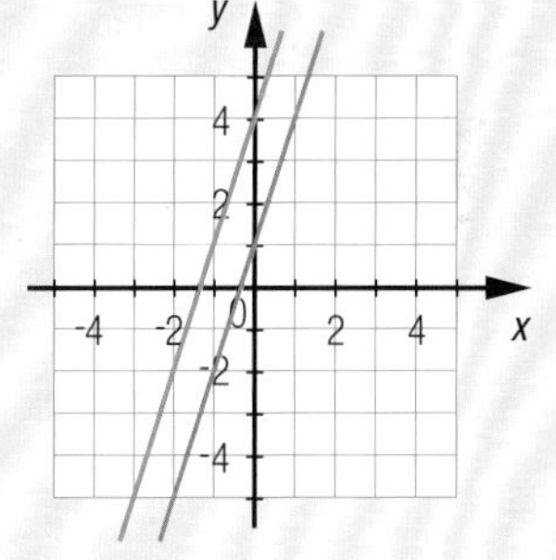

Coinciding parallel lines

Two parallel lines are called coinciding when their equations have the same slope and the same y-intercept. Solving this type of system algebraically leads to a true equality and has an infinite number of solutions.

E.g.

The solution to the system $\begin{matrix} -4x + 2y = -4 \\ y = -2x - 2 \end{matrix}$ leads to a true equality and has an infinite number of solutions.

$4x + 2(-2x - 2) = -4$

$4x - 4x - 4 = -4$

$\mathbf{-4 = -4}$

practice 3.3

1 Use the most appropriate method to solve the equations below.

a) $2x + y = 7$
$x - y = -1$

b) $y = 3x + 1$
$2x + y = 6$

c) $y = 3x - 7$
$y = 2x - 5$

d) $8x - 12y + 4 = 0$
$x + 4y = 5$

e) $2x + 2y + 10 = 0$
$-4x + 3y + 7 = 0$

f) $y = 2x - 4$
$-8x + y = -16$

g) $0.7x - 0.4y = 0.1$
$2x - 0.5y = 0.35$

h) $x + y = 2365$
$1.5x + 0.9y = 2586$

i) $y = 9 - (3x - 7)$
$y = \frac{4(2x - 5)}{3}$

2 For each case, do the following:

1) Identify the unknowns and represent them using different variables.
2) Express the situation as a system of equations.
3) Find the solution.

a) Garo has a choice between two plumbers for his renovations. The first plumber charges \$25/h and a \$50 travelling cost. The second charges \$35/h and a \$20 travelling cost. After how many hours of work will the amount charged by either plumber be the same?

b) Thea places two orders with an electrical supplier. Her first order consists of 50 resistors and 75 condensers and costs \$90. Her second order consists of 125 resistors and 90 condensers and costs \$135. Determine the price of one resistor and one condenser.

c) A school rented a 54-passenger bus for a ski trip. When the bus was completely full, there were twice as many skiers as there were snowboarders. Find the number of skiers and the number of snowboarders that attended the trip.

d) A tray on Scale **A** holds 2 bottles and 5 glasses, and a tray on Scale **B** holds 3 bottles and 3 glasses. In grams, what is the mass of one glass if the total mass of Scale **A** is 440 g and the total mass of Scale **B** is 534 g. All bottles are identical and all glasses are identical.

3 The front view of a house is represented in the adjacent Cartesian plane. Find the coordinates of the rooftop of the house.

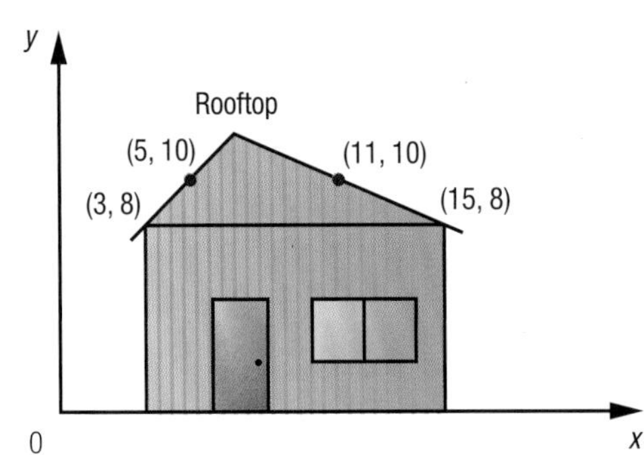

4 Considering that the scales below are balanced, do the following:

1) Express the situation as a system of equations.
2) Determine the mass of each object.

a) **Scale A**

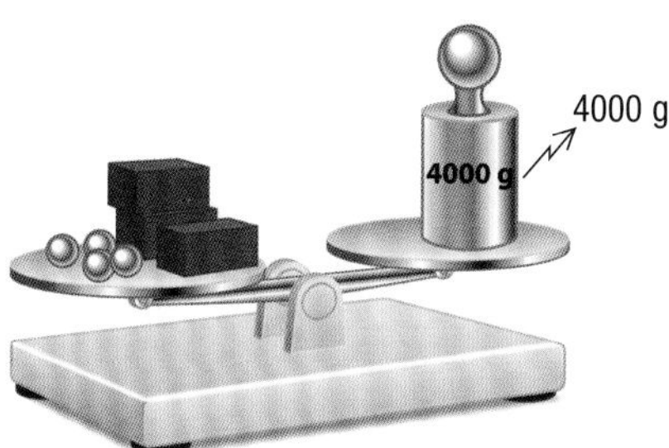

Scale B

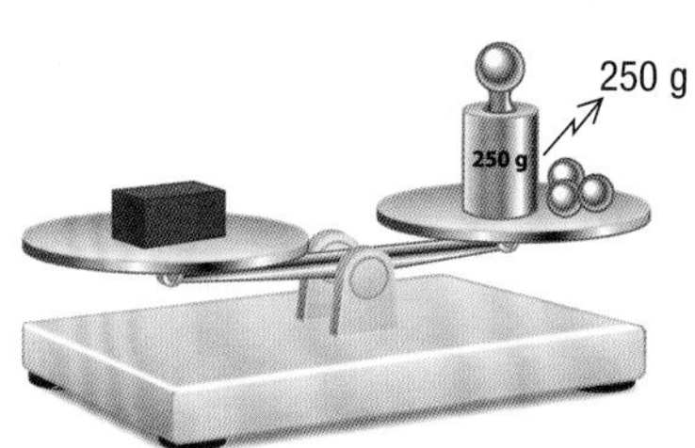

b) **Scale A**

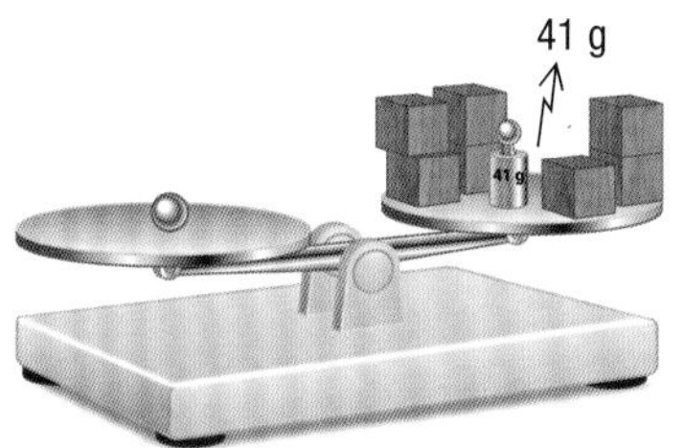

Scale B

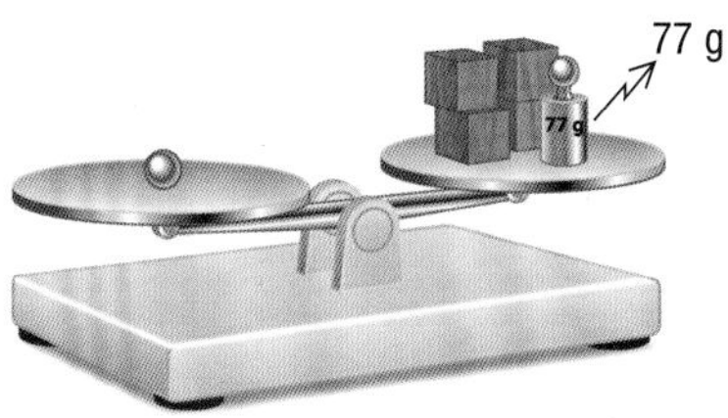

c) **Scale A**

Scale B

5 Determine the equation of a line whose graphical representation is parallel to:

a) a non-coinciding line with equation $^-5x + 2y - 7 = 0$

b) a coinciding line with equation $y = 12x + 5$

6 **MERCURY** The diameter of the planet Mercury corresponds to $\frac{1220}{3189}$ that of Earth. The sum of three times the diameter of Earth and double the diameter of Mercury is 48 028 km. What is the diameter of Mercury?

7 A straight line l_1 whose slope is $\frac{3}{4}$ and whose x-intercept is $^-16$, intersects and is perpendicular to line l_2 whose y-intercept is $^-6.75$. Determine the coordinates of the point where lines l_1 and l_2 intersect.

8 In each case, determine the solution to the system of equations.

a) System of equations ①

x	-1.75	-1.5	-1.25	-1	-0.75
y_1	-37.5	-35	-32.5	-30	-27.5
y_2	-29.5	-31	-32.5	-34	-35.5

b) System of equations ②

x	y_1	y_2
50	480	460
60	580	560
70	680	660
80	780	760
90	880	860

c) System of equations ③

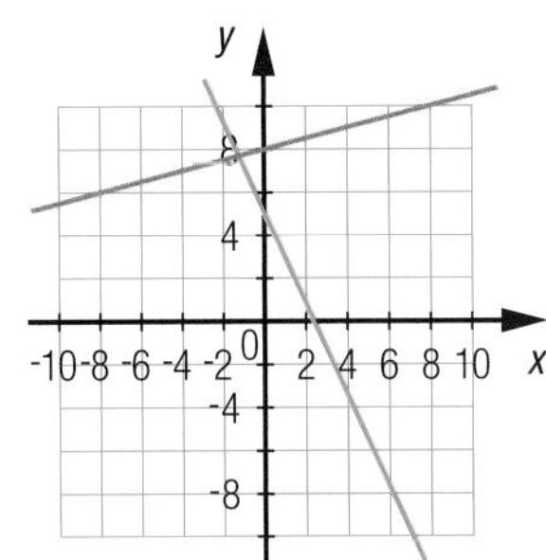

d) System of equations ④

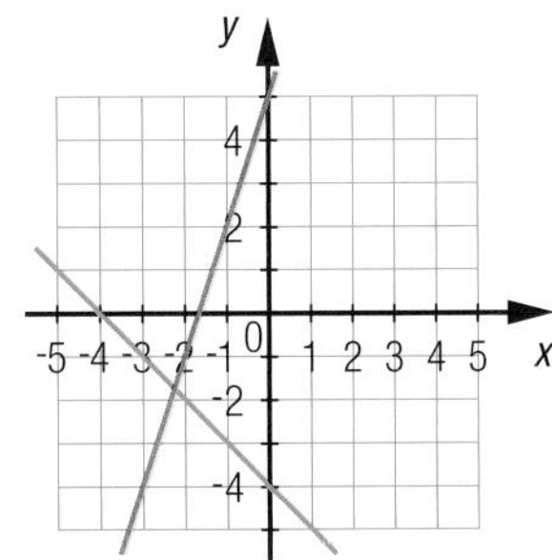

9 Over the past month, 8365 people have visited the exhibit at the Civilization Museum. The price of admission is $12 for an adult and $7 for a child. Considering that the box office received an amount of $77,880, determine the number of adults and the number of children who visited the exhibit.

10 A company specializes in the manufacturing of floor screws. When two machines operate simultaneously for 10 hours, they produce 23 000 screws. If one machine operates for 15 hours and the other for 20 hours, 41 000 screws are produced. How many floor screws does each machine produce per hour?

11 The rules below represent the assets A (in $) of three companies where m corresponds to the number of months elapsed since the beginning of the year.

Company ①	**Company ②**	**Company ③**
$A = 1000m + 12\ 000$	$A = -1000m + 28\ 000$	$A = 2000m + 4000$

Which of the three companies has the best financial situation? Explain your answer.

12 **METEOROLOGY** Fahrenheit measurements in degrees can be converted to Celsius measurements in degrees using the equation $C = \frac{5}{9}(F - 32)$ where C corresponds to the temperature in Celsius and F to the temperature in Fahrenheit. At what temperature are both Celsius and Fahrenheit measurements identical?

1975 was the first year the temperature in Canada was measured in Celsius rather than in Fahrenheit.

13 A graduating student from a Marketing program analyzes two employment offers in sales.

Employer A	Employer B
Base annual salary of \$30,000 plus a commission of 2% of total sales	Base annual salary of \$20,000 plus a commission of 3% of total sales

At what amount of total sales does Employer **B**'s offer become more appealing to the student?

14 In the system of equations $\begin{array}{c} 3x - 2y = 9 \\ 9x = 6y + k \end{array}$:

a) for what value of k is there are an infinite number of solutions for this system?

b) for what value of k is there are no solutions to this system?

15 The adjacent diagram is a representation of the triangular garden Christopher set up behind his house. The scale is in metres.

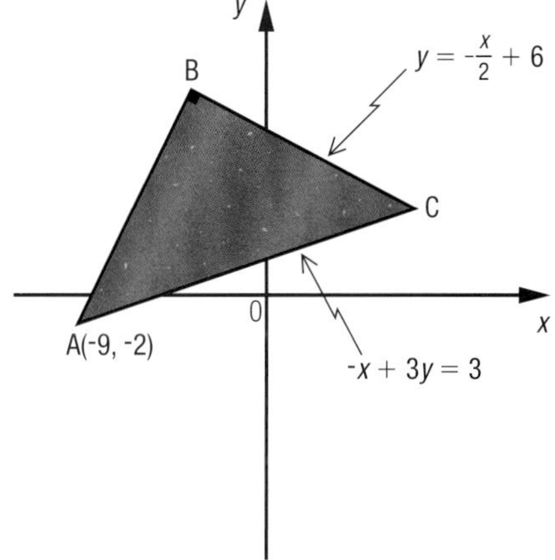

a) What are the coordinates of the points corresponding to the automatic sprinklers installed at the vertices B and C?

b) How much soil must Christopher buy in order to spread a 25 cm thick layer over his garden?

16 For experimental purposes, a cup of water filled to capacity with water at a temperature of 100°C is placed in a freezer. At the same time, another cup is filled to capacity with water at a temperature of 60°C and is placed on a counter. Thirty minutes later, the temperature of the water that was placed in the freezer is 22°C while the temperature of the water placed on the counter is 30°C.

If, in both cases, the change in temperature is constant, when will the temperature of the water in both cups be the same?

17 **SNOWBIRDS** The Snowbirds are Canada's aerial acrobatic team. During air shows, the planes fly a few metres apart. The table below offers some information on the altitude of two of the planes as a function of time during a demonstration:

Time (s)	0	15	30	45
Altitude of plane A (m)	100	150	100	275
Altitude of plane B (m)	200	50	150	175

Considering that these two planes follow linear paths that alternate between ascending and descending manoeuvres every 15 seconds, determine at what moments they will be flying at the same altitude.

SECTION 3.4 Half-planes in the Cartesian plane

This section is related to LES 6.

PROBLEM Internet use

Since 2000, the popularity of the Internet has continued to grow. It is estimated that nearly 80% of Québec families have an Internet service provider. When downloading from the Internet, the megabyte is the main unit of measure. One megabyte is equal to one million bytes.

Below is some information on the Internet use of a family of five:

Internet use in November

User	Time online (min)	Number of bytes downloaded per hour
Andrew	300	5 000 000
Joanne	450	6 000 000
Camille	150	2 000 000
Florence	600	10 000 000
Joey	960	25 000 000

The rule $C = 0.06n + 15.5$ represents the monthly Internet service cost C (in \$) according to the number n of megabytes downloaded during the month of November.

Will this family stay within its \$50/month Internet-service budget?

Binary language is based on two digits, 0 and 1. Depending on their position in a series the 0 and 1 have a specific value: 0 has a value of 0; 1 has a value of 1, but the sequence 10 has a value of 2, and the sequence 11 has a value of 3, etc. Computers are programmed to understand this language and use it to process data which they do very quickly since this language is based on simple concepts.
A byte is a sequence of eight binary digits, either 0 or 1, that can represent 256 different values.

ACTIVITY 1 One situation, many solutions

Everyday situations can often be described with the help of inequalities.

Below are four situations:

Situation ①

Last month, the number of houses h built was more than triple the number of duplexes d built.

$h > 3d$

Situation ②

At my last show, the number of green spotlights g and red spotlights r was at least 80.

$g + r \geq 80$

Situation ③

The degree of precision c of a compass is four times less than the degree of precision g of a GPS.

$4c < g$

Situation ④

Last season, the difference between the number of passes p made and the number of goals g scored was less than or equal to 10.

$p - g \leq 10$

a. In Situation ①, is it possible that 15 houses and 5 duplexes were built during the last month? Explain your answer.

b. In Situation ②, how many green spotlights could there have been at the last show if there were 50 red spotlights?

c. In Situation ③, what is the GPS's lowest degree of precision if the compass' degree of precision is at 15%?

d. In Situation ④, could this player have made 45 passes and scored 30 goals? Explain your answer.

e. How many possible solutions are there for each of these situations?

ACTIVITY 2 Demolition

Demolition engineers generally use a technique called implosion to demolish large buildings. To avoid being injured by falling debris, engineers must follow strict safety rules whenever a fuse is lit. For safety reasons, the distance that separates the engineers from the building must be greater than $\frac{10}{3}$ the height of the building.

An implosion is an "explosion" that is directed inward. In demolition, implosions are used to demolish buildings by making them collapse onto themselves.

a. If x corresponds to the height of the building to be demolished and y corresponds to the required distance that must be respected during an implosion, express this situation as an inequality.

b. If the height of the building being demolished is 54 m, is a person safe at a distance of:

1) 100 m from the building?
2) 150 m from the building?
3) 180 m from the building?
4) 200 m from the building?

This situation is represented in the adjacent graph.

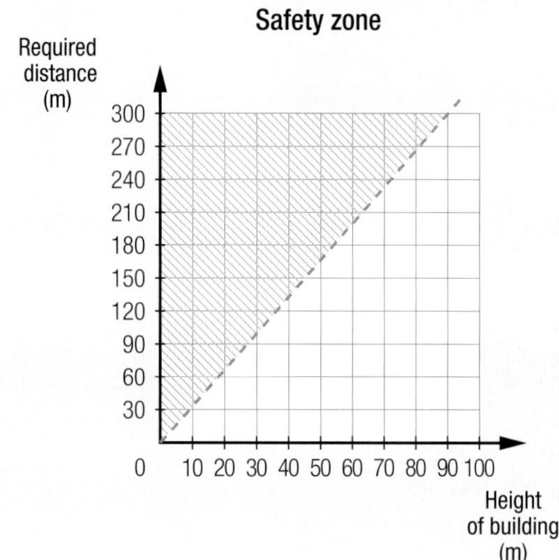

c. What is the relationship between:

1) the coloured half-plane and the symbol used to represent the inequality in this situation?
2) the dotted line and the symbol used to represent the inequality in this situation?

d. Are the following statements true?

1) The coordinates of the points below the dotted line are solutions to this situation. Explain your answer.
2) The coordinates of the points above the dotted line are solutions to this situation. Explain your answer.
3) The coordinates of the points on the dotted line are solutions to this situation. Explain your answer.

After studying the situation, specialists concluded that, for safety reasons, the distance separating the engineers from the building must be greater than or equal to $\frac{10}{3}$ of the height of the building.

e. What effect does this change have:

1) on the inequality that represents this situation?
2) on the graphical representation?
3) on the solution set?

Techno math

A graphing calculator allows you to view the solution set of an inequality on a Cartesian plane.

This screen allows you to enter and edit the equations of one or more curves. The type of line used to draw the curves can also be modified.

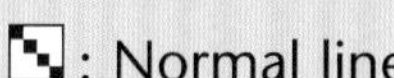: Normal line

: Dotted line

: Thick line

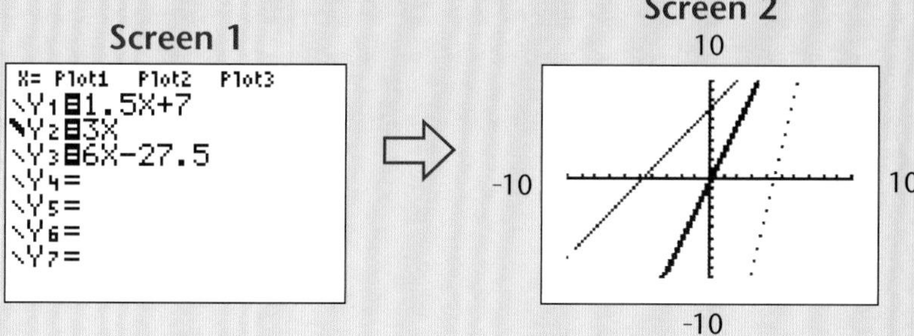

It is possible to represent the solution to an inequality by shading the appropriate region on a Cartesian plane that is bound by a curve.

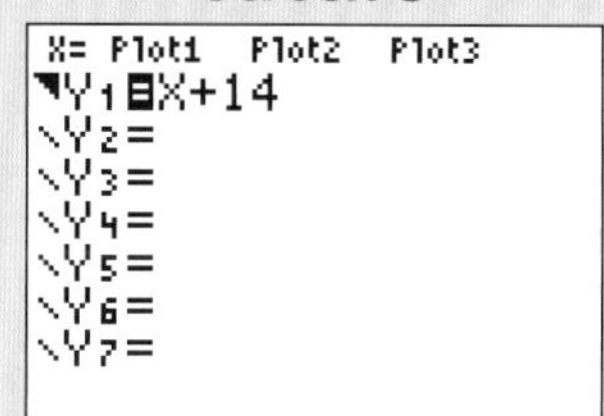

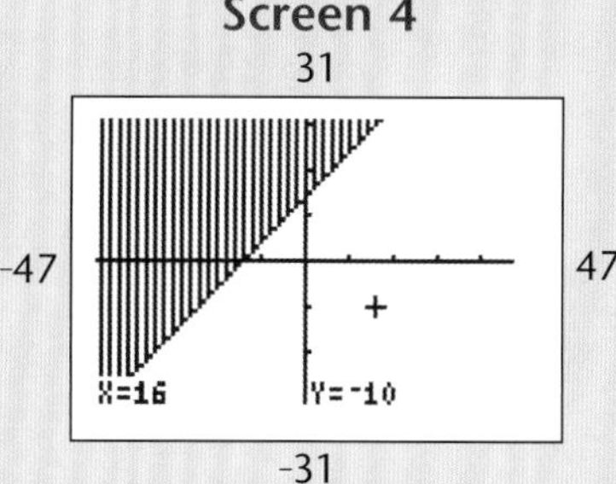

By moving the cursor on the graph screen, it is possible to view the coordinates of the ordered pairs that may, or may not be part of the solution set.

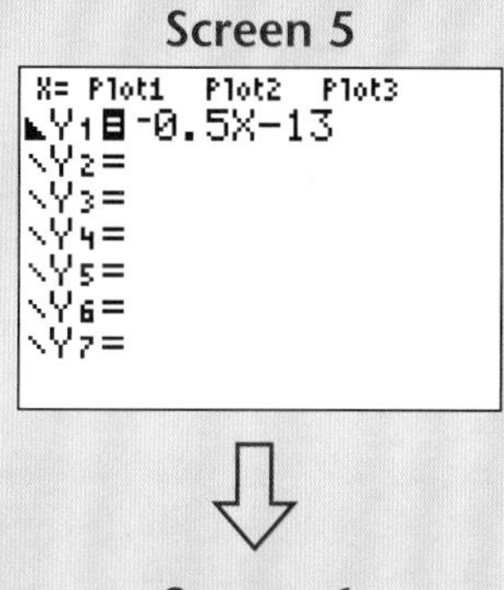

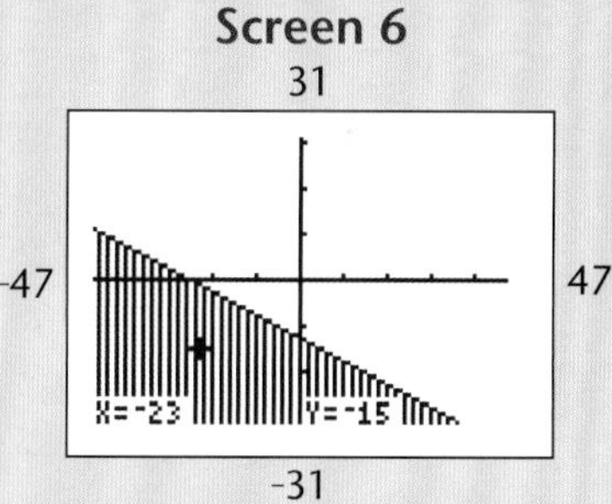

a. What inequality is represented in:

1) Screens **3** and **4**?

2) Screens **5** and **6**?

b. Use algebra to show that the coordinates of the point shown:

1) on Screen **4** do not belong to the solution set of the inequality

2) on Screen **6** do belong to the solution set of the inequality

c. Find the coordinates of a point located:

1) in the 3rd quadrant of Screen **4** and not in the shaded region

2) in the 2nd quadrant of Screen **6** and in the shaded region

d. Using a graphing calculator, graph the solution set of each of the following inequalities:

1) $y \geq 2x + 3$

2) $y \leq -0.25x + 1$

3) $y \geq -3x + 8$

knowledge 3.4

FIRST-DEGREE INEQUALITY IN TWO VARIABLES

To translate a situation into a first-degree inequality in two variables, proceed as follows:

1. Identify the unknown variable(s) in the situation.	E.g. In a store, Type **A** cellphones are sold for \$55 and Type **B** cellphones for \$85. If the total monthly sales for these two types of phones is at most \$3,540, how many cellphones of each type have been sold? The variables are: • the number of Type **A** phones: x • the number of Type **B** phones: y
2. Express the situation algebraically.	The expressions defining this situation are: • the total monthly sales for these two types of phones: $55x + 85y$ • the maximum sales: \$3,540
3. Write the inequality using an appropriate symbol. Once the inequality is written, it can be validated by replacing the variable(s) with numerical values.	Inequality: $55x + 85y \leq 3540$ Validation: For example, 8 phones can be sold for \$55 and 14 phones at \$85. By substituting 8 for x and 14 for y, you get $55(8) + 85(14) \leq 3540$, which results in a true statement: $1630 \leq 3540$.

The solution to an inequality in two variables corresponds to ordered pairs that satisfy the inequality. The complete set of ordered pairs that satisfy an inequality is called the solution set.

Half-plane

It is possible to graphically represent an inequality in two variables on a Cartesian plane.

- All the points whose coordinates satisfy the inequality are found on the same side of the line that corresponds to the equation associated with this inequality. All these points form a **half-plane** which represents the solution set to the inequality. This half-plane is usually coloured or shaded.
- The **boundary line** of a half-plane is a **solid line** when the equality is included in the inequality ($\leq$ or $\geq$) and is a **dotted line** when the equality is not included ($<$ or $>$).

E.g. 1) Solution set for inequality $y \geq 2x - 3$.

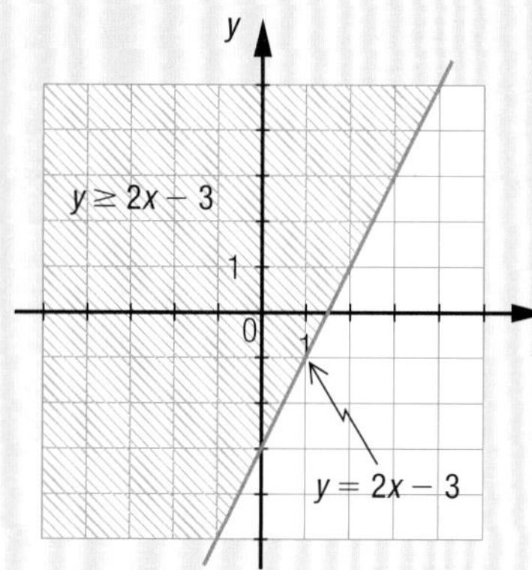

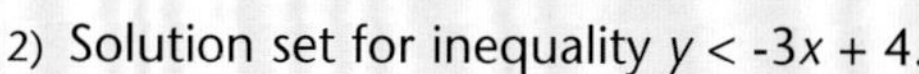
2) Solution set for inequality $y < -3x + 4$.

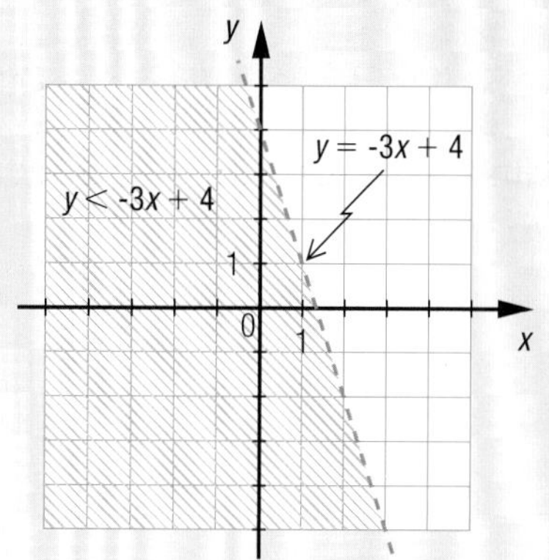

To graphically represent the solution set of a first-degree inequality in two variables, proceed as follows.

1. Write the inequality using the form $y < ax + b$, $y > ax + b$, $y \leq ax + b$ or $y \geq ax + b$.	E.g. Represent the solution set of the inequality $2y < 4x + 7$ on a Cartesian plane. $2y < 4x + 7$ $y < 2x + 3.5$
2. Draw the boundary line of equation $y = ax + b$ using a solid or dotted line, depending on whether or not the equality is included in the inequality.	The equation of the boundary line is $y = 2x + 3.5$. 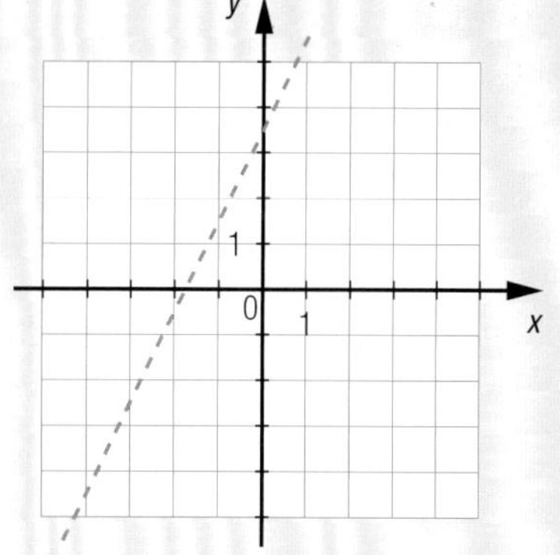
3. Colour or shade the resulting half-plane below the boundary line if the symbol is $<$ or $\leq$, or above the boundary line if the symbol is $>$ or $\geq$.	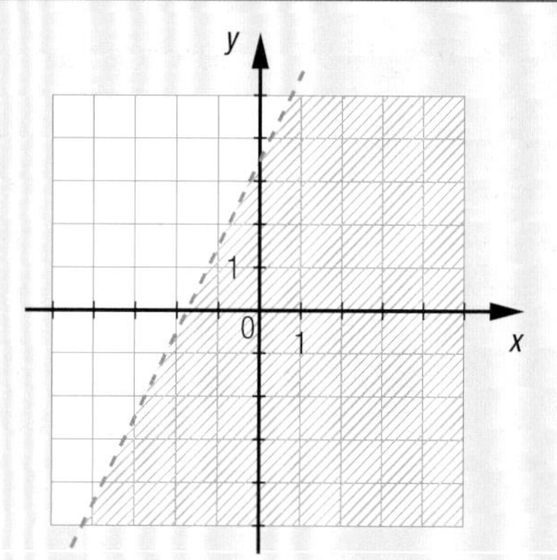

practice 3·4

1 Express each of the following statements as a first-degree inequality in two variables.

a) There are x five dollar bills and y ten dollar bills in a wallet. The total amount of money in the wallet is less than $300.

b) The x number of watts, consumed by a radio is three times more than the y number of watts consumed by a telephone.

c) At a dealership, the x number of cars is at least twice the y number of trucks.

d) On an airplane, the x number of seats in economy class is at least five times greater than the y number of seats in business class.

2 Represent the following inequalities using the form $y \leq ax + b$ or $y \geq ax + b$.

a) $2x + 3y \leq 4$ b) $5x \geq 22 - 2y$ c) $6 - 4y \leq 12x + 26$

d) $\frac{x}{4} + \frac{y}{5} \geq 1$ e) $0.5y + 5 \leq 0.25x$ f) $2(3x - 4) > 3(6 - 2y)$

g) $\frac{2}{3}x + \frac{3}{5}y \geq 3$ h) $\frac{4 - x}{4} + \frac{y - 7}{3} \geq \frac{3}{2}$ i) $\frac{1}{x + y} \leq 2$

3 Considering that the boundry line is included in the solution, associate each of the adjacent equations with one of the calculator screens shown below.

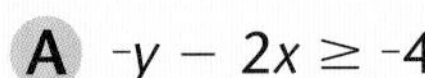

A $^{-}y - 2x \geq {}^{-}4$ **B** $y \geq {}^{-}2x + 4$

C $y - 2x - 4 \geq 0$ **D** $y - 2x \leq 4$

1

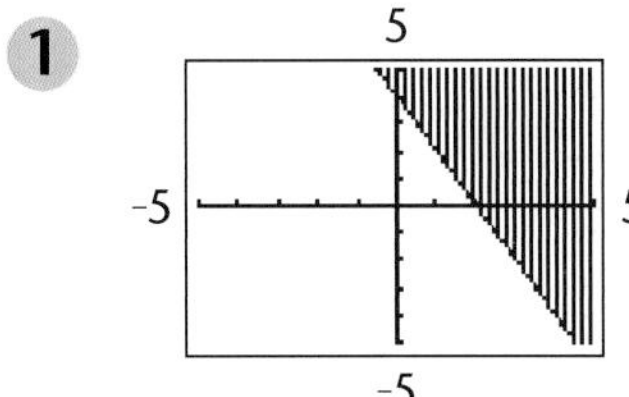

2

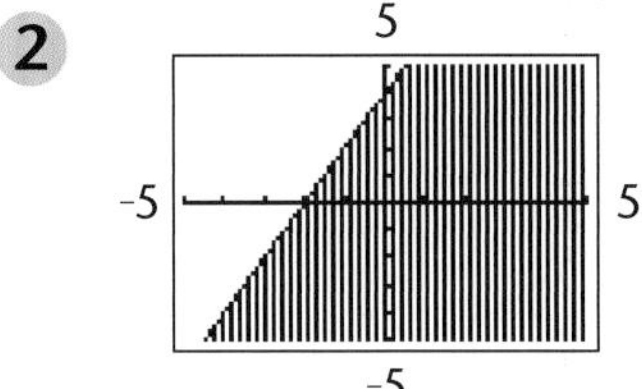

3

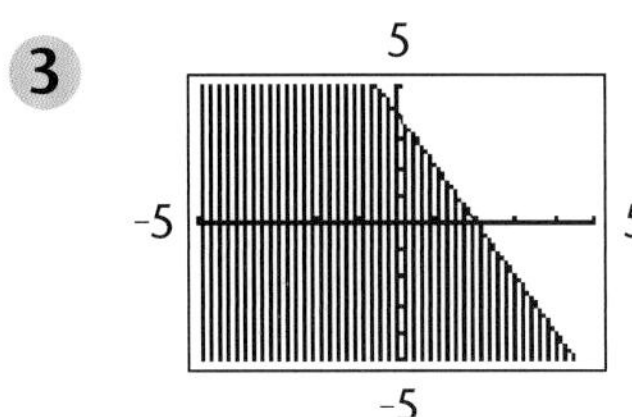

4

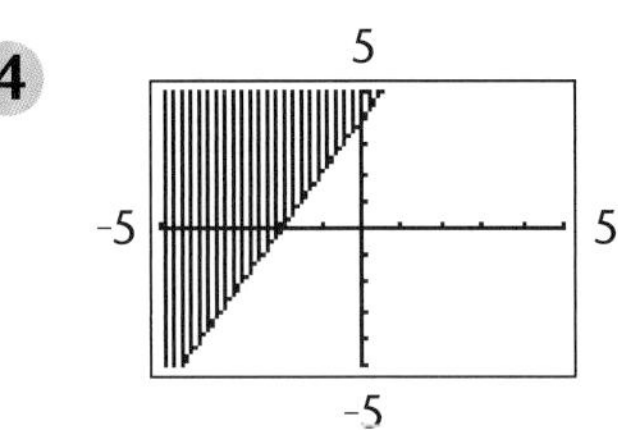

4 Graphically represent the solution set of the inequalities.

a) $x \geq 3$ b) $y < 7$ c) $3x + y \leq 8$

d) $15x - 30y - 60 > 0$ e) $12x - 3y < 15$ f) $\frac{x - y}{3} \geq {}^{-}1$

g) $6x + 3y \leq 9$ h) ${}^{-}0.2y + 0.4x > {}^{-}0.3$ i) $\frac{3x - 4}{2} \geq \frac{5 - 2y}{3}$

5 Describe each situation with an inequality.

a) 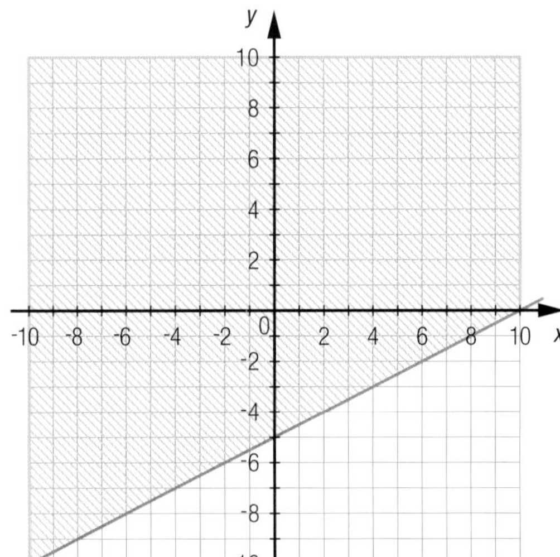

b) 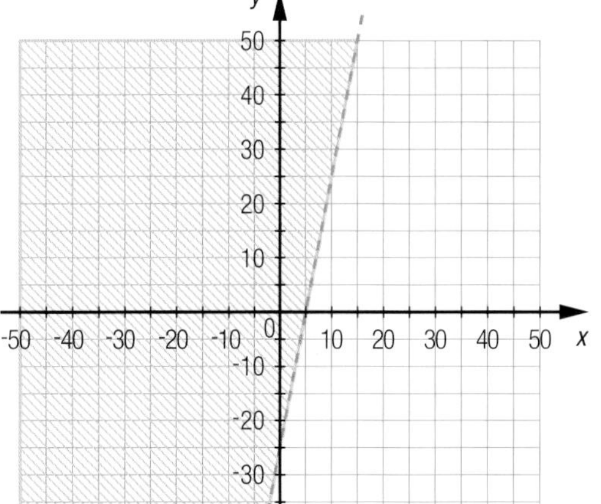

c) 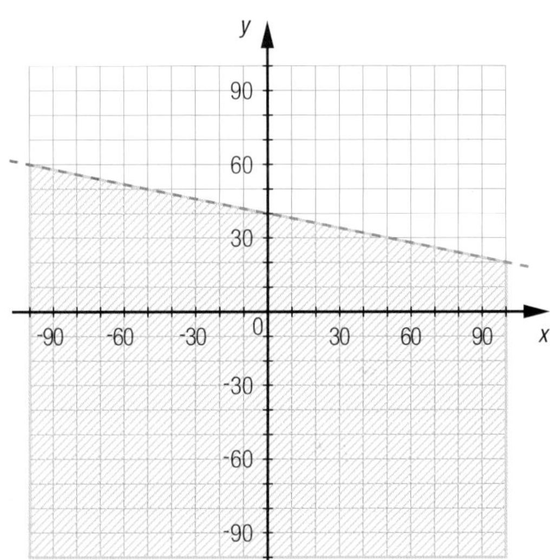

d) 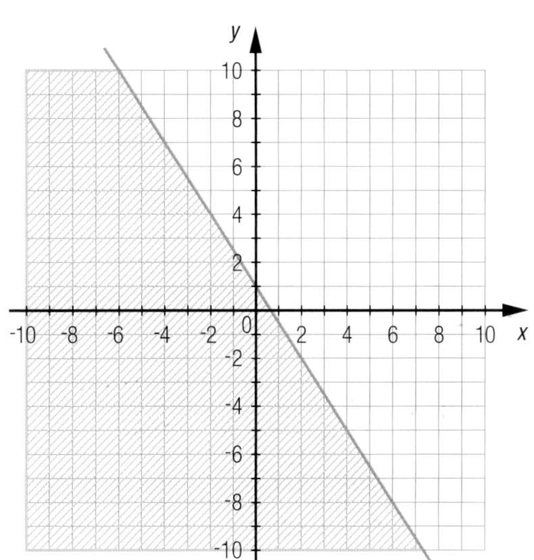

6 For each of the cases below, do the following:

1) Graphically represent the solution set of the inequality.
2) Determine the area of the region that corresponds to the solution set defined by the axes and each of the boundary lines.

a) $y \leq 0.5x + 5$ b) $y - 8x \geq -8$ c) $-4y - 3x - 24 \leq 0$

7 In each of the following cases, identify an inequality having, as part of its solution set, the given ordered pairs.

a) (1, 2), (3, 8), (4, 1), (3, 3)
b) (-1, 3), (-3, 0), (2, 1), (5, 1)
c) (-9, 3), (-2, 3), (3, 3), (7, 3)
d) (-4, 3), (-5, -1), $\left(\frac{1}{2}, 3\right)$, (0, 0)

8 Determine the number of solutions for the inequality $4x + 2y - 8 \leq 0$ where the coordinates of the points are exclusively prime numbers.

9 In each of the cases below, identify the coordinates that are included in the solution set.

a) $y > -3$

b) $x + y - 1 < 0$

c) $y \geq \frac{7x - 9}{3}$

d) $-3x - 4y - 5 \leq 0$

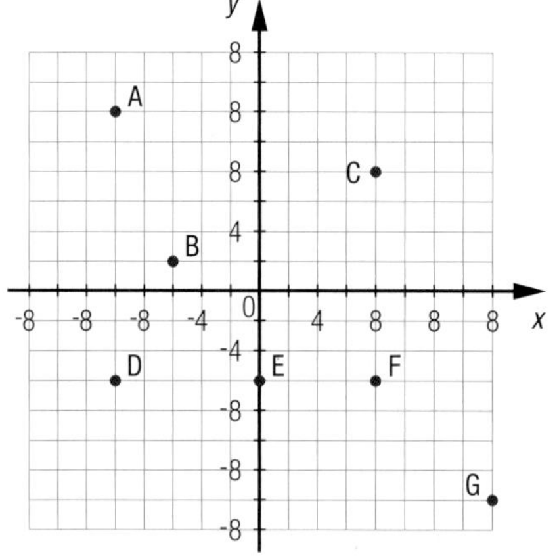

10 On a football team, the sum of the x number of offensive players and the y number of defensive players must be between 24 and 45.

a) Express this situation using two inequalities.

b) What does the blue region in the adjacent graph represent?

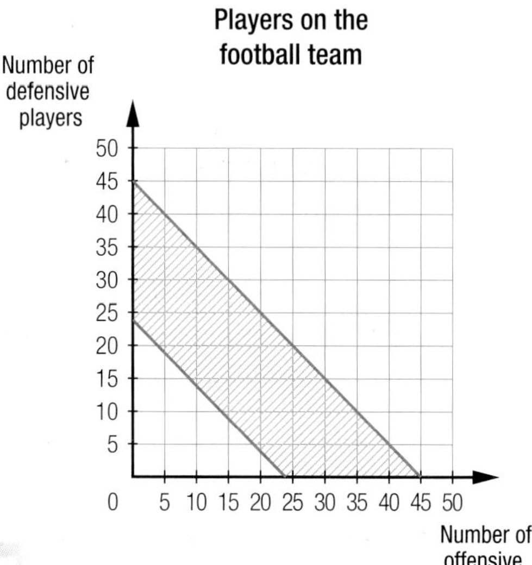

Canadian and American football originated from rugby which has been played in England since the 1800s.

11 A hockey player has graphically represented her summer training schedule in the adjacent graph.

a) Express the situation with an inequality.

b) What does the orange region represent?

c) Is it possible for this player to train for:

1) 15 h?

2) 18 h?

3) 20 h?

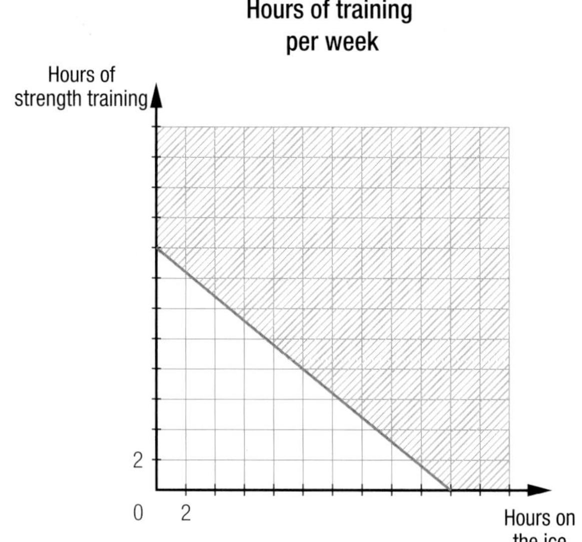

12 An *a* amount of money is invested at an annual interest rate of 5%. A second *b* amount is invested at an annual interest rate of 6%. The combined annual interest of both investments is at most $207. The adjacent graph represents this situation:

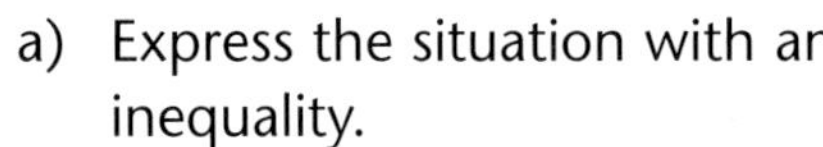

a) Express the situation with an inequality.

b) What is the equation of the boundary line corresponding to the inequality in **a)**?

c) Are the points on the boundary line included in the solution set? Explain your answer.

d) For this situation which Region, **A** or **B**, represents the solution set for the inequality?

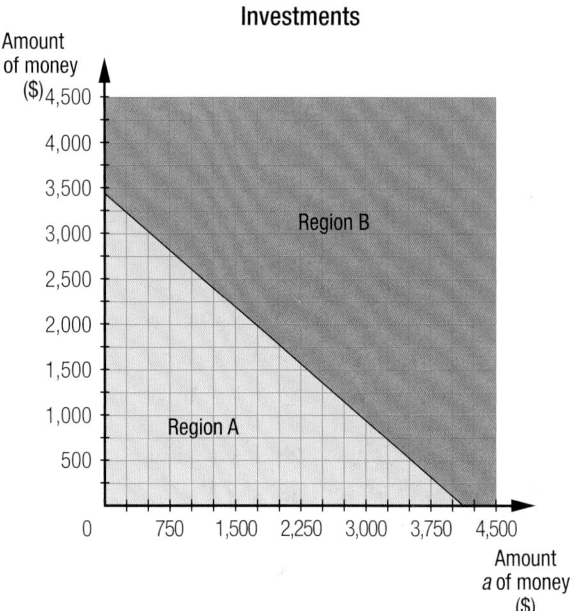

13 A farmer wants to put a fence around his new field. To do this, he must respect certain restrictions. The width of the field must be a minimum of 40 m and a maximum 100 m. The perimeter of the field must be more than 1000 m and less than 2000 m.

a) Identify the two unknowns in this situation, and represent them using different variables.

b) Determine the inequality that corresponds to:

1) the minimum width of the field 2) the minimum length of the field

c) Determine the inequality that corresponds to:

1) the smallest perimeter of this field 2) the largest perimeter of this field

14 An industrial park is marked off in a Cartesian plane by the following inequalities:

$y \geq 4$ $x \geq 3$ $2x + 3y \leq 24$

Find the area of this industrial park. The scale of the Cartesian plane is in kilometres.

Chronicle of the past René Descartes

His life

René Descartes

French philosopher, physician and mathematician, René Descartes made his mark in each of these disciplines. In mathematics he made several discoveries in geometry and algebra. An important law of optical physics bears his name, and in philosophy, he is the author of a famous sentence.

Born March 31, 1596 in Descartes, France; died at age 53 on February 11, 1650 in Stockholm, Sweden.

DISCOURS
DE LA METHODE
Pour bien conduire sa raison, & chercher la verité dans les sciences.
PLUS
LA DIOPTRIQVE.
LES METEORES.
ET
LA GEOMETRIE.
Qui sont des essais de cete METHODE.

A LEYDE
De l'Imprimerie de IAN MAIRE.
cIↄ Iↄ c XXXVII.
Auec Priuilege.

His *Discourse on Method* (1668) is his most famous work. He qualified it himself as his "discourse on how to think correctly and to seek the truth through science."

Algebraic and exponential notation

In Descartes' era, equations such as $4x^2 + 3x = 16$ were written as 4Aq+3A equals 16, where A was the unknown and q stood for quadratus which meant "squared." The notation "4cc+3c.16" was used where c was the unknown and the period represented the equals sign.

René Decartes was the first to propose a simplified notation in which unknowns were expressed as the last few letters of the alphabet and known quantities as the first few. He developed a more practical notation for powers: exponents which considerably simplified written equations.

Using geometry for arithmetic operations

René Descartes demonstrated his creativity by using geometric constructions to find the square roots of numbers and to carry out mathematical operations such as multiplication and division.

Multiplication and division

Descartes used a scale diagram similar to the one shown below to determine the product or the quotient of two numbers. In the diagram below, $\overline{BD} \parallel \overline{CE}$ and $m\,\overline{AD} \times m\,\overline{AC} = m\,\overline{AB} \times m\,\overline{AE}$. Assuming the length of $\overline{AB}$ to be 1, you can conclude that $m\,\overline{AC} \times m\,\overline{AD} = m\,\overline{AE}$, that $\frac{m\,\overline{AE}}{m\,\overline{AD}} = m\,\overline{AC}$ and that $\frac{m\,\overline{AE}}{m\,\overline{AC}} = m\,\overline{AD}$.

For example, Descartes could find the product of 3 and 2 with the help of the adjacent diagram.

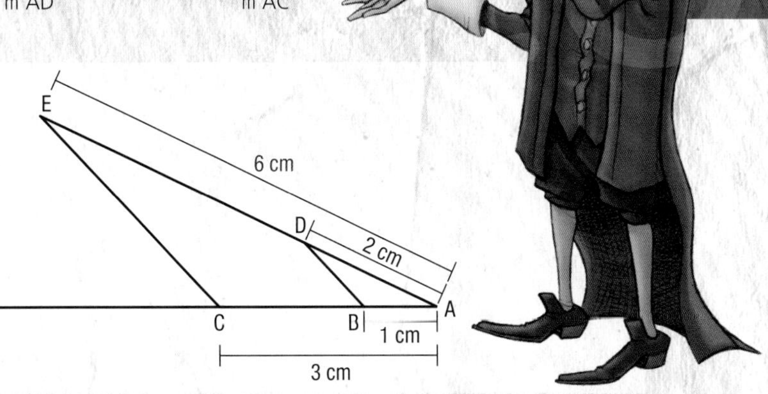

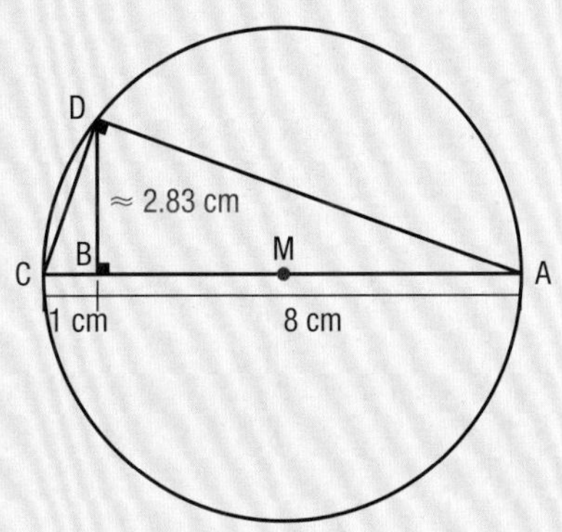

Finding square roots

Descartes used a method based on the adjacent diagram to find square roots. In this scale diagram, $\overline{AC}$ is the diameter of a circle with centre M and $\frac{m\,\overline{BC}}{m\,\overline{BD}} = \frac{m\,\overline{BD}}{m\,\overline{AB}}$. By letting the length of $\overline{BC}$ be equal to one unit, you can deduce that $(m\,\overline{BD})^2 = m\,\overline{AB}$; therefore $\overline{BD} = \sqrt{m\,\overline{AB}}$.

Thus Descartes could, for example, determine the square root of 8 with the help of this diagram.

The Cartesian plane

In some situations, Descartes used algebra to develop geometric concepts. He used a grid to describe equations of lines. Legend has it that the idea came to him from watching a fly walk across the square panes of a window. He realized that he could define the position of the fly with the help of the squares which resulted in the creation of the coordinate system that is used today.

1. Rewrite and simplify this equation using modern notation. Let the unknown value be *x*.

6Aq – 15A – 3 – 9Aq + 17A + 5Aq + 7 equals 0.

2. Rewrite this equation using the notation from the 16th century.

$$9x^2 + 5x = 5$$

3. Using Descartes' geometric method, do the following:

a) Show that $2 \times 4 = 8$.

b) Show that $10 \div 4 = 2.5$.

c) Find the square root of 10.

In the workplace Power linepersons

The profession

The main role of power linepersons is to build power distribution networks. Their tasks include climbing electric towers and installing high-tension lines that carry the electricity produced by hydroelectric dams to distribution stations. Power linepersons sometimes have to do maintenance work on systems to keep them in proper working order. They may also be asked to connect homes and other buildings to the power distribution networks and to install telecommunication cables.

Kilometres of wiring

The electric power system of Québec includes tens of thousands of kilometres of overhead cable. The diameter of this aluminum wire is usually somewhere between 25 mm and 30 mm. While performing maintenance tasks on high-tension, tower-supported lines, these linepersons use a type of basket that is suspended from the lines. The basket can be moved along the lines like a trolley, so the linepersons can inspect them. Sometimes, the basket is suspended under a helicopter and the lineperson is deposited directly on the tower to be repaired.

Situation ① lists the materials linepersons need when they are installing a new high-tension line.

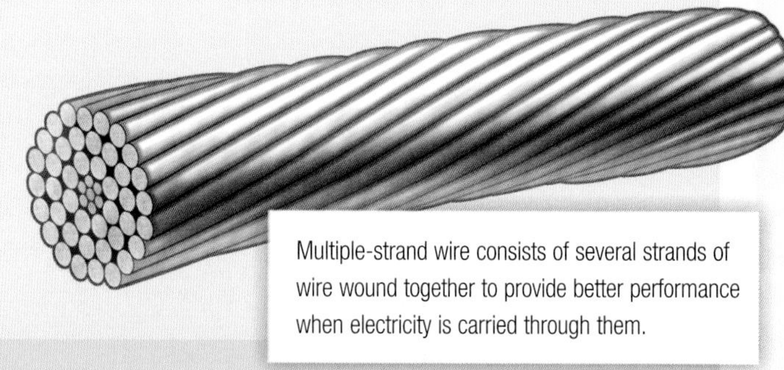

Multiple-strand wire consists of several strands of wire wound together to provide better performance when electricity is carried through them.

Situation ①

- at least 30 km of wire with a 30 mm diameter
- at least 20 km of wire with a 25 mm diameter
- at most 60 km of wire with both of these diameters

The Manic-5 Hydroelectric Dam, the largest multiple arch and buttress dam in the world, has a capacity of 1528 MW.

A calculated risk

This type of work is governed by strict safety standards because power linepersons sometimes work near wires of up to 735 000 V, so they have to make sure they are working in a perfectly insulated environment. Sometimes, this type of work is done on very rough terrain such as when workers install new high-tension lines that will carry electricity from hydroelectric dams to distribution stations.

The adjacent graph represents a high-tension line that has to be installed to provide a distribution station with enough electricity for two different cities. The scale is in kilometres.

Diagram ①
Electric power supply system

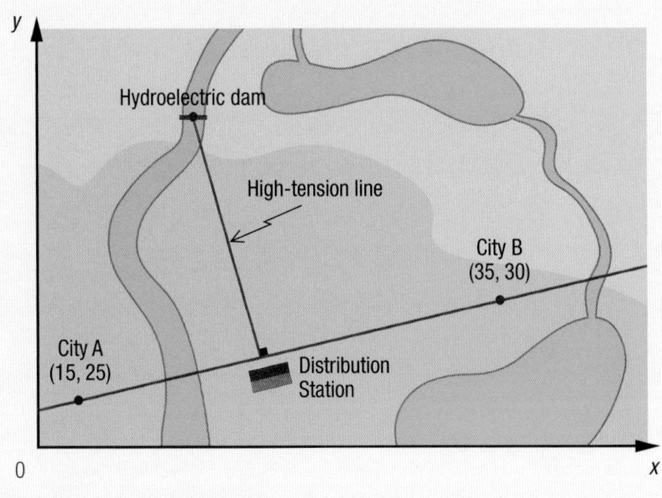

Responding to emergencies

Power linepersons may have to work at all hours of the day or night to respond to certain types of emergencies. The power system may sustain major damage during a natural disaster such as a violent storm. When fire breaks out, the linepersons help firefighters by cutting the power supply to burning buildings.

In the days following the infamous ice storm, linepersons had an enormous task ahead of them.

1. a) Express Situation ① using inequalities.

b) Represent these inequalities on the Cartesian plane.

c) What does the region between the boundary lines of these inequalities represent?

2. a) In Diagram ①, the distribution station is located at $\frac{2}{5}$ of the distance between City **A** and City **B**. Determine the equation of the line corresponding to the high-tension line.

b) What is the equation of the line passing through City **A** and City **B**?

c) What is the distance between City **A** and City **B**?

d) If the distance between the dam and the distribution station is $\sqrt{153}$ km, find the exact coordinates representing the location of the dam.

overview

1 Solve the following systems of equations.

a) $y = 3.5x$
$y = 5x - 1$

b) $-3x + 2y = 4$
$4x - 18y = 10$

c) $-3x - y + 9 = 0$
$y = -x$

d) $4x + 6y = 12$
$-2x + y = 3$

e) $y = -x - 5$
$y = 3x - 21$

f) $-x + 0.1y = 2.8$
$-x - 10.7y = -25.28$

g) $8x + y = -100$
$x = 3y$

h) $y = -0.75x + 10$
$y = 2.25x + 2$

i) $2x + 5.2y = 18$
$x = 2.2y$

2 For each of the polygons below, find:

1) the slope of side AB
2) the perimeter
3) the name of the polygon

a)

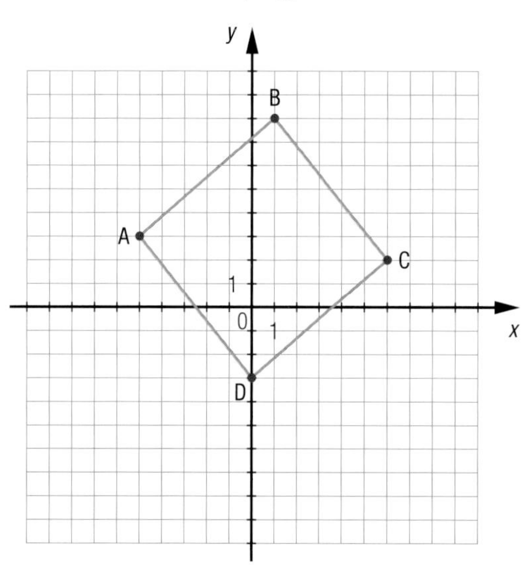

b)

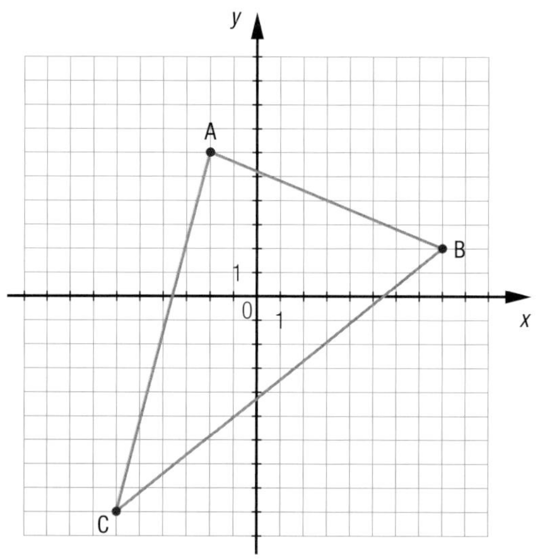

c)

Graph ③

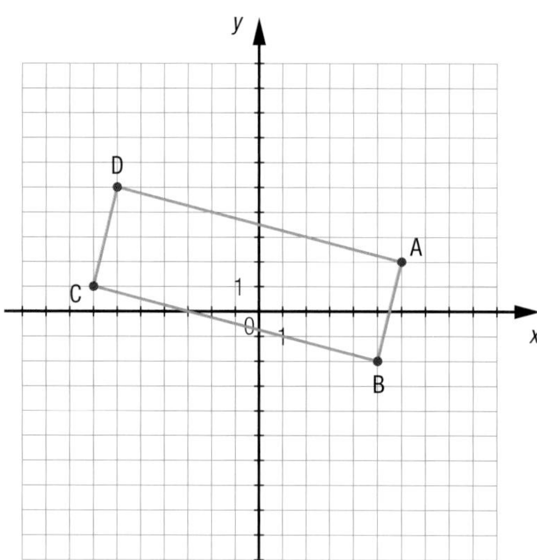

d)

Graph ④

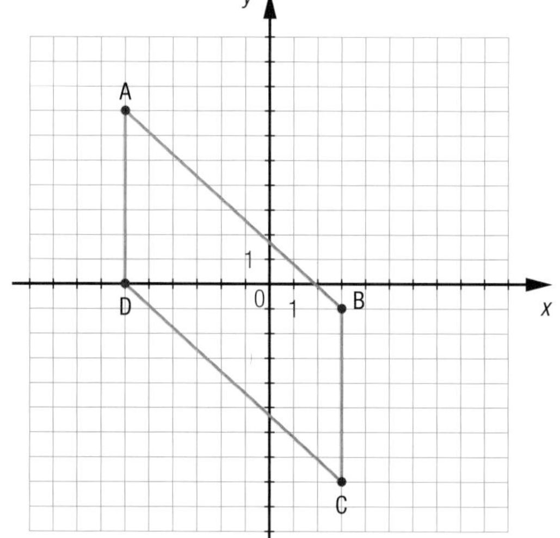

3 Two positive whole numbers are chosen at random such that the sum of the first number and twice the second number is less than 120.

a) If x corresponds to the first number and y to the second, create an inequality that represents this situation.

b) In each of the cases below, determine whether or not the ordered pair is a solution to the inequality found in **a)**.

1) (0, 120) 2) (40, 40) 3) (60, 60) 4) (0, 0) 5) (60, 0) 6) (-70, 180)

4 Graphically represent the solution set of each of the following inequalities.

a) $x \geq 4$ b) $y < 9$ c) $y > -x + 5$

d) $-15x + 3y \leq -120$ e) $\frac{x - y}{2} \leq 5$ f) $\frac{x}{3} + \frac{y}{5} - 2 > 0$

g) $0.4x + 0.2y \geq 1.2$ h) $y < 2x + 3$ i) $3.5x - 2y \leq 0$

5 Determine the inequality associated with each of the solution sets below.

a)

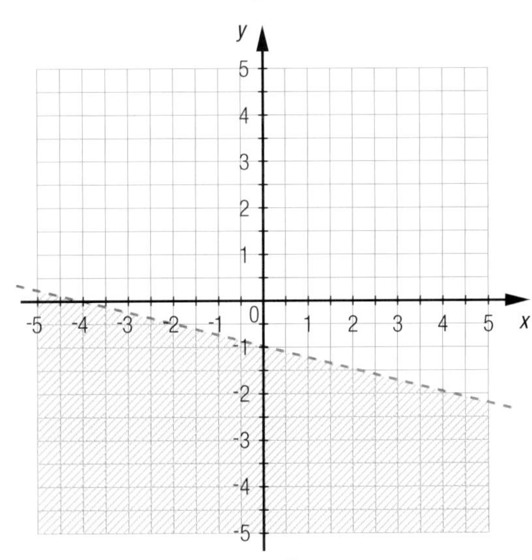

b)

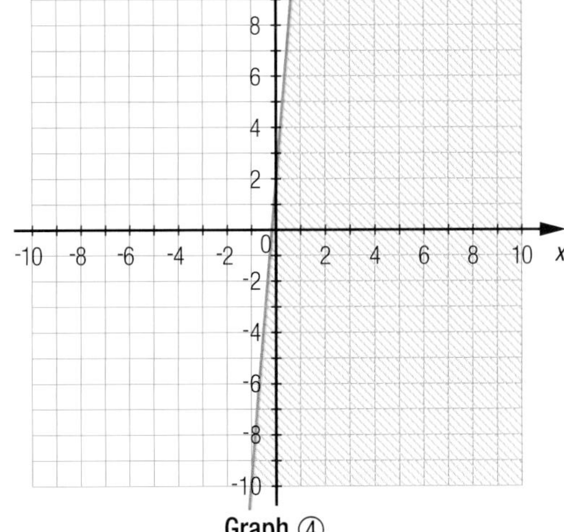

c)

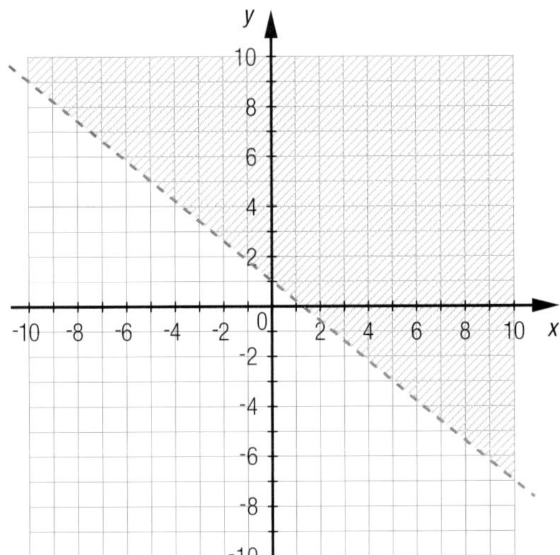

d)

Graph ④

y
10
8
6
4
2
0
-2
-4
-6
-8
-10
-10 -8 -6 -4 -2 2 4 6 8 10 x

6 Determine the coordinates of the point:

a) located $\frac{3}{5}$ of the length of segment CD whose endpoints are C(3, 10) and D(18, 20)

b) located $\frac{2}{3}$ of the length of segment FE whose endpoints are E(-3, 5) and F(15, 26)

c) that divides segment AB whose endpoints are A(8, 7) and B(-16, 2) in a ratio of 2:2

d) that divides segment CB whose endpoints are B(-1, -5) and C(83, 58) in a ratio of 2:5

7 Without manipulating the following equations, describe the position of the two lines in relation to each other. Explain your reasoning.

a) $y = 3x - 8$
$y = 3x + 8$

b) $y = -2x + 5$
$y + 2x - 5 = 0$

c) $y = 4x + 3$
$y = 0.25x + 3$

d) $y = -4x - 1$
$y = 4x - 3$

8 Using the information provided, determine the equation of each line.

a) Lines l_1 and l_2 pass through point P(8, 4). The slope of l_1 is 2, and it is perpendicular to l_2.

b) Line l_3 passes through points A(1, 3) and B(9, 12), and line l_4 passes through points C(9, 12) and D(–6, 4).

c) The slope of line l_5 is negative and forms a 45° angle with the *y*-axis. Line l_6 passes through the origin and forms a 30° angle with the *x*-axis. The two lines intersect at point E $\left(\frac{\sqrt{3}}{2}, \frac{1}{2}\right)$.

9 Based on segment AB, find the possible coordinates of point A considering the following:

a) the coordinates of point B are (2, 5) and m $\overline{AB}$ = 5 cm

b) the coordinates of point B are (-5, -7) and m $\overline{AB}$ = 41 cm

c) the coordinates of point B are (0, 8) and m $\overline{AB}$ = $\sqrt{13}$ cm

d) the coordinates of point B are (6, -3) and m $\overline{AB}$ = $\sqrt{242}$ cm

10 Determine the equation of a line that is:

a) parallel to the line with equation $y = 3x - 12$ and passes through point (2, 1)

b) perpendicular to the line with equation $2x + 2y - 6 = 0$ and has a *y*-intercept of 8

11 For each of the situations below, do the following:

1) Identify the unknowns and represent them using different variables.
2) Express the situation as an inequality.

a) The rhythm of a metronome set to *vivace* is at least 6 beats per minute faster than double the beats per minute of a metronome set to *andante*.

b) A heliostat captures a maximum of 70% more solar energy than fixed solar panels.

c) A carpenter's compass allows him to take measurements at least twice as fast as with an ordinary ruler.

d) The images cast by an overhead projector are at least 15 times bigger than those on an original document.

A heliostat consists of a mirror or set of mirrors that track the movement of the sun and reflect its rays in a single, fixed direction. The energy from these rays can be used to produce electricity or provide natural light in a dim location.

12 Considering that the triangles below are right triangles, find the coordinates of point B in each of the situations below.

a) Point D divides segment AC in a ratio of 4:9.

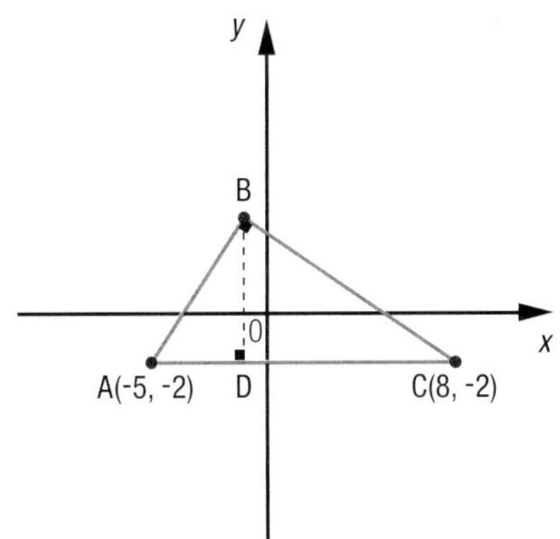

b) Point D divides segment AC in a ratio of 4:16.

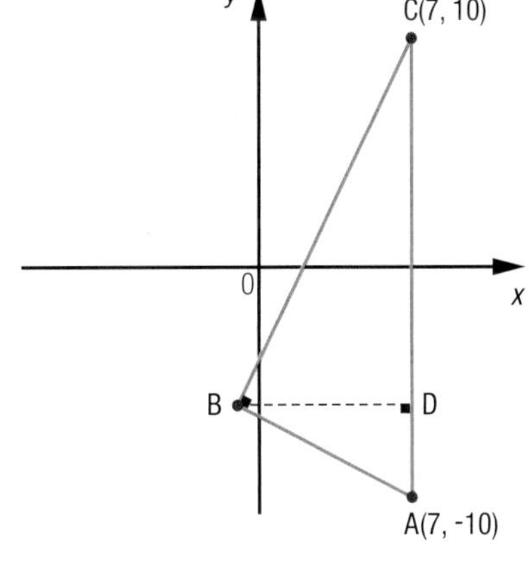

c)

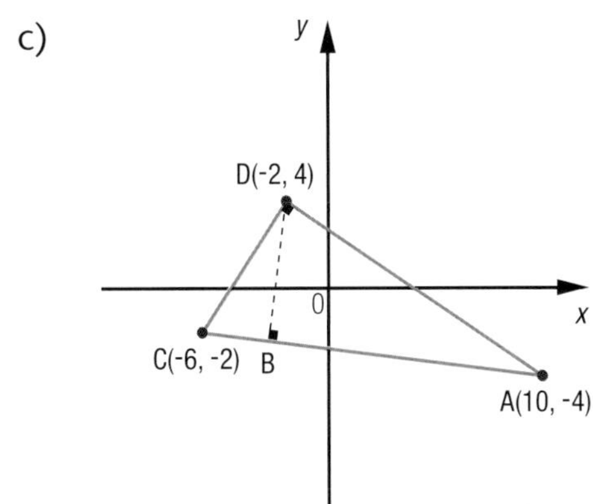

d)

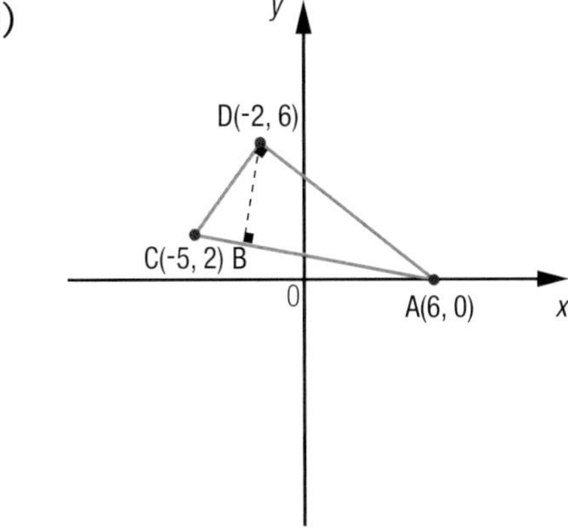

13 **AMBULANCE TRANSPORTATION** The following is information on the flight plan of two helicopter ambulances.

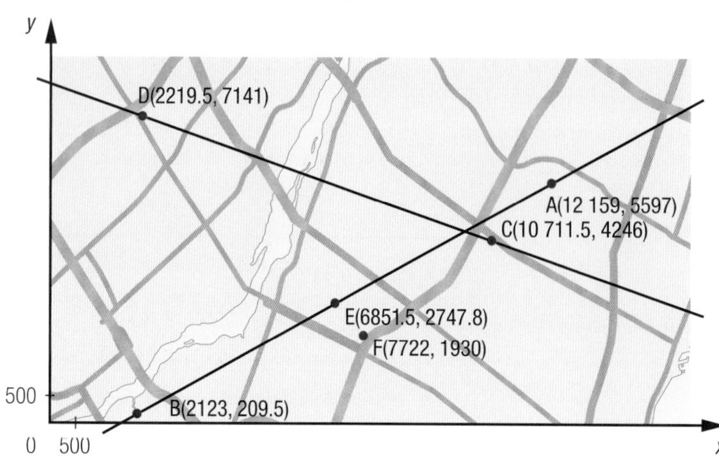

Helicopter ①	Helicopter ②
Departure: Louis-Hyppolite-Lafontaine Hospital (point A)	Departure: Maisonneuve-Rosemont Hospital (point C)
Arrival: Sacré-Coeur Hospital in Montréal (point B)	Arrival: Cité de la Santé in Laval (point D)

a) Considering that the scale is in metres, what distance must Helicopter ① cover?

b) At a certain location, Helicopter ① flies directly over Helicopter ②. What are the coordinates of the point where this takes place?

c) 1) Helicopter ① lands at point E to pick up a patient. What portion of the total flight plan does this represent?

2) When Helicopter ① takes off again, its route must be altered to transport the patient to Cité de la Santé in Laval. What is the equation associated with this new route?

d) The new route charted for Helicopter ③ is parallel to the initial route charted for Helicopter ①. Find the equation associated with this helicopter's route if it passes through point F.

14 **HYDROCHLORIC ACID** Concentrated hydrochloric acid is a colourless or yellowish vaporous liquid used to clean and strip metals, process ore, and manufacture pharmaceutical, photographic and food products. Determine the quantity of 5% hydrochloric acid and the quantity of 20% hydrochloric acid needed to prepare 10 mL of 12.5% hydrochloric acid?

15 The adjacent graph represents the location of three surveying posts. The urban surveyors who placed them there would like to install a fire hydrant at the intersection of the median from vertex A and the altitude from vertex C. Where will the fire hydrant be located?

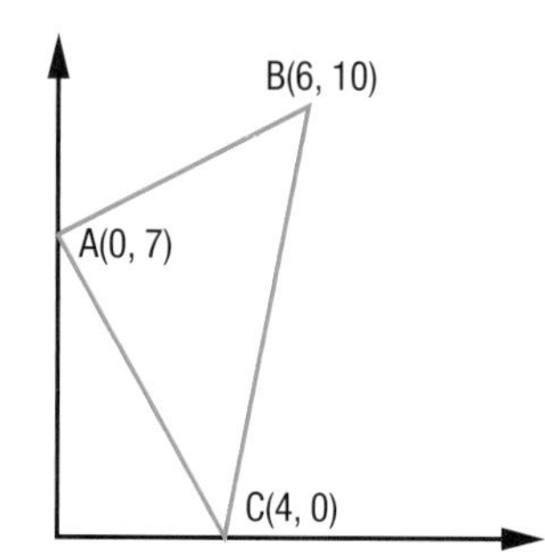

16 Determine the largest area of each of the following rectangular lots, considering the following:

a) The length of the lot is four times its width and its perimeter is 140 m.

b) The perimeter of the lot is 160 m, and twice the length of the lot is 40 m more than double its width.

c) The length of the lot is 30 times its width, and its perimeter is 610 m more than its width.

17 Considering that the centre of gravity is the point where the steel plate should be placed to balance on a pin, determine the coordinates of the centre of gravity of each of the steel plates below.

a) Graph ①

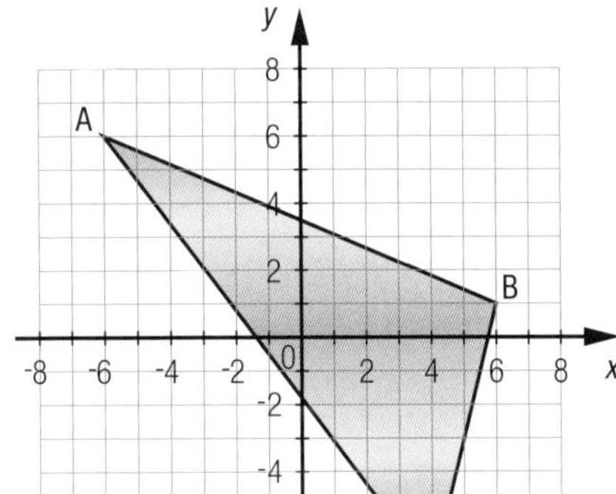

b) Graph ②

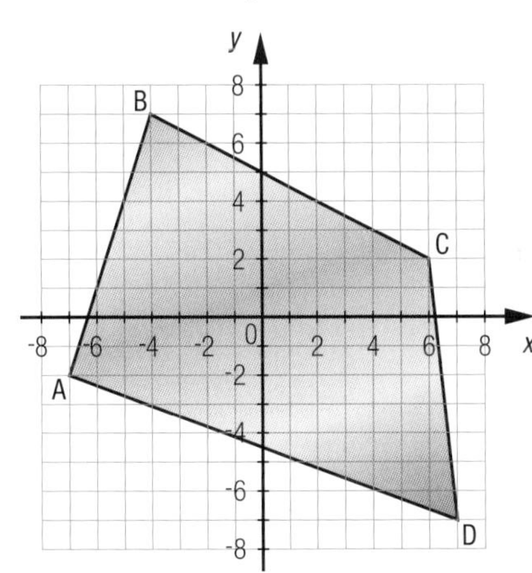

c) Graph ③

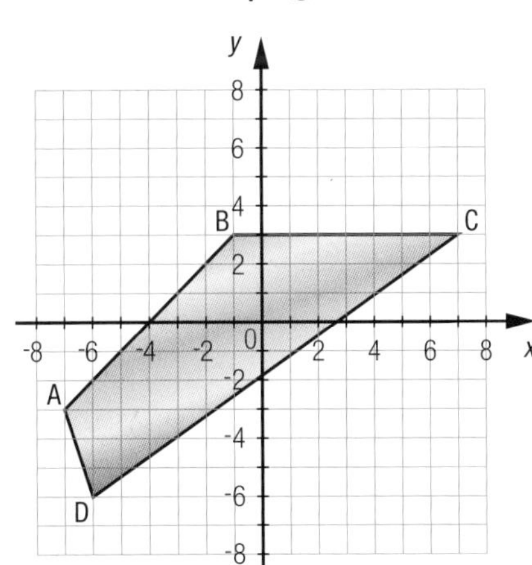

d) Graph ④

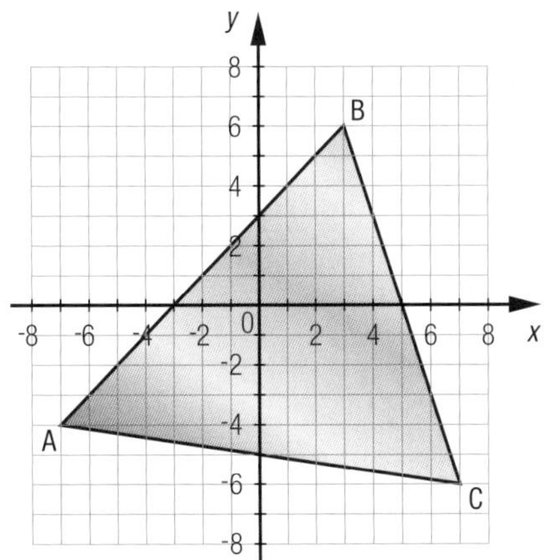

18 An energy company produces oil and natural gas. The company expects that over the next month three times the amount of oil produced, in kL, in addition to twice the amount of natural gas produced, in kL, will total 400 kL. The company's total production must be 175 kL. Expressed in kL, how much oil and how much natural gas will be produced over the next month?

19 **THE PARTHENON** The Parthenon is a structure that is an integral part of Athen's Acropolis. It was built by a military alliance to safeguard the silver confiscated from its enemies. Below is an aerial view of the rectangular floor plan of this structure. The scale is in metres.

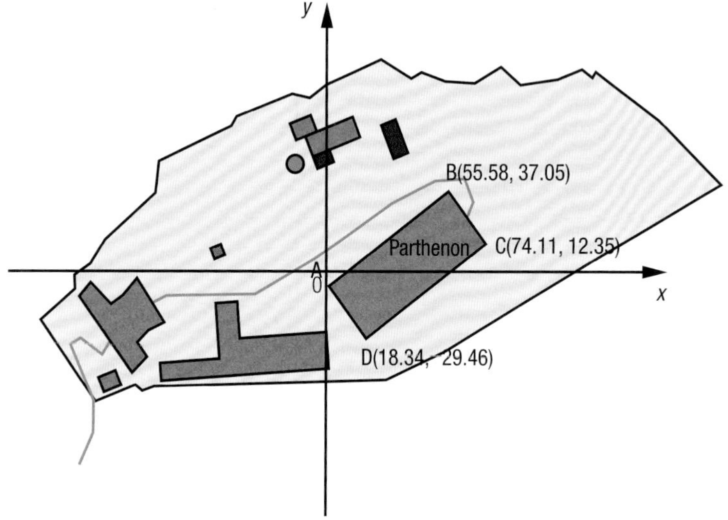

The Parthenon was built in the 5th century BCE. Time, vandalism, pollution, and even fire have caused a great deal of damage to the building, and it has undergone restoration since 1975.

There is a plan to restore the Parthenon by replacing the marble floor slabs. How much would it cost per square metre to restore this floor if the total cost to restore the marble floor is $536,540?

20 As shown in the adjacent map, a navigator has plotted the flight plan of a plane transporting supplies and medication to South Africa. The scale is in kilometres.

a) Determine the distance between:
   1) Johannesburg and Pietersburg
   2) Pietersburg and Pietermaritzburg
   3) Pietermaritzburg and Umtata
   4) Umtata and Johannesburg

b) Are the opposite sides of this quadrilateral parallel? Justify your answer.

c) If the flight plan determined by the officer had been Johannesburg–Pietermaritzburg–Pietersburg–Umtata–Johannesburg, would the pilot's route have been longer or shorter than the original route? Justify your answer.

y
x
(178, 290) Pietersburg
(0, 0) Johannesburg 0
Pietermariztburg (246, -456)
Umtata (64, -706)

Navigators coordinate and direct tactical support missions to help ensure the success of various military air operations such as search and rescue missions, anti-drug operations and humanitarian assistance.

bank of problems

21 The graph below represents an aerial view of the underground passages used by theatre technicians. During a show, someone can enter from underneath the stage through a trap door located at point H. Following is some additional information on these passages:

- Point B is located $\frac{2}{3}$ of the length of $\overline{AC}$.
- $\overline{EH}$ is situated on the perpendicular bisector of $\overline{DF}$.

Considering that the scale is in metres, find the coordinates of the trap door.

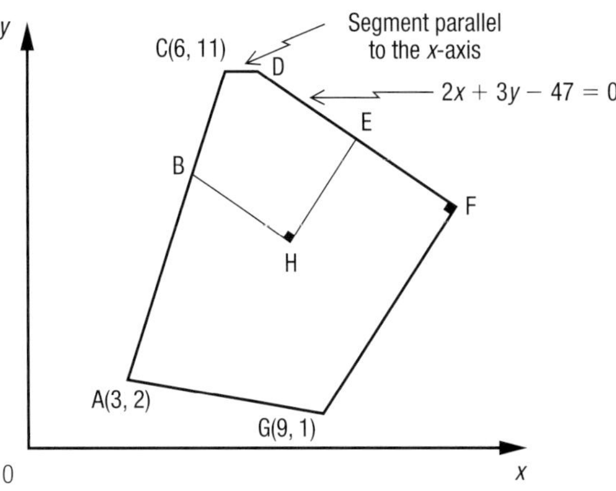

The origins of Western theatre stem from the religious ceremonies of ancient Greece. During these ceremonies, the actors, all of whom were male, put on tragedies in open-air amphitheatres like the one in the adjacent photograph.

22 A graphic designer has to reproduce a shape on a computer using the following information:

- Point B is located $\frac{5}{19}$ of the length of the segment with endpoints A(-8, 10) and H(11, -9).
- Point B is the midpoint of the segment with endpoints C and K(-8, 0).
- Segment CD is parallel to the *x*-axis.
- The perpendicular bisector of segment CD passes through point I(6, -4).
- Point I divides the segment with endpoints F(20, 10) and J in a ratio of 14:5.
- Point E is on segment FJ.
- The angle formed by segment DE and segment FJ measures 90°.
- When points B, C, D, E and I are connected, the desired shape is obtained.

Draw this shape on a Cartesian plane.

23 In a laboratory, two different solutions are used to produce a product called Cecoluse. The first solution contains a 50 mL dose and costs a total of \$12.50; the second solution contains a 350 mL dose and costs \$11. The price of producing 550 mL of Cecoluse is \$60. How much of each solution is needed to prepare 20 L of Cecoluse?

24 A pilot wants to land his plane at an airport and has received landing instructions from the air traffic controller. The diagram below displays his landing instructions. To begin his descent, he must first follow the path represented by line l_1. At a certain point, he has to change course by making a 90° counter-clockwise turn to get to the beginning of the runway located at point A where he will make another 90° counter-clockwise turn to reach his destination at point B.

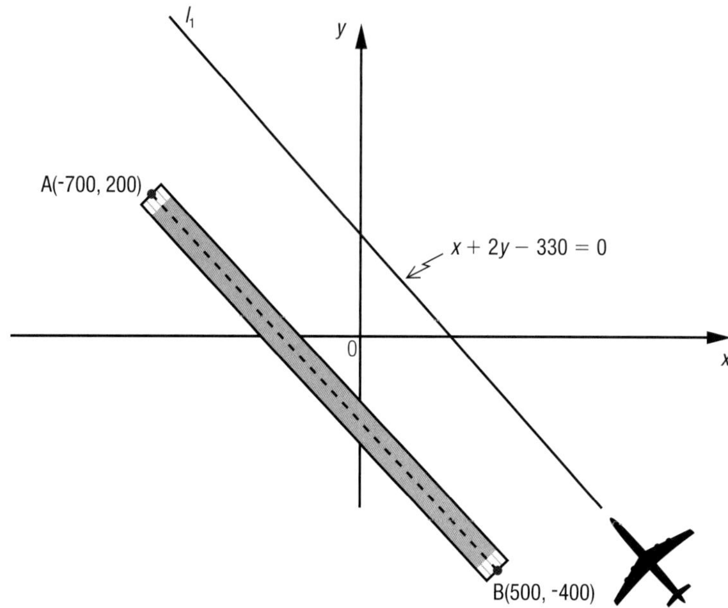

Considering that line l_1 and the runway are parallel and that the scale is in metres, calculate the distance travelled by the plane from the time the pilot made the first turn until the plane reaches point B.

25 Archaeologists search the bottom of the sea in hopes of finding artifacts from the ship *Empress of Ireland* which sank in the St. Lawrence River off the coast of Sainte-Luce near Rimouski on May 29, 1914. To avoid searching in the same place more than once, they marked off the searched territory on a Cartesian plane whose scale is in metres using the inequalities below. On this Cartesian plane, each point whose coordinates are whole numbers, corresponds to one of the locations the archaeologists plan to explore.

On a foggy night, the *Empress of Ireland,* collided with the freighter, the *Storstad.* Hit dead centre, the *Empress of Ireland* sank in only 14 minutes, and 1014 of its 1477 passengers perished. In comparison, approximately 1500 people died when the Titanic sank.

$y \leq \frac{-x + 64}{5}$

$y > 5$

$5x - y - 60 \leq 0$

$y < x + 2$

How many different locations will the archaeologists explore?

LEARNING AND EVALUATION SITUATIONS

TABLE OF CONTENTS

VISION 1

Sustainable development 204

LES 1 A plan for a healthier living environment 205

LES 2 Conserve and save 207

VISION 2

The logging industry 209

LES 3 Reforestation 210

LES 4 Minimizing travel 211

VISION 3

GPS [Global Positioning System] 212

LES 5 Air traffic control 213

LES 6 A precision tool 214

VISION 1

Sustainable development

Learning context

Throughout the ages, people have developed habitats without much concern for the environment. However, in light of current environmental data, we are now being forced to review our habits as consumers, builders and producers. Everything we do from now on must be rooted in an effort to create sustainable development.

A smog-polluted street in Mexico.

Environmental protection must not only include natural habitats and their inhabitants but also focus on our urban environment. We must have an integrated vision for our urban environment and take into account land-use planning, architectural development, materials used, and pollution control measures in terms of atmospheric, environmental, light, and sound pollution.

To develop our habitat in a healthy manner, we need to use technology to ensure that the results of our actions are in harmony with our environment and that we leave a minimal ecological footprint. Our individual health and the health of our entire planet are at stake.

This amazing dwelling, called *Hundertwasserhaus*, was designed by Friedenshreich Hundertwasser and drawn by Joseph Krawina. Its floors are irregular and 250 trees and shrubs are integrated into its architecture.

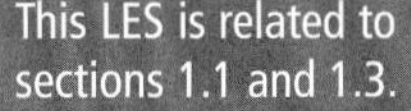

LES 1

C3

A plan for a healthier living environment

The increasing population in large cities has led urban planners to develop new zones for housing construction such as lots near high-power lines and along highways. However, they have a duty to ensure that such locations are not hazardous to people's health.

An urban planner wants to know the development potential of the lot shown in the adjacent diagram. It is located next to a highway, and he wants to turn it into a residential district.

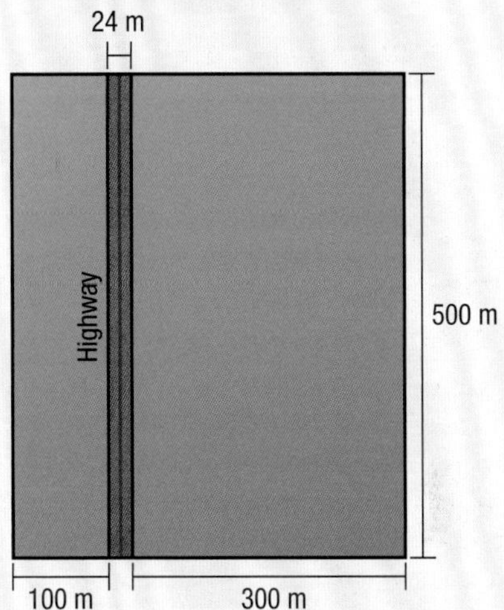

Consider the following information concerning sounds detected by the human ear:

- The human ear captures sound at frequencies from 16 000 Hz up to 20 000 Hz.
- The human ear does not have the same sensitivity at all frequencies. Depending on the frequency output, the sound's intensity will be adjusted as follows.

Adjustment relative to ear sensitivity

Frequency (Hz)	63	125	250	500	1000	2000	4000	8000	16 000
Perceived intensity (dB)	-26	-16	-9	-3	0	1	1	-6	-10

- In the case of highway traffic in which all audible frequencies are present, those frequencies measuring in the vicinity of 750 Hz are the most prevalent.
- For health reasons, it has been established that the mean intensity of sound measured around a house should be no more than 60 dB.

Below is some information regarding sound intensity as measured under these circumstances:

Sound intensity

Distance from the highway (m)	1	2	4	8	16	32	100	250
Intensity (dB)	80	77	74	71	68	65	60	57

Below are the acoustic measurements for three sound barriers:

- A vegetation barrier absorbs a maximum of 16 dB when the wave frequency is 8000 Hz.

Sound absorption of a vegetation barrier

Frequency (kHz)	4	6	8	10	12	14	16
Sound reduction (dB)	13.1	15.2	16	15.1	13.0	9.3	6

- A concrete-panelled barrier reduces sound by 10 dB for all audible frequencies.
- The graph below provides information on a barrier made of recycled material.

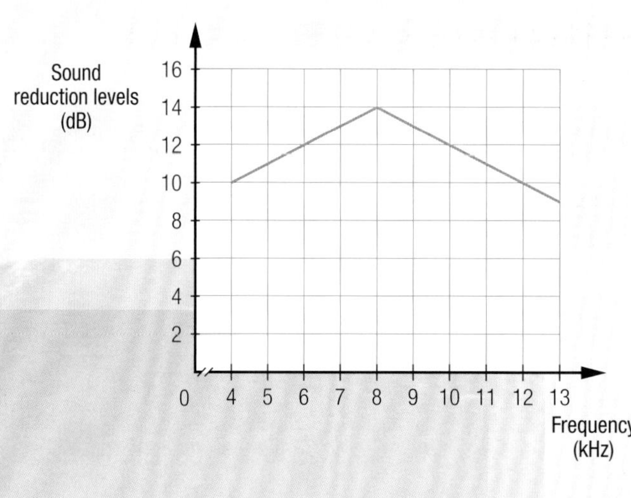

Your task is to draw a noise map for the lot presented on the previous page. Your map will show the high-risk zones (more than 60 dB), the moderate risk zones (55 to 60 dB), and the low-risk zones (less than 55 dB). You must also recommend the installation of one of the three types of sound barriers along the highway.

This LES is related to sections 1.2 and 1.3.

LES 2

C2

Conserve and save

Controlling the inside temperature of our dwellings is essential in a country where the outside temperature varies from -50°C to 35°C. Many companies are working to improve heating and air conditioning systems in order to minimize the power and operation time of the systems as well as to improve their performance. Electricity consumption is most often measured in kilowatt-hours (kWh). One kilowatt-hour is equivalent to 1000 watts consumed in one hour.

Programs such as ecoENERGY Retrofit in Canada, or Rénoclimat in Québec, have been set up to help homeowners who want to improve the energy performance of their homes.

An air conditioning contractor presents the benefits of a new type of air conditioner with a variable-speed compressor by comparing it to the standard on/off model.

The graphs below provide information on the performance of a standard on/off air conditioner:

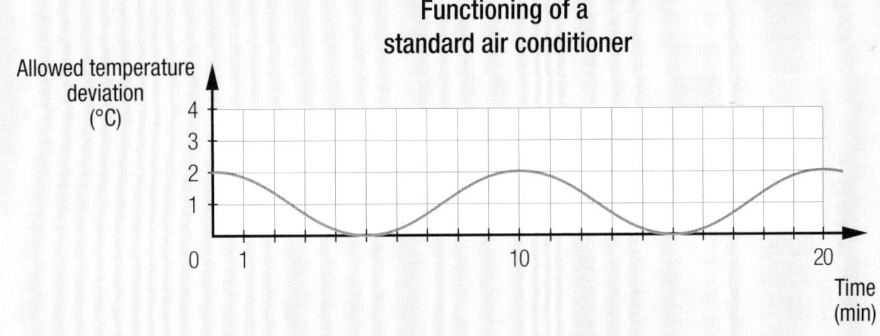

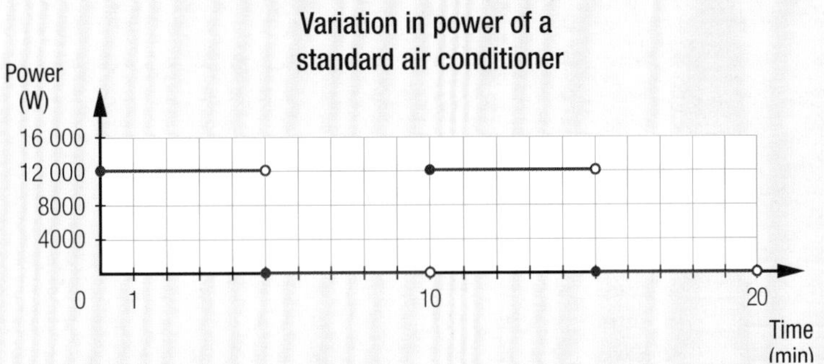

According to the contractor's claims, the new type of air conditioner with a variable-speed compressor has the following features.

- For better comfort, temperature adjustments occur five times more often with the variable-speed compressor type than with the standard model.
- Allowed temperature deviation is five times less with the variable-speed compressor model than with the standard type.
- Compared to the standard model, the variable-speed compressor model cuts power consumption by half.

The information below indicates the performance of the variable-speed compressor model as provided by the manufacturer:

Air conditioner with variable-speed compressor

Time (s)	10	20	30	40	50	60	70	80	90	100	110	120
Allowed temperature deviation (°C)	0.17	0.09	0	−0.1	0.17	−0.21	−0.16	−0.9	0.02	0.1	0.17	0.2

Variation in power of an air conditioner with variable-speed compressor

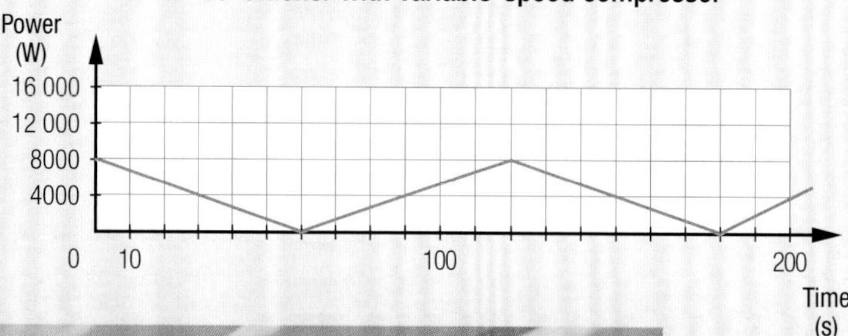

Considering that the energy required to operate each type of air conditioner corresponds to the product of the power of the unit and the time of operation, confirm or refute the air conditioning contractor's claims.

VISION 2

The logging industry

Learning context

Québec has some of the richest forests on Earth. In addition to the economic benefits these forests generate, they provide an essential environment for its inhabitants. Over the centuries, the forest ecosystem has been destabilized by human activity. Forests the world over are threatened by forest fires, the migration of harmful insects and excessive logging practices. We need to try to minimize the negative impacts on all wooded territories.

Over the last century, logging has proved to be an industry in which over-consumption prevails. From an economic perspective, wood is an important resource used for heating, construction, and the pulp and paper industry. Since the 1990s, Canada has established more stringent regulations and developed a diversification of logging methods to protect this natural resource from abusive practices and to provide for forest regeneration. Reforestation and paper recycling are two effective means of controlling deforestation. Québec society is faced with major challenges in the logging industry, and only through a collective effort can we ensure the future of this precious resource.

This LES is related to sections 2.2 and 2.5.

Reforestation

Before logging companies can begin to operate in a given territory, they must promise to protect the forest and the habitats of the local animal species. Companies must also agree to reforest and maintain the wooded territory to ensure resource sustainability. The diagram below represents a territory earmarked for reforestation.

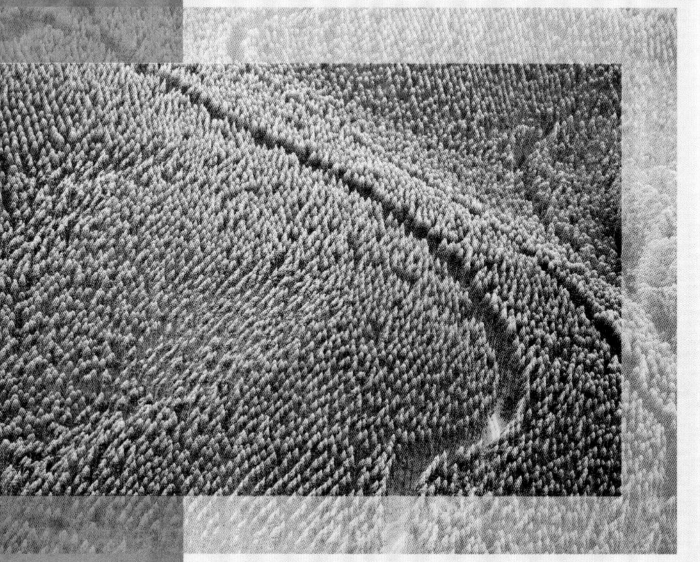

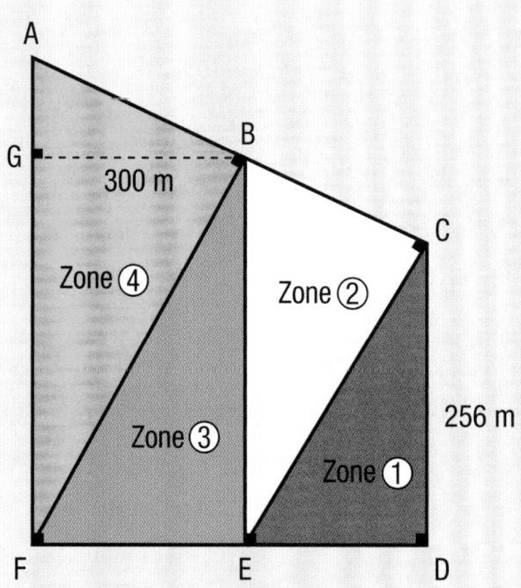

Consider the following information regarding this reforestation project:

- A minimum of 25 000 trees must be planted in the entire territory.
- The ratio of the areas of Zone ② with respect to Zone ③ is 16:25.
- Each zone must be entirely reforested with the same type of tree.
- If two zones are adjacent to each other, one of them must be reforested in hardwood and the other in softwood.

Type of hardwood	Cost per seedling ($)	Space required per seedling (m^2)	Quantity of available seedlings	Type of softwood	Cost per seedling ($)	Space required per seedling (m^2)	Quantity of available seedlings
Maple	3	12	4000	Cedar	1.5	4	25000
Birch	2	6	6500	Fir	2	6	5000
Oak	4	12	8000	Pine	2.5	8	8000
Ash	2.5	4	10000	Spruce	3	10	3000

Calculate the best possible price for the project. You must also provide a diagram of the territory indicating the type of tree to be planted in each zone.

This LES is related to sections 2.1 to 2.4.

LES 4

Minimizing travel

To build logging roads, companies must, among other things, take into account production costs. This economic constraint leads logging companies to set up road networks that reduce transportation to a minimum.

The diagram below represents the road plan for a logging territory. On this plan, the entrance to the logging territory is located off the main road so that the sum of the lengths of Roads **S10** and **S14** is minimal.

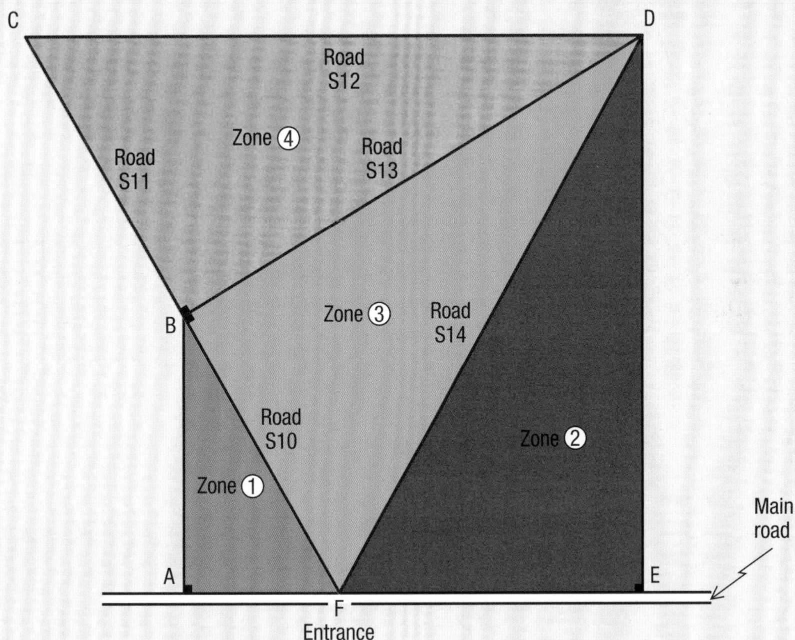

Below is important information about this logging territory:

- Road **S13** corresponds to the bisector of angle CDF.
- Road **S12** is parallel to the main road.
- The area of Zone ① is $(xy + 1.5x)$ km^2.
- The area of Zone ② is $(30y + 45 - 2xy - 3x)$ km^2.

Prove that the length of Road **S13** is $10\sqrt{3}$ km.

GPS [Global Positioning System]

Learning context

In a world where technology manages to change and grow at the speed of light, a pervasive pressure has overcast our society over the past few years. Many people and businesses are making efforts to be more efficient, more effective, and more productive. Now, there are tools that help us to work more quickly and be more precise. Computers and robotic devices, among others, make mass production possible in record time. The GPS system, initially developed for military purposes, is now being used in many other fields. Whether it's for a car trip, to survey an area, to support emergency vehicles in reaching their destinations, or to use on a construction site, this new technology has changed the lives of many people and the reality of many businesses.

Used in the U.S. since 1998, the JDAM bomb is GPS-guided.

GPS is an acronym for *Global Positioning System* which means "worldwide tracking system." The GPS system uses signals from various satellites to find a specific location by calculating the distance between that location and the satellites.

This LES is related to sections 3.1 to 3.3.

LES 5

Air traffic control

A GPS system helps air traffic controllers locate the exact position of planes flying in their observation zone. Using this system and depending on the number of planes in the zone, controllers recommend different air corridors to prevent mid-air collisions.

Below is a graphical representation of the aerial view of three air corridors.

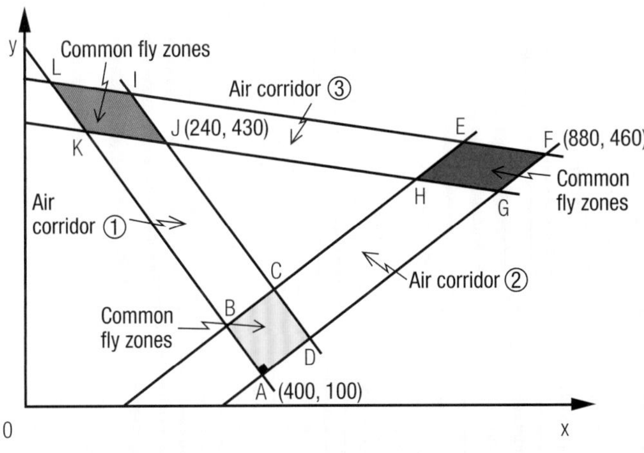

In this representation, note the following:

- Point D is located at $\frac{1}{6}$ the length of segment AF.
- Point J divides segment DI in a ratio of 4:1.
- Each air corridor is marked off by two parallel lines.
- Air corridors ① and ② have the same width.
- The scale is in metres.

Determine:
- the equation of the lines that mark off each air corridor
- the coordinates of the vertices for each of the three polygons corresponding to the common fly zones
- the area of each common fly zones

This LES is related to sections 3.2 and 3.3.

LES 6

C3

A precision tool

A GPS system helps roadwork to be done with greater precision. These devices are sometimes installed on tractors to help calculate the inclination of a road under construction.

Below is a diagram of a cross-section of a road construction project:

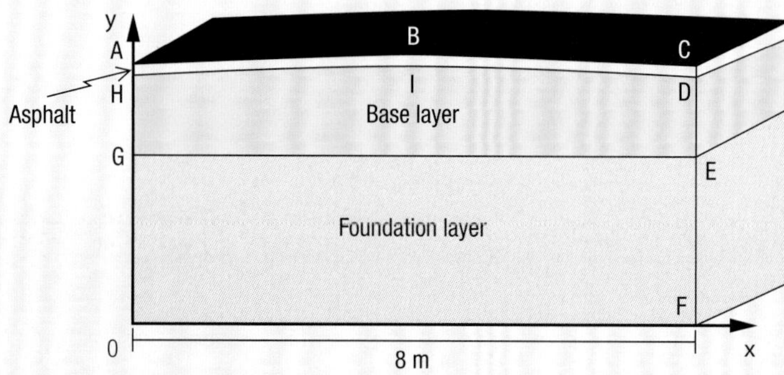

- The thickness of the asphalt layer must be between 20 cm and 25 cm.
- The base layer is $\frac{2}{7}$ of the total volume of building materials.
- The foundation layer is a minimum of 80 cm thick.
- The inclination of segments AB and BC is 2%.

Cost of the building materials

Asphalt : $80/m^3	Foundation layer: $5/m^3	Base layer: $10/m^3

Prepare a bid to build 1 km of road.

Your bid must include:

- a diagram of the cross-section of the road on which the inequalities defining each layer are clearly indicated
- the coordinates of all points from A to H
- the total cost of the project

REFERENCE
TABLE OF CONTENTS

Technology **216**
- Graphing calculator **216**
- Spreadsheet **218**
- Dynamic geometry software **220**

Knowledge **222**
- Notations and symbols **222**
- Geometric statements **224**
- Glossary **228**

Graphing calculator

Sample Calculations

It is possible to perform scientific calculations and to evaluate both algebraic and logical expressions.

Scientific calculations

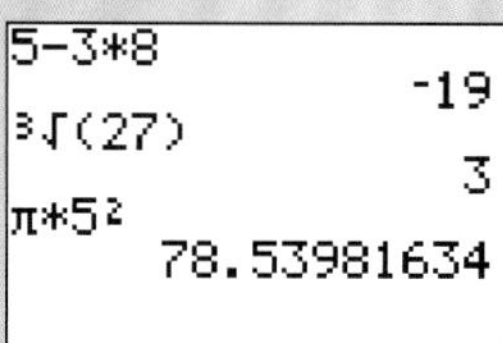

Logical expressions

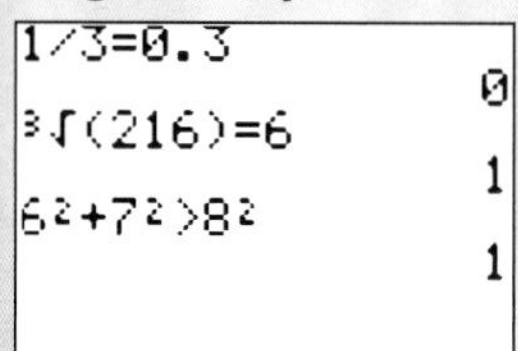

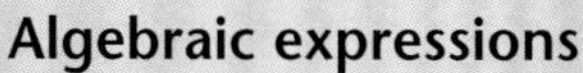

Algebraic expressions

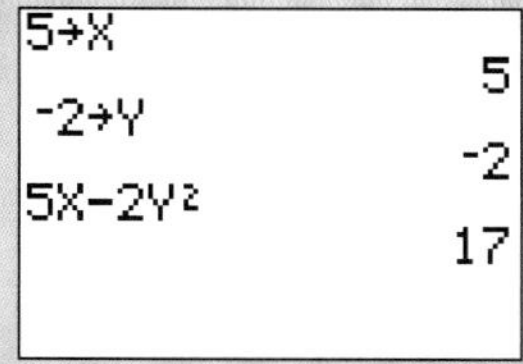

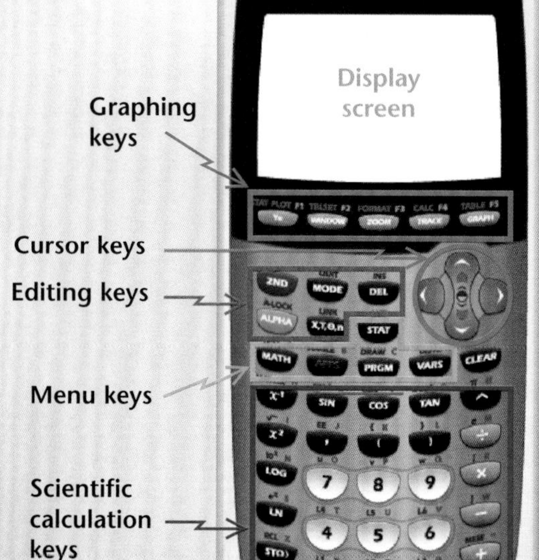

Probability

1. Display the probability menu.

```
MATH NUM CPX PRB
1:rand
2:nPr
3:nCr
4:!
5:randInt(
6:randNorm(
7:randBin(
```

- Among other things, this menu allows the simulation of random experiments. The fifth option generates a series of random whole numbers. Syntax: randInt (minimum value, maximum value, number of repetitions).

2. Display calculations and results.

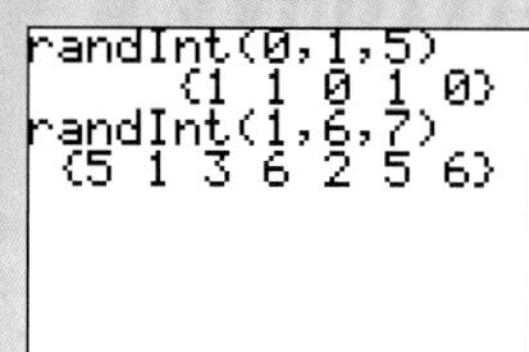

- The first example simulates flipping a coin 5 times where 0 represents tails and 1 represents heads. The second example simulates seven rolls of a die with 6 faces.

Display a table of values

1. Define the rules.

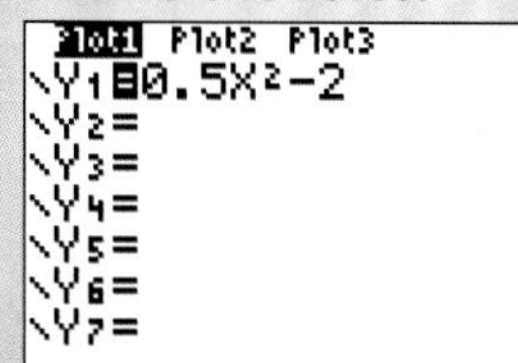

- This screen allows you to enter and edit the rule for one or more functions where Y is the dependent variable and X is the independent variable.

2. Define the viewing window.

- This screen allows you to define the viewing window for a table of values indicating the starting value of X and the step size for the variation of X.

3. Display the table of values.

X	Y1	Y2
0	1	0
1	2	.5
2	4	2
3	8	4.5
4	16	8
5	32	12.5
6	64	18

X=0

- This screen allows you to display the table of values of the rules defined.

Display a graphical representation

1. Define the rules.

```
Plot1 Plot2 Plot3
\Y1=0.5X+1
\Y2=4X-3
\Y3=
\Y4=
\Y5=
\Y6=
\Y7=
```

- If desired, the thickness of the curve (E.g. normal, thick or dotted) can be adjusted for each rule.

2. Define the viewing window.

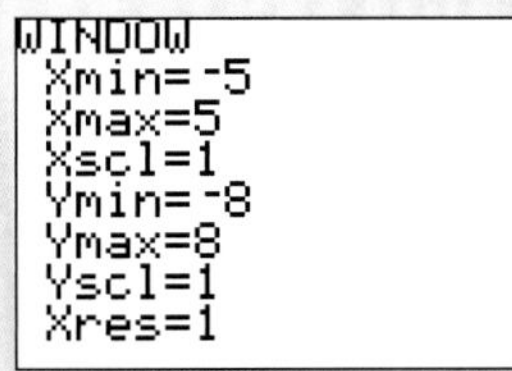

- This screen allows you to define the viewing window by limiting the Cartesian plane: `Xscl` and `Yscl` correspond to the step value on the respective axes.

3. Display the graph.

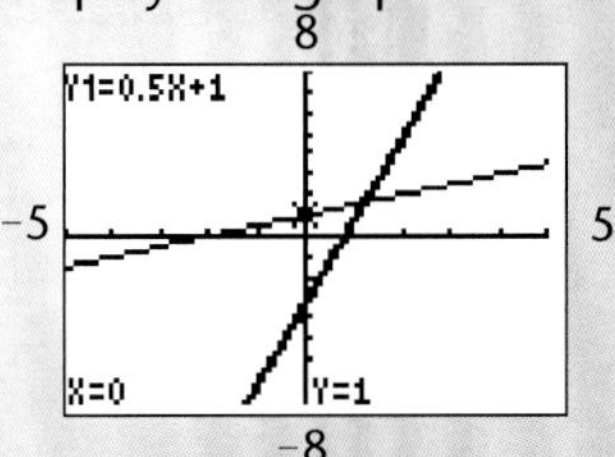

- This screen allows you to display the graphical representation of the rules previously defined. If desired, the cursor can be moved along the curves and the coordinates displayed.

Displaying a scatter plot and statistical calculations

1. Enter the data.

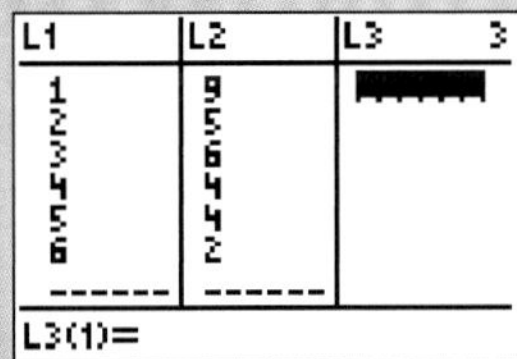

- This screen allows you to enter the data from a distribution. For a two-variable distribution, data entry is done in two columns.

2. Select the mode of representation.

- This screen allows you to choose the type of statistical diagram.

: scatter plot
: broken-line graph
: histogram
: box and whisker plot

3. Display the graph.

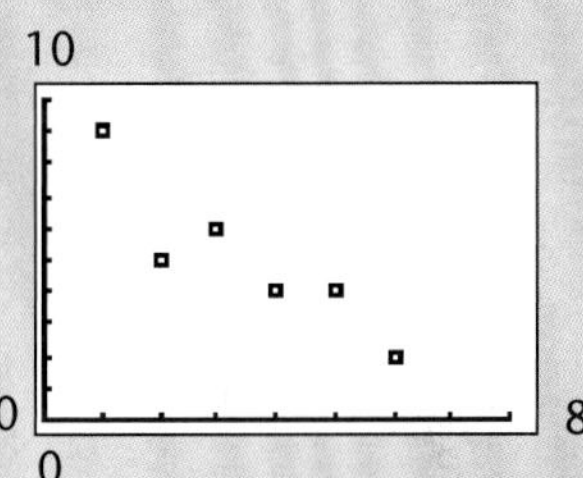

- This screen allows you to displays the scatter plot.

4. Perform statistical calculations.

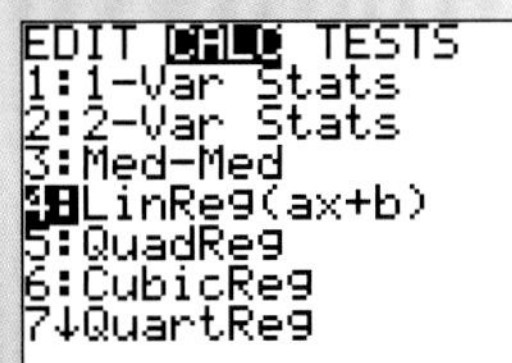

- This menu allows you to access different statistical calculations, in particular that of the linear regression.

5. Determine the regression and correlation.

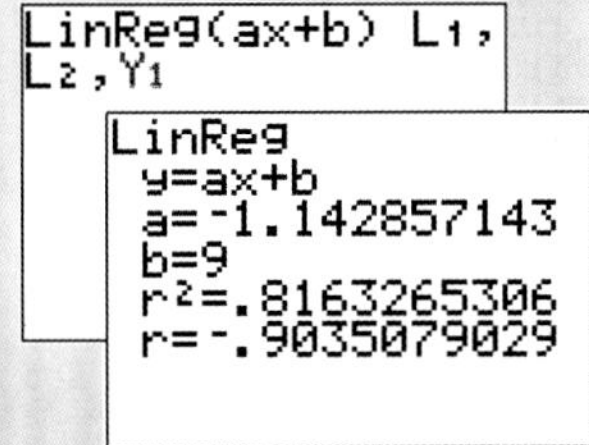

- These screens allow you to obtain the equation of the regression line and the value of the correlation coefficient.

6. Display the line.

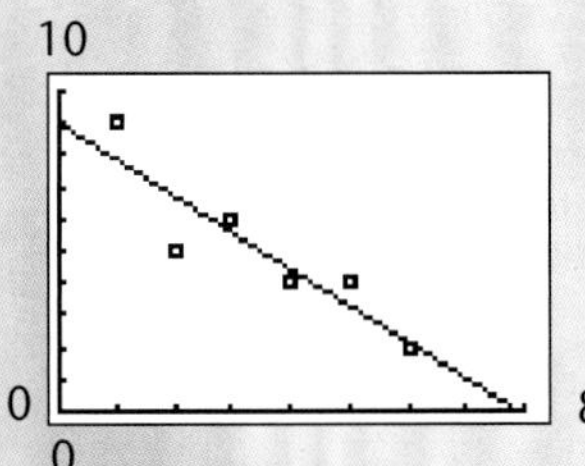

- The regression line can be displayed on the scatter plot.

Spreadsheet

TECHNOLOGY

A spreadsheet is a software that allows you to perform calculations on numbers entered into cells. It is used mainly to perform calculations on large amounts of data, to construct tables and to draw graphs.

Spreadsheet Interface

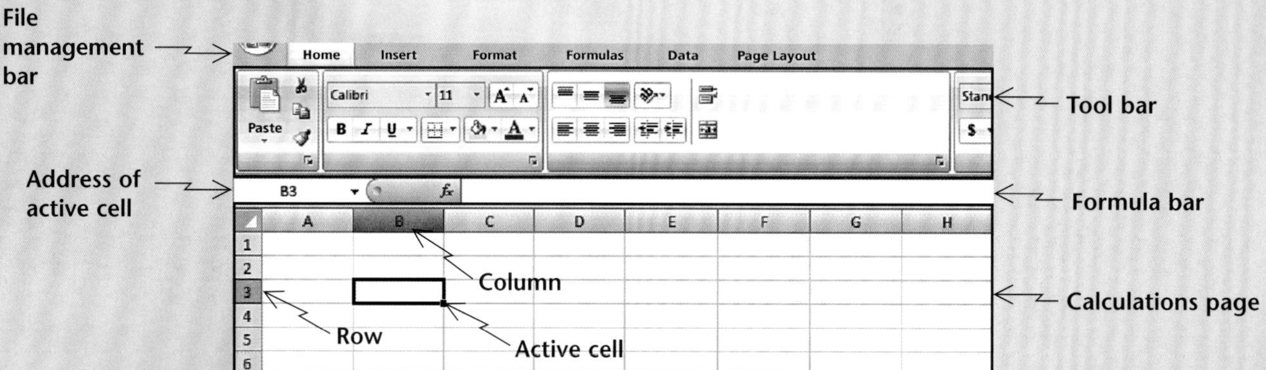

What is a cell?

A cell is the intersection of a column and a row. A column is identified by a letter and a row is identified by a number. Thus, the first cell in the upper right hand corner is identified as A1.

Entry of numbers, text and formulas in the cells

You can enter a number, text or a formula in a cell after clicking on it. Formulas allow you to perform calculations on numbers already entered in the cells. To enter a formula in a cell, just select it and begin by entering the "=" symbol.

Example:
Column **A** contains the data to be used in the calculations.

In the spreadsheet, certain functions are predefined to calculate the sum, the minimum, the maximum, the mode, the median, the mean and the mean deviation of a set of data.

◇	A	B	C	
1	Results			
2	27.4	Number of data	17	→ =COUNT(A2:A18)
3	30.15			
4	15	Sum	527	→ =SUM(A2:A18)
5	33.8			
6	12.3	Minimum	12.3	→ =MIN(A2:A18)
7	52.6			
8	28.75	Maximum	52.6	→ =MAX(A2:A18)
9	38.25			
10	21.8	Mode	33.8	→ =MODE(A2:A18)
11	35			
12	29.5	Median	30.15	→ =MEDIAN(A2:A18)
13	27.55			
14	33.8	Mean	31	→ =MEAN(A2:A18)
15	15			
16	33.8	Mean deviation	8.417647059	→ =MEAN DEVIATION (A2:A18)
17	50			
18	42.3	Standard deviation	11.2543325	→ =STANDARD DEVIATION (A2:A18)
19				

How to construct a graph

Below is a procedure for drawing a graph using a spreadsheet:

1) Select the range of data.

	A	B
1	Length of a femur (cm)	Height of a person (cm)
2	36	144
3	37	146
4	40	153
5	42	158
6	43.5	162
7	45	165
8	46.5	168
9	46.8	169
10	47	170
11	47.5	171

2) Select from the graph assistant.

Cells...
Rows
Columns
Worksheet
Chart...
List...
Page Break
Function...
Name
Comment
Picture
Movie...
Object...
Hyperlink... ⌘K

3) Choose the graph type.

Graph Wizard – Step 1 of 4 – Graph Type

4) Confirm the data for the graph.

Graph Wizard – Step 2 of 4 – Graph Source Data

5) Choose the graph options.

Graph Wizard – Step 3 of 4 – Graph Options

6) Choose the location of the graph.

Graph Wizard – Step 4 of 4 – Graph Location

7) Draw the graph.

Height of a person compared to length of femur

Height (cm): 140, 145, 150, 155, 160, 165, 170, 175

Length of femur (cm): 30, 32, 34, 36, 38, 40, 42, 44, 46, 48, 50

After drawing the graph you can modify different elements by double-clicking on the element to be changed: title, scale, legend, grid, type of graph, etc.

Below are different types of graphs you can create using a spreadsheet:

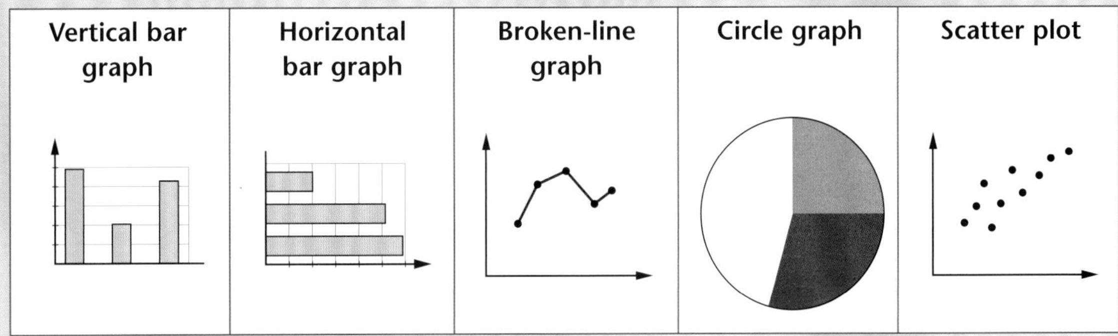

Dynamic geometry software

Dymanic geometry software allows you to draw and move objects in a workspace. The dynamic aspect of this type of software allows you to explore and verify geometric properties and to validate constructions.

The workspace and the tools

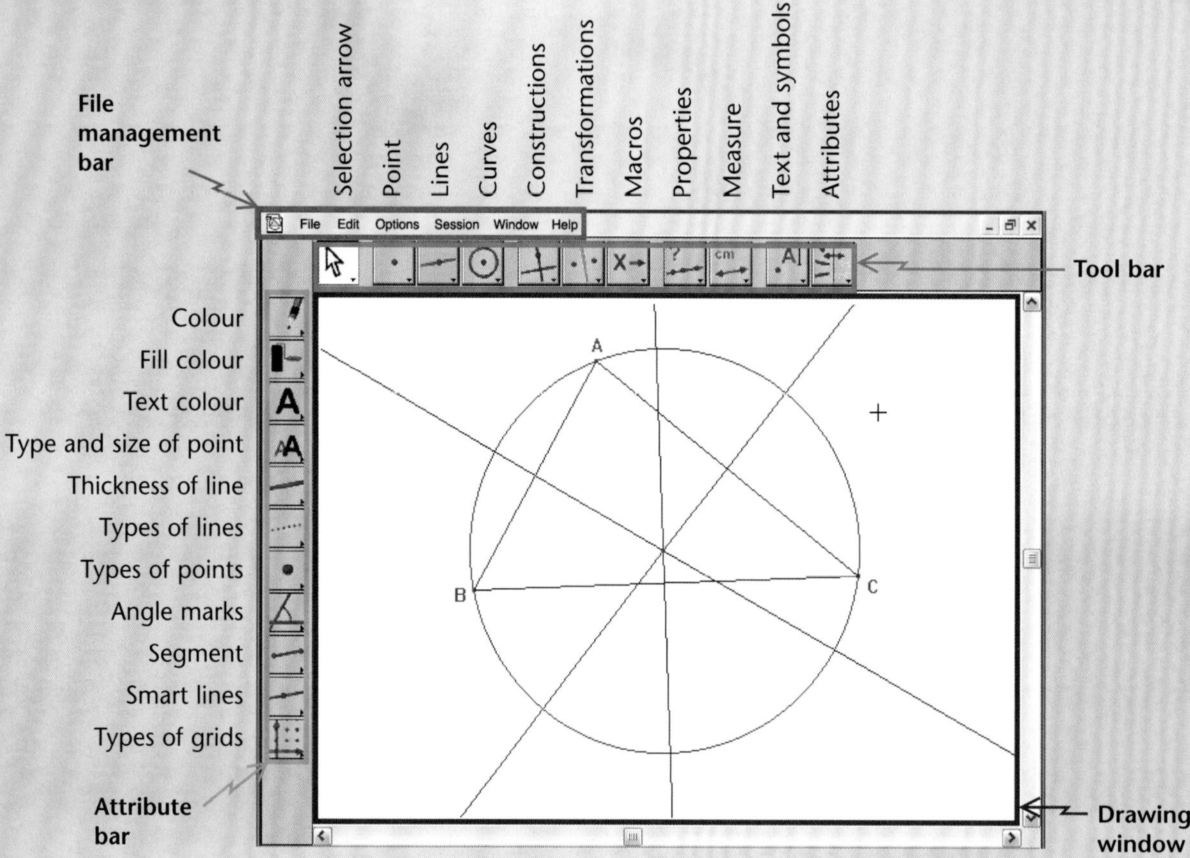

Cursors and their interpretations

	Cursor used when moving in the drawing window.
	Cursor used when drawing an object.
What object?	Cursor used when there are several objects.
	Cursor used when tracing an object.
	Cursor used to indicate that movement of an object is possible.
	Cursor used when working in the file management bar and in the tool bar.
	Cursor used when filling an object with a colour.
	Cursor used to change the attribute of the selected object.

Geometric explorations

1) A median separates a triangle into two other triangles. In order to explore the properties of these two triangles, perform the following construction. To verify that triangles ABD and ADC have the same area, calculate the area of each triangle. By moving the points A, B and C, notice that the areas of the two triangles are always the same.

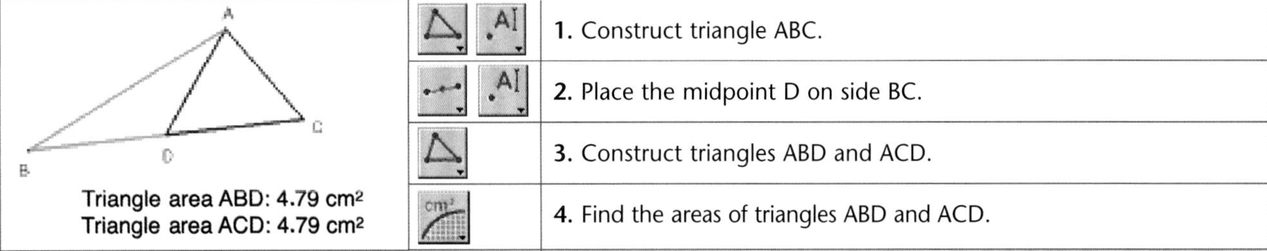

1. Construct triangle ABC.	
2. Place the midpoint D on side BC.	
3. Construct triangles ABD and ACD.	
4. Find the areas of triangles ABD and ACD.	

2) In order to determine the relation between the position of the midpoint of the hypotenuse in a right triangle and the three vertices of the triangle, perform the construction below. By moving points A, B, and C, note that the midpoint of the hypotenuse of a right triangle is equidistant from its three vertices.

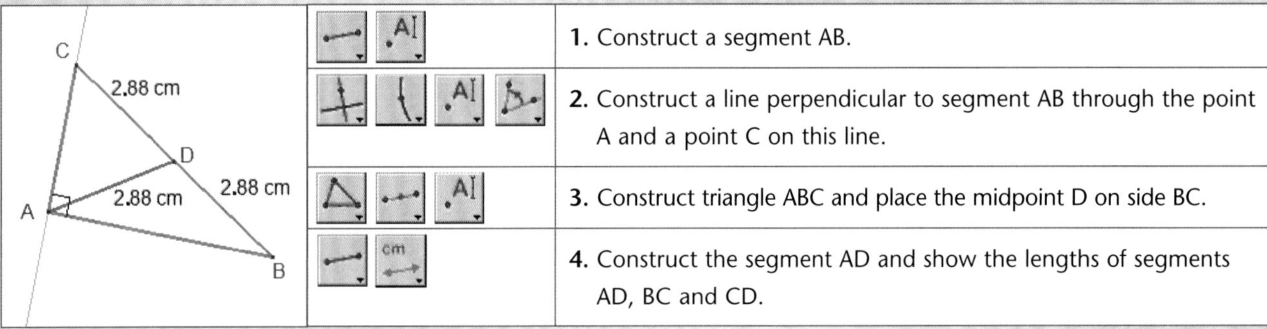

1. Construct a segment AB.	
2. Construct a line perpendicular to segment AB through the point A and a point C on this line.	
3. Construct triangle ABC and place the midpoint D on side BC.	
4. Construct the segment AD and show the lengths of segments AD, BC and CD.	

Graphical exploration

In order to discover the relation between the slopes of two perpendicular lines in the Cartesian plane, perform the construction below. By showing the product of the slopes and modifying the inclination of one of the lines, note a particular property of these slopes: the product of the slopes of two perpendicular lines is -1.

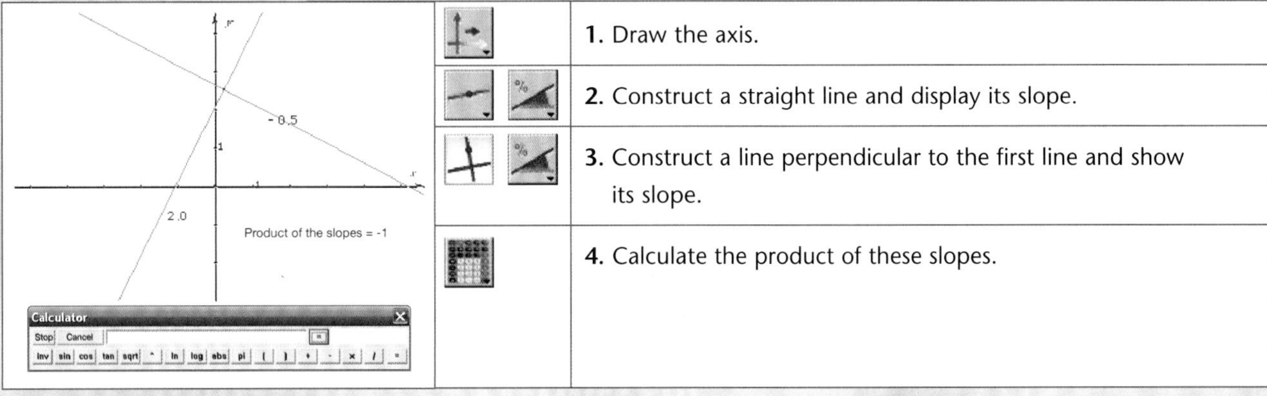

1. Draw the axis.	
2. Construct a straight line and display its slope.	
3. Construct a line perpendicular to the first line and show its slope.	
4. Calculate the product of these slopes.	

Notations and symbols

KNOWLEDGE

Notation & symbols	Meaning
{ }	Brace brackets, used to list the elements in a set
$\mathbb{N}$	The set of Natural numbers
$\mathbb{Z}$	The set of Integers
$\mathbb{Q}$	The set of Rational numbers
$\mathbb{Q}'$	The set of Irrational numbers
$\mathbb{R}$	The set of Real numbers
$\cup$	The union of sets
$\cap$	The intersection of sets
Ω	Read "omega," it represents the sample space in a random experiment
$\varnothing$ or { }	The empty set (or the null set)
$=$	…is equal to…
$\neq$	…is not equal to…or…is different from…
$\approx$	…is approximately equal to…
$<$	…is less than…
$>$	…is greater than…
$\leq$	…is less than or equal to…
$\geq$	…is greater than or equal to…
$[a, b]$	Interval, including a and b
$[a, b[$	Interval, including a but excluding b
$]a, b]$	Interval, excluding a but including b
$]a, b[$	Interval, excluding both a and b
∞	Infinity
(a, b)	The ordered pair a and b
$f(x)$	Is read as f of x or the value (image) of the function f at x
Δx	Variation or growth of x

Notation & symbols	Meaning
( )	Parentheses show which operation to perform first
$-a$	The opposite of a
$\frac{1}{a}$ or a^{-1}	The reciprocal of a
a^2	The second power of a or a squared
a^3	The third power of a or a cubed
$\sqrt{a}$	The square root of a
$\sqrt[3]{a}$	The cube root of a
$\lvert a \rvert$	The absolute value of a
%	Percent
$a : b$	The ratio of a to b
π	Read "pi," it is approximately equal to 3.1416
$^\circ$	Degree, unit of angle measure
$\overline{AB}$	Segment AB
m $\overline{AB}$	Measure of segment AB
$\angle$	Angle
m $\angle$	The measure of an angle
$\overset{\frown}{AB}$	The arc of the circle AB
m $\overset{\frown}{AB}$	The measure of arc of the cirle AB
//	… is parallel to…
$\perp$	… is perpendicular to …
∟	Indicates a right angle in a geometric plane figure
$\triangle$	Triangle
$\cong$	…is congruent to…
$\sim$	…is similar to…
$\triangleq$	…corresponds to…
$P(E)$	The probability of event E
Med	The median of a distribution

Geometric statements

KNOWLEDGE

	Statement	Example
1.	If two lines are parallel to a third line, then they are all parallel to each other.	If $l_1 // l_2$ and $l_2 // l_3$, then $l_1 // l_3$. 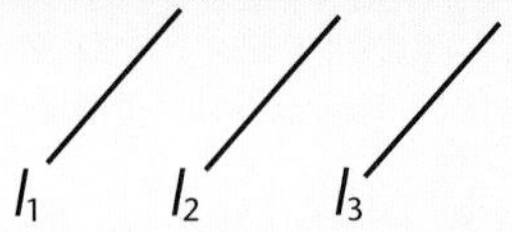
2.	If two lines are perpendicular to a third line, then the two lines are parallel to each other.	If $l_1 \perp l_3$ and $l_2 \perp l_3$, then $l_1 // l_2$. 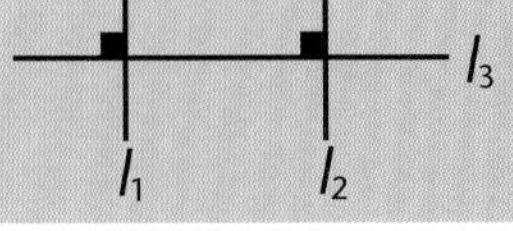
3.	If two lines are parallel, then every line perpendicular to one of these lines is perpendicular to the other.	If $l_1 // l_2$ and $l_3 \perp l_2$, then $l_3 \perp l_1$. 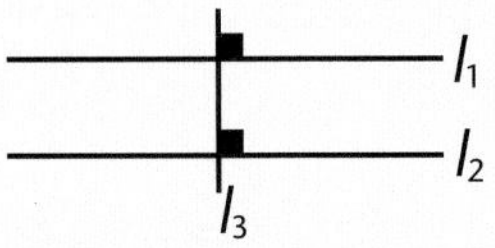
4.	If the exterior arms of two adjacent angles are collinear, then the angles are supplementary.	Points A, B and D are collinear. ∠ ABC and ∠ CBD are adjacent and supplementary. 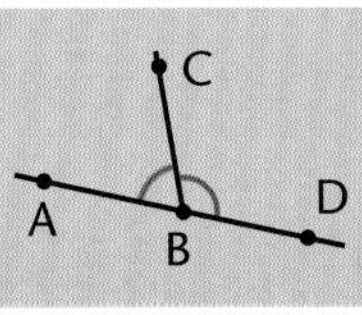
5.	If the exterior arms of two adjacent angles are perpendicular, then the angles are complementary.	$\overline{AB} \perp \overline{BD}$ ∠ ABC and ∠ CBD are adjacent and complementary. 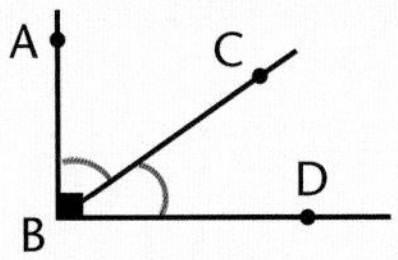
6.	Vertically opposite angles are congruent.	∠ 1 ≅ ∠ 3 ∠ 2 ≅ ∠ 4 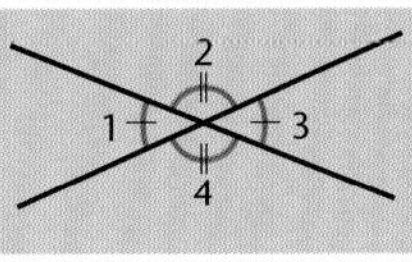
7.	If a transversal intersects two parallel lines, then the alternate interior, alternate exterior and corresponding angles are respectively congruent.	If $l_1 // l_2$, then angles 1, 3, 5 and 7 are congruent as are angles 2, 4, 6 and 8. 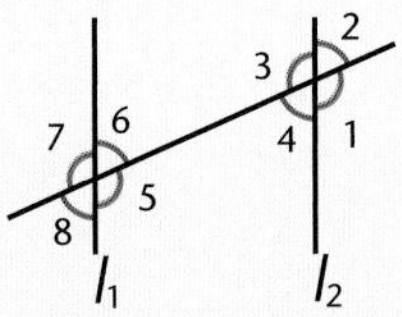
8.	If a transversal intersects two lines resulting in congruent corresponding angles (or alternate interior angles or alternate exterior angles), then those two lines are parallel.	In the figure for statement 7, if the angles 1, 3, 5 and 7 are congruent and the angles 2, 4, 6 and 8 are congruent, then $l_1 // l_2$.
9.	If a transversal intersects two parallel lines, then the interior angles on the same side of the transversal are supplementary.	If $l_1 // l_2$, then m ∠ 1 + m ∠ 2 = 180° and m ∠ 3 + m ∠ 4 = 180°. 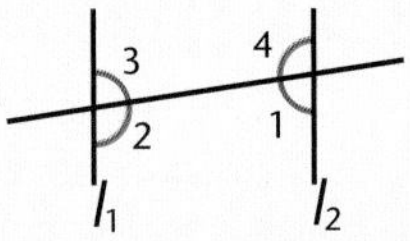

	Statement	Example
10.	The sum of the measures of the interior angles of a triangle is 180°.	$m\angle 1 + m\angle 2 + m\angle 3 = 180°$
11.	Corresponding elements of congruent plane or solid figures have the same measurements.	$\overline{AD} \cong \overline{A'D'}$, $\overline{CD} \cong \overline{C'D'}$, $\overline{BC} \cong \overline{B'C'}$, $\overline{AB} \cong \overline{A'B'}$ $\angle A \cong \angle A'$, $\angle B \cong \angle B'$, $\angle C \cong \angle C'$, $\angle D \cong \angle D'$
12.	In an isosceles triangle, the angles opposite the congruent sides are congruent.	In the isosceles triangle ABC: $\overline{AB} \cong \overline{AC}$ $\angle C \cong \angle B$
13.	The axis of symmetry of an isosceles triangle represents a median, a perpendicular bisector, an angle bisector and an altitude of the triangle.	Axis of symmetry of triangle ABC. Median from point A Perpendicular bisector of the side BC Bisector of angle A Altitude of the triangle
14.	The opposite sides of a parallelogram are congruent.	In the parallelogram ABCD: $\overline{AB} \cong \overline{CD}$ and $\overline{AD} \cong \overline{BC}$
15.	The diagonals of a parallelogram bisect each other.	In the parallelogram ABCD: $\overline{AE} \cong \overline{EC}$ and $\overline{DE} \cong \overline{EB}$
16.	The opposite angles of a parallelogram are congruent.	In the parallelogram ABCD: $\angle A \cong \angle C$ and $\angle B \cong \angle D$
17.	In a parallelogram, the sum of the measures of two consecutive angles is 180°.	In the parallelogram ABCD: $m\angle 1 + m\angle 2 = 180°$ $m\angle 2 + m\angle 3 = 180°$ $m\angle 3 + m\angle 4 = 180°$ $m\angle 4 + m\angle 1 = 180°$
18.	The diagonals of a rectangle are congruent.	In the rectangle ABCD: $\overline{AC} \cong \overline{BD}$
19.	The diagonals of a rhombus are perpendicular.	In the rhombus ABCD: $\overline{AC} \perp \overline{BD}$
20.	The measure of an exterior angle of a triangle is equal to the sum of the measures of the interior angles at the other two vertices.	$m\angle 3 = m\angle 1 + m\angle 2$

	Statement	Example
21.	In a triangle, the longest side is opposite the largest angle.	In triangle ABC, the largest angle is A; therefore the longest side is BC.
22.	In a triangle, the smallest angle is opposite the smallest side.	In triangle ABC, the smallest angle is B; therefore the smallest side is AC.
23.	The sum of the measures of two sides in a triangle is larger than the measure of the third side.	$2 + 5 > 4$ $2 + 4 > 5$ $4 + 5 > 2$ 2 cm, 5 cm, 4 cm
24.	The sum of the measures of the interior angles of a quadrilateral is 360°.	$m\angle 1 + m\angle 2 + m\angle 3 + m\angle 4 = 360°$
25.	The sum of the measures of the interior angles of a polygon with n sides is $n \times 180° - 360°$ or $(n - 2) \times 180°$.	$n \times 180° - 360°$ or $(n - 2) \times 180°$
26.	The sum of the measures of the exterior angles (one at each vertex) of a convex polygon is 360°.	$m\angle 1 + m\angle 2 + m\angle 3 +$ $m\angle 4 + m\angle 5 + m\angle 6 = 360°$
27.	The corresponding angles of similar plane figures or of similar solids are congruent, and the measures of the corresponding sides are proportional.	The triangle ABC is similar to triangle A'B'C': $\angle A \cong \angle A'$ $\angle B \cong \angle B'$ $\angle C \cong \angle C'$ $\frac{m\,\overline{A'B'}}{m\,\overline{AB}} = \frac{m\,\overline{B'C'}}{m\,\overline{BC}} = \frac{m\,\overline{A'C'}}{m\,\overline{AC}}$
28.	In similar plane figures, the ratio of the areas is equal to the square of the ratio of similarity.	In the above figures: $\frac{m\,\overline{A'B'}}{m\,\overline{AB}} = \frac{m\,\overline{B'C'}}{m\,\overline{BC}} = \frac{m\,\overline{A'C'}}{m\,\overline{AC}} = k$ ← Ratio of similarity $\frac{\text{area of triangle A'B'C'}}{\text{area of triangle ABC}} = k^2$
29.	Three non-collinear points define one and only one circle.	There is only one circle which contains the points A, B and C.
30.	The perpendicular bisectors of any chords in a circle intersect at the centre of the circle.	l_1 and l_2 are the perpendicular bisectors of the chords AB and CD. The point of intersection M of these perpendicular bisectors is the centre of the circle.

	Statement	Example
31.	All the diameters of a circle are congruent.	$\overline{AD}$, $\overline{BE}$ and $\overline{CF}$ are diameters of the circle with centre O. $\overline{AD} \cong \overline{BE} \cong \overline{CF}$ 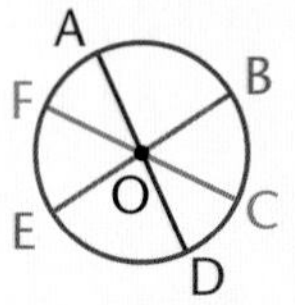
32.	In a circle, the measure of the radius is one-half the measure of the diameter.	$\overline{AB}$ is a diameter of the circle with centre O. $m\ \overline{OA} = \frac{1}{2} m\ \overline{AB}$ 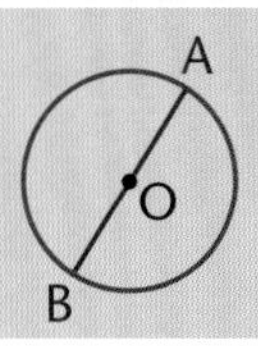
33.	In a circle, the ratio of the circumference to the diameter is a constant represented by π.	$\frac{C}{d} = \pi$ 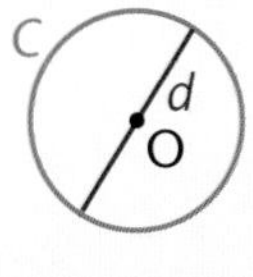
34.	In a circle, a central angle has the same degree measure as the arc contained between its sides.	In the circle with centre O, $m\ \angle\ AOB = m\ \overparen{AB}$ is stated in degrees. 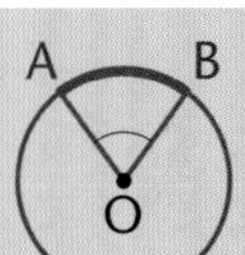
35.	In a circle, the ratio of the measures of two central angles is equal to the ratio of the arcs intercepted by their sides.	$\frac{m\angle\ AOB}{m\ \angle COD} = \frac{m\ \overparen{AB}}{m\ \overparen{CD}}$ 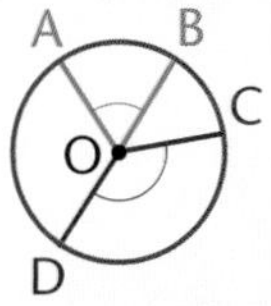
36.	In a circle, the ratio of the areas of two sectors is equal to the ratio of the measures of the angles at the centre of these sectors.	$\frac{\text{Area of the sector AOB}}{\text{Area of the sector COD}} = \frac{m\angle\ AOB}{m\ \angle COD}$ 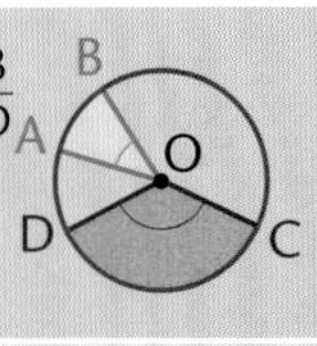

Glossary

KNOWLEDGE

A

Algebraic term - see Term.

Altitude of a triangle
Segment from one vertex of a triangle, perpendicular to the line containing the opposite side. The length of such a segment is also called a height of the triangle.

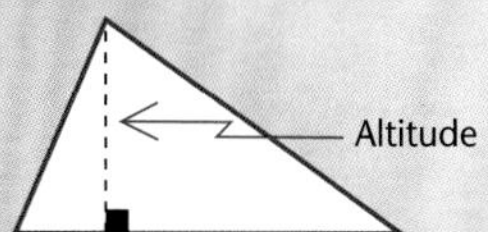

Angle

Classification of angles according to their measure

Name	Measure	Representation
Zero	0°	
Acute	Between 0° & 90°	
Right	90°	
Obtuse	Between 90° & 180°	
Straight	180°	
Reflex	Between 180° & 360°	
Perigon	360°	

Angles

Alternate interior angles, p. 65

Alternate exterior angles, p. 65

Corresponding angles, p. 65

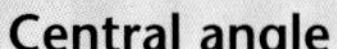

Central angle
Angle formed by two radii in a circle. The vertex of the angle is the centre of the circle.

Centre angle

Apothem of a regular polygon
Segment (or length of segment) from the centre of the regular polygon perpendicular to any of its sides. It is determined by the centre of the regular polygon and the midpoint of any side.

Apothem

Centre of a regular polygon

Arc of a circle
Part of a circle defined by two points on the circle.

Area
The surface of a figure. Area is expressed in square units.

Area of a circle, p. 64

Area of a parallelogram, p. 64

Area of a rectangle, p. 64

Area of a regular polygon, p. 64

Area of a rhombus, p. 64

Area of a right circular cone

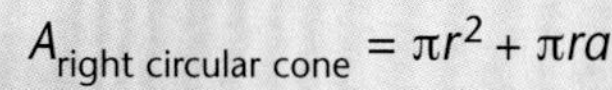

$A_{\text{right circular cone}} = \pi r^2 + \pi ra$

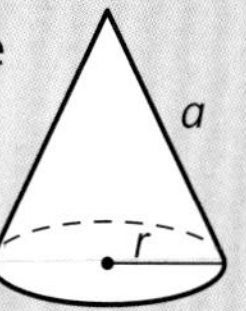

Area of a sector

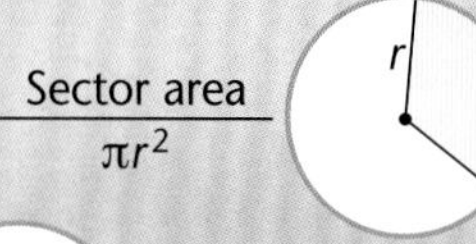

$$\frac{\left(\text{Measure of the central angle of a sector}\right)}{360°} = \frac{\text{Sector area}}{\pi r^2}$$

Area of a sphere

$A_{\text{sphere}} = 4\pi r^2$

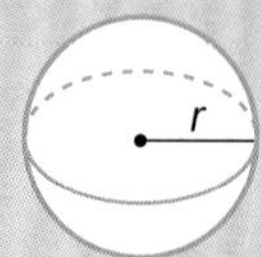

Area of a square, p. 64

Area of a trapezoid, p. 64

Area of a triangle, p. 64

C

Capacity
Volume of a fluid which a solid can contain.

Cartesian plane
A plane formed by two scaled perpendicular lines. Each point is located by its distance from each of these lines respectively.

Central angle - see Angles.

Change on the axes, p. 148

Circle
The set of all points in a plane at an equal distance from a given point called the centre.

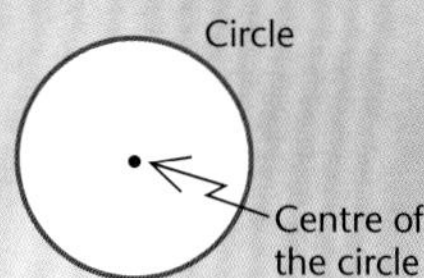

Circumference
The perimeter of a circle. In a circle whose circumference is C, diameter is d and radius is r:
$C = \pi d$ and $C = 2\pi r$.

Conjecture, p. 73

Coordinates of a point
Each of the two numbers used to describe the position of a point in a Cartesian plane.

Counter-example, p. 73

Cube root
The inverse of the operation which consists of cubing a number is called finding the cube root. The symbol for this operation is $\sqrt[3]{\ }$.

E.g. 1) $\sqrt[3]{125} = 5$
2) $\sqrt[3]{-8} = -2$

D

Degree of a monomial
The sum of the exponents of the monomial.
E.g. 1) The degree of the monomial 9 is 0.
2) The degree of the monomial $-7xy$ is 2.
3) The degree of the monomial $15a^2$ is 2.

Degree of a polynomial in one variable
The largest exponent of that variable in the polynomial.
E.g. The degree of the polynomial $7x^3 - x^2 + 4$ is 3.

Diameter
Segment (or length of segment) which is determined by two points on a circle passing through the centre of the circle.

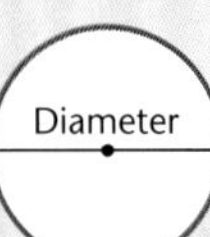

Dilatation
A geometric transformation defined by a centre, initial point, corresponding image and scale factor. A dilatation results in an enlargement or reduction of the initial figure.

Distance between two points, p. 148

Distance from a point to a line, p. 162

Domain of a function, p. 6

E

Edge
Segment formed by the intersection of any two faces of a solid.

Equation
Mathematical statement of equality involving one or more variables.
E.g. $4x - 8 = 4$

Equation of a line, p. 161

Equivalent equations
Equations having the same solution.
E.g. $2x = 10$ and $3x = 15$ are equivalent equations, because the solution of each is 5.

Exponentiation
Operation which consists of raising a base to an exponent.
E.g. In 5^8, the base is 5 and the exponent is 8.

Extrema: minimum and maximum, p. 6

F

Face
Plane or curved surface bound by edges.

Factoring, p. 65, 84, 85

Families of functions, p. 15

Function, p. 5

Direct variation function
A function in which a constant change in the independent variable results in a constant, non-zero change to the dependent variable. Its graph is an oblique line through the origin of the Cartesian plane.

Types of functions, p. 39
- Periodic
- Piecewise
- Step

First-degree polynomial function
A function whose rule can be written as a first-degree polynomial.
E.g. $f(x) = 7.1x + 195$

Inverse, p. 5

Inverse variation function
A function that represents an inversely proportional situation. The product of each ordered pair is a constant and non-zero. The graphical representation is a curve whose extremities gradually approach the axes but never touches them.

Partial variation function
A function in which a constant change in the independent variable results in a constant, non-zero change to the dependent variable. Its graph is an oblique line which does not pass through the origin of the Cartesian plane.

Polynomial function
A function whose rule can be written as a polynomial.
E.g. $f(x) = 3x^2 + 7$

Zero-degree polynomial function (constant function)
A function in which a constant change in the independent variable results in no change in the dependent variable. Its graph is a horizontal line parallel to the x-axis.
E.g. $f(x) = -5$

Half-plane, p. 184

Hypotenuse
The side opposite the right angle in a right triangle. It is the longest side of a right triangle.

Hypotenuse

Image
In geometry, figure obtained by a geometric transformation performed on an initial figure.

Independent variable - see Variable

Inequality
A mathematical statement which compares two numerical expressions with an inequality symbol (which may include variables).
E.g. 1) $4 < 4.2$
2) $-10 \leq -5$
3) $4a > 100$

First-degree inequality
in one variable, p. 139
in two variables, p. 184, 185

Initial figure
Figure on which a geometric transformation is performed.

Initial value (see Function), p. 7

Integer
A number belonging to the set
$\mathbb{Z} = \{...,-2, -1, 0, 1, 2, ...\}$.

Interval
A set of all the real numbers between two given numbers called the endpoints. Each endpoint can be either included or excluded in the interval.
E.g. The interval of real numbers from -2 included to 9 excluded is [-2, 9[.

Irrational number
A number which cannot be expressed as a ratio of two integers, and whose decimal representation is non-periodic and non-terminating.

Laws of Exponents

Law	
Product of powers:	$a^m \times a^n = a^{m+n}$
Quotient of powers: For $a \neq 0$	$\frac{a^m}{a^n} = a^{m-}a^n$
Power of a product:	$(ab)^m = a^m b^m$
Power of a power:	$(a^m)^n = a^{mn}$
Power of a quotient: $b \neq 0$	$\left(\frac{a}{b}\right)^m = \frac{a^m}{b^m}$

Legs (or arms) of a right triangle
The sides that form the right angle in a right triangle.

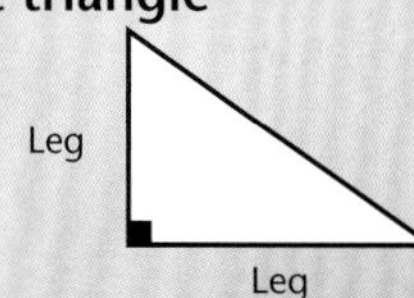

Like terms - see Terms.

Lines
parallel, p. 162, 175
perpendicular, p. 162

M

Mathematical model, p. 16

Maximum of a function, p. 6

Median of a triangle
Segment determined by a vertex and the midpoint of the opposite side.
E.g. The segments AE, BF and CD are the medians of triangle ABC.

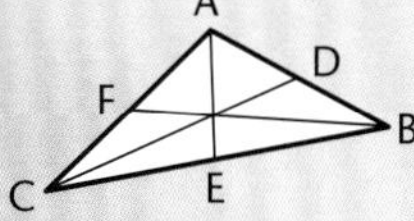

Metric relations in a right triangle, p. 114

Minimum of a function, p. 6

Monomial
Algebraic expression formed by one number or a product of numbers and variables.
E.g. 9, $-5x^2$ and $4xy$ are monomials.

Natural number
Any number belonging to the set
$\mathbb{N} = \{0, 1, 2, 3, ...\}$.

Numerical coefficient of a term
Numerical value multiplied by the variable or variables of a term.
E.g. In the algebraic expression $x + 6xy - 4.7y$, the numerical coefficients of the first, second and third terms are 1, 6 and -4.7 respectively.

Optimizing a distance, p. 106

Origin of a Cartesian plane
The point of intersection of the two axes in a Cartesian plane. The coordinates of the origin are (0, 0).

Parallelogram, p. 63

Parameter, p. 26, 27

Perimeter
The length of the boundary of a closed figure. It is expressed in units of length.

Perpendicular bisector, p. 162

Point of division, p. 149

Polygon
A closed plane figure with three of more sides.

Polygons

Number of sides	Name of polygon
3	Triangle
4	Quadrilateral
5	Pentagon
6	Hexagon
7	Heptagon
8	Octagon
9	Nonagon
10	Decagon
11	Undecagon
12	Dodecagon

Polygon (regular), p. 64

Polyhedron
A solid determined by plane polygonal faces.
E.g.

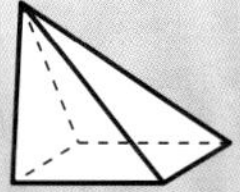

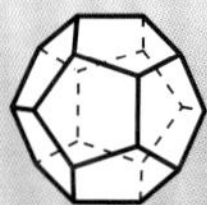

Polynomial
An algebraic expression containing one or more terms. E.g. $x^3 + 4x^2 - 18$

Prism
A polyhedron with two congruent parallel faces called "bases." The parallelograms defined by the corresponding sides of these bases are called the "lateral faces."

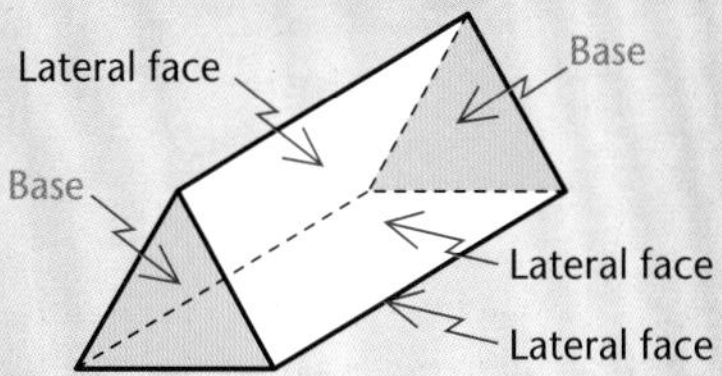

Regular prism
A prism whose bases are regular polygons.
E.g. A regular heptagonal prism.

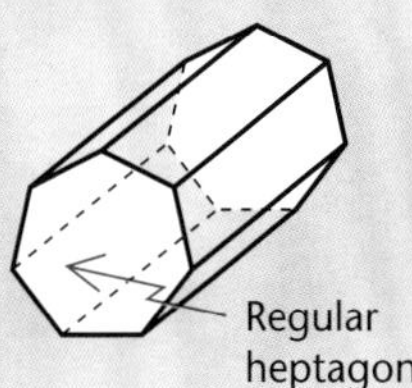

Right prism
A prism whose lateral faces are rectangles.
E.g. A right trapezoidal prism

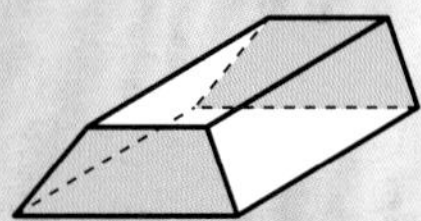

Proof, p. 73

Proportion
A statement of equality between two ratios or two rates.
E.g. 1) 3:11 = 12:44
2) $\frac{7}{5} = \frac{14}{10}$

Pyramid
A polyhedron with one polygonal base, whose lateral faces are triangles with a common vertex called the apex.
E.g. Octagonal pyramid

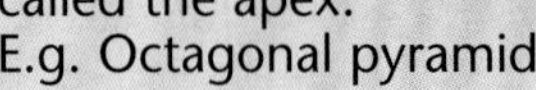

Regular pyramid
A pyramid whose base is a regular polygon.
E.g. A regular hexagonal pyramid

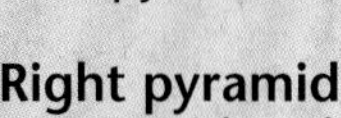

Right pyramid
A pyramid such that the segment from the apex, perpendicular to the base, intersects it at the centre of the polygonal base.
E.g. A right rectangular pyramid

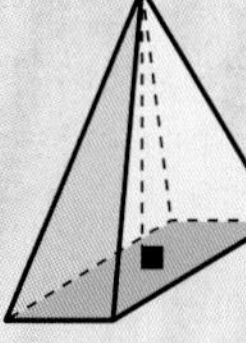

Pythagorean theorem
In a right triangle, the sum of the squares of the legs is equal to the square of the hypotenuse.

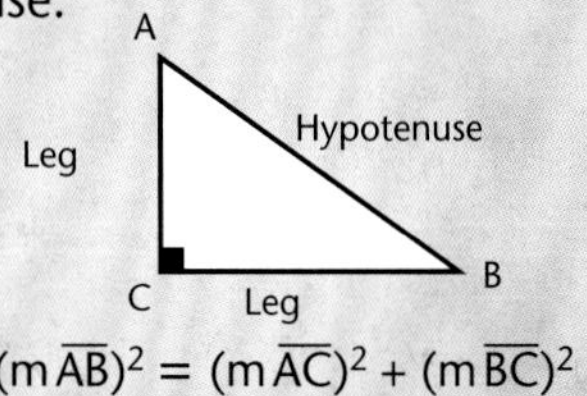

$(m\overline{AB})^2 = (m\overline{AC})^2 + (m\overline{BC})^2$

Quadrant
Each of the four regions defined by the axis of a Cartesian plane. The quadrants are numbered 1 to 4.

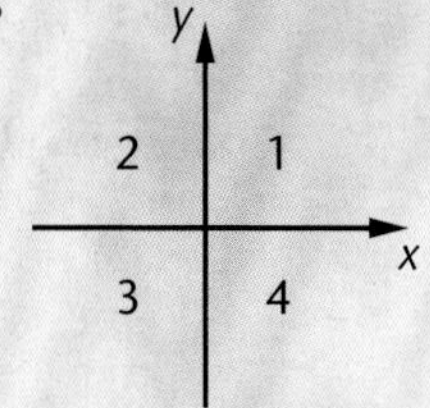

Radius
A radius is a segment (or length of a segment) which is determined by the centre of a circle and any point on the circle.

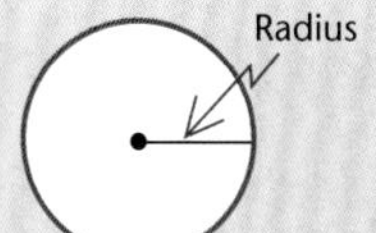

Range, p. 6

Rate
A way of comparing two quantities or two sizes expressed in different units and which requires division.

Rate of change
In a relation between two variables, a comparison between two corresponding variations.

$$\text{Rate of change} = \frac{\left(\text{variation of the dependent variable}\right)}{\left(\text{variation of the independent variable}\right)}$$

Ratio
A way of comparing two quantities or two sizes expressed in the same units and which requires division.

Rational number
A number that can be written as the quotient of two integers where the denominator is not zero. Its decimal representation can be terminating or non-terminating and periodic.

Ratio of similarity
Ratio of corresponding segments resulting from a dilatation.

Real number
A number belonging to the union of the set of rational numbers and the set of irrational numbers.

Rectangle, p. 65

Reflection
A geometric transformation which maps an initial point to an image point such that a given line (called the reflection line) is the perpendicular bisector of the segment determined by the point and its image. The reflection of a figure is the reflection of all of its points.

Relation, p. 5

Rhombus, p. 65

Right circular cone
Solid made of two faces, a circle and a sector. The circle is the base and the sector forms the lateral face.

Right circular cylinder
Solid made of three faces, two congruent circles and a rectangle. The circles form the bases and the rectangle forms the lateral face.

Rotation
A geometric transformation which maps an object to an image using a centre, an angle and a direction of rotation.

Rule
An equation which translates a relationship between variables.

Rules for transforming equations
Rules that allow you to obtain equivalent equations.
You can preserve the value of the equation:
- by adding the same amount to both sides of the equation
- by subtracting the same amount from both sides of the equation
- by multiplying both sides of the equation by the same amount or
- by dividing both sides of the equation by the same non-zero amount

Rules for transforming inequalities
Rules that allow you to obtain equivalent inequalities.
The value of an inequalities is preserved:
- by adding the same amount to both sides of the inequality
- by subtracting the same amount from both sides of the inequality
- by multiplying or dividing both sides of the inequalities by the same strictly positive amount

The direction of an inequality is reversed:
- by multiplying or dividing both sides of the inequality by the same strictly negative amount

Scientific notation
A notation which facilitates the reading and writing of numbers which are very large or very small.

E.g. 1) $56\ 000\ 000 = 5.6 \times 10^7$
2) $0.000\ 000\ 008 = 8 \times 10^{-9}$

Sector
Part of a circle defined by two radii.

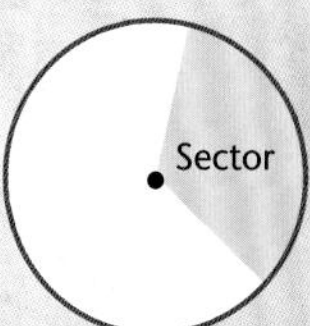

Similar figures
Two figures are similar if and only if a dilation enlargement or reduction of one results in a figure congruent to the other.

Sign of a function, p. 7

Slant height of a right circular cone
Segment (or length of a segment) defined by the apex and any point on the edge of the base.
E.g.

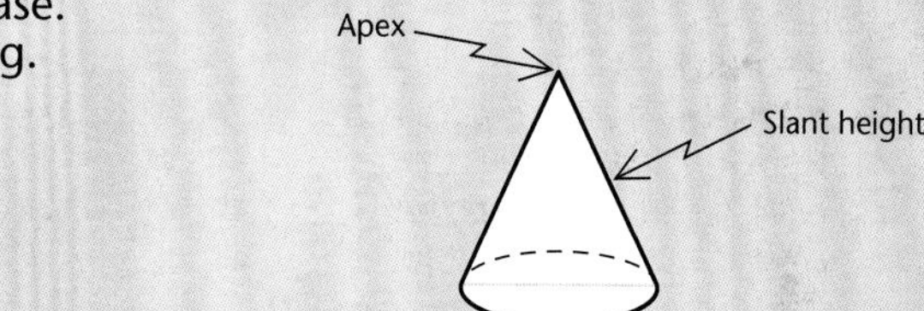

Slant height of a regular pyramid
Segment from the apex perpendicular to any side of the polygon forming the base of the pyramid. It corresponds to the altitude of a triangle which forms a lateral face.

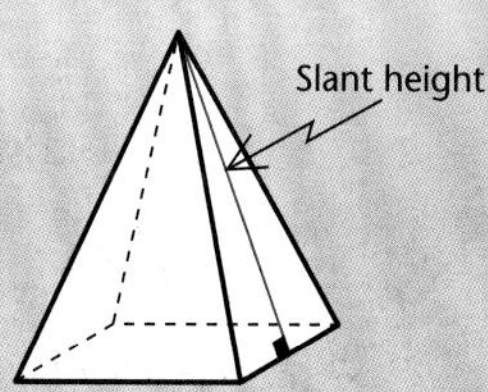

Solid
Portion of space bounded by a closed surface.
E.g.

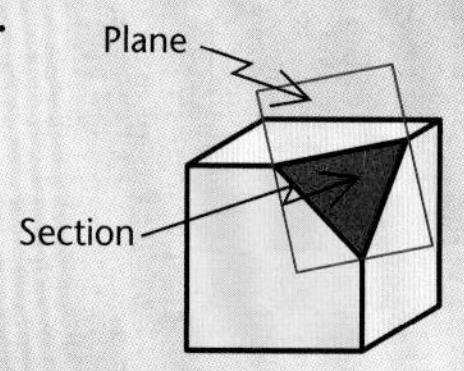

The section obtained by the intersection of this plane and this cube is a triangle.

Sphere
The set of all points in space at a given distance (radius) from a given point (centre).

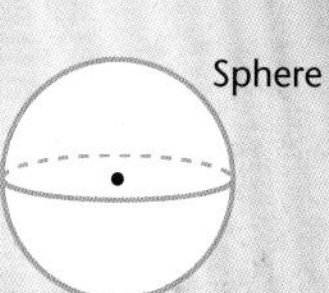

Square, p. 64

Square root
The inverse of the operation which consists of squaring a positive number is called finding the square root. The symbol for this operation is $\sqrt{\ }$.
E.g. The square root of 25, written $\sqrt{25}$, is 5.

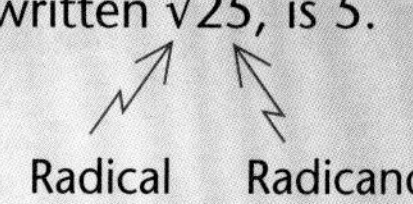

Surface area - see Area.

Solving system of equations
by comparison, p. 138, 139
by elimination, p. 174, 175
by substitution, p. 174

System of equations, p. 138, 139, 174, 175

T

Terms

Like terms
Terms composed of the same Constants or terms composed of the same variables raised to the same exponents.
E.g. 1) $8ax^2$ and ax^2 are like terms.
2) 8 and 17 are like terms.

Algebraic term
A term can be composed of one number or of a product of numbers and variables.
E.g. 9, x and $3xy^2$ are terms.

Theorem, p. 73

Translation
A geometric transformation which maps an initial point to an image point given a specified direction and length.

Trapezoid, p. 63

Triangles, p. 63
congruent, p. 72
similar, p. 84

Units of area
The square metre is the basic unit of area in the metric system (SI).

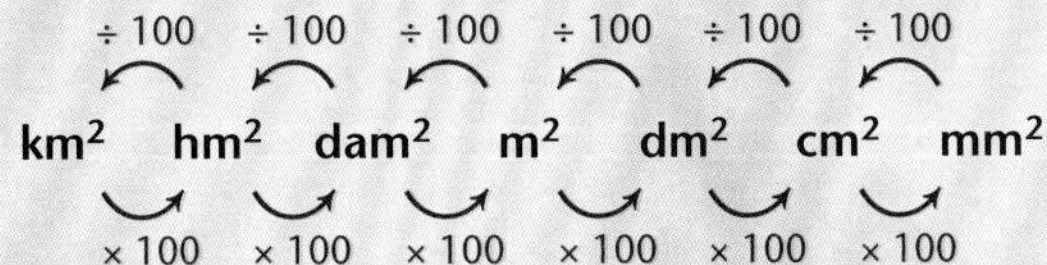

Units of capacity
The litre is the basic unit of capacity in the metric system (SI).

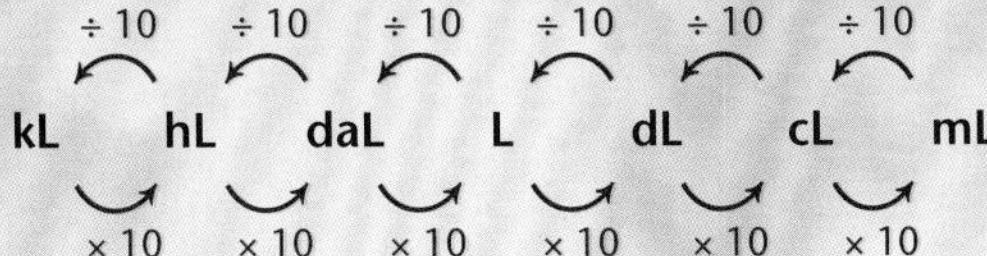

Units of length
The metre is the basic unit of length in the metric system (SI).

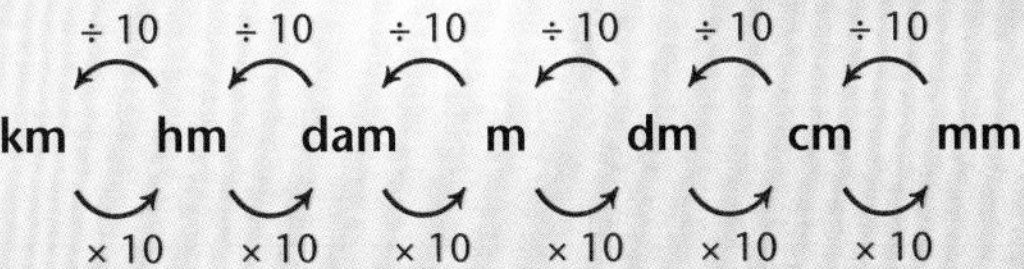

Units of volume
The cubic metre is the basic unit of volume in the metric system (SI).

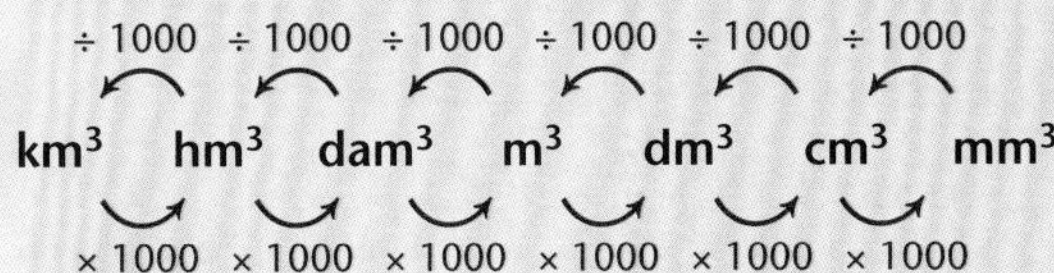

V

Variable
A symbol (generally a letter) which can take different values.

Dependent variable, p. 5

Independent variable, p. 5

Vertex of a solid
In geometry, a point common to at least two edges of a solid.

Volume
A measure of the space occupied by a solid, volume is expressed in cubic units.

Volume of a right circular cone

$$V_{cone} = \frac{(\text{area of the base}) \times (\text{height})}{3}$$

Volume of a right circular cylinder

$$V_{right\ circular\ cylinder} = (\text{area of the base}) \times (\text{height})$$

Volume of a right prism

$$V_{right\ prism} = (\text{area of the base}) \times (\text{height})$$

Volume of a pyramid

$$V_{pyramid} = \frac{(\text{area of the base}) \times (\text{height})}{3}$$

Volume of a sphere

$$V_{sphere} = \frac{4\pi r^3}{3}$$

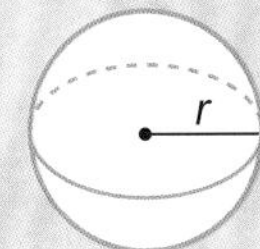

X

x-axis (horizontal)
A scaled line which allows you to determine the x-value (abscissa) of any point in the Cartesian plane.

x-intercept (zero)
In a Cartesian plane, an x-intercept is the x-value (abscissa) of a intersection point of a curve with the x-axis.

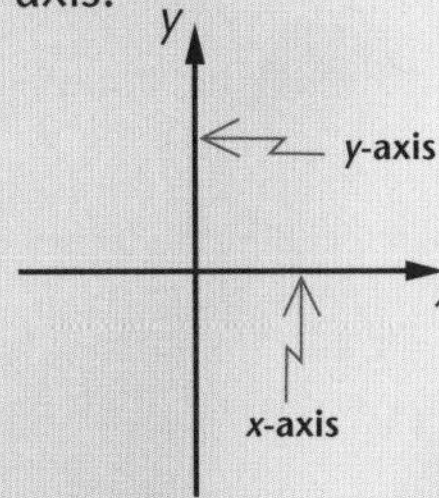

x-value (abscissa)
The first coordinate of a point in the Cartesian plane.
E.g. The x-value (abscissa) of the point (5, -2) is 5.

y-axis (vertical)
A scaled line which allows you to determine the y-value (ordinate) of any point in the Cartesian plane.

y-intercept (initial value)
In a Cartesian plane, the y-value (ordinate) of the point of intersection of a curve with the y-axis.

y-value (ordinate)
The second coordinate of a point in the Cartesian plane.
E.g. The y-value (ordinate) of the point (5, -2) is -2.

Photography Credits

T Top **B** Bottom **L** Left **R** Right **C** Centre **BG** Background

Cover

(1) © Shutterstock

Vision 1

3 TL © Damir Fabijanic/Grand Tour/Corbis **TR** © Bettmann/Corbis **CL** © David Fokos/Corbis **CR** © Scot Frei/ Corbis **4 CR** © Dominique Boivin **BR** 36915390 © 2008 Jupiter Images et ses représentants **5 TR** © TDLG **BL** 34610754 © 2008 Jupiter Images et ses représentants **9 BC** © Henrik Trygg/Corbis **10 BR** © Underwood & Underwood/Corbis **11 TR** Image Source/La Presse Canadienne **BL** © Otto Rogge/Corbis **BC** © Jon Hicks/Corbis **BR** © Richard T. Nowitz/Corbis **12 TR** 36907872 © 2008 Jupiter Images et ses représentants **CR** © Shutterstock **13 TL** © Chris Hellier/Corbis **TR** 36968988 © 2008 Jupiter Images et ses représentants **17 CR** © Shutterstock **18 TR** © Réunion des Musées Nationaux/Art Resource, NY **19 TL** © Martin Bernetti/AFP/Getty Image **21 BL** © Shutterstock **22 CR** © Collection/Publiphoto **BR** © Keith Delong/Flickr **23 TR** © cgb.fr **BR** © Peter Johnson/Corbis **24 TL** © Bettmann/Corbis **31 CL** 24726671© 2008 Jupiter Images et ses représentants **BR** © Shutterstock **32 BL** © iStockphoto **33 BL** © Shutterstock **34 TR** © louisetanguay **BR** 34727186 © 2008 Jupiter Images et ses représentants **36 BR** © Andre Forget/La Presse Canadienne **37 TR** © Jacques Boissinot/La Presse Canadienne **42 BR** © SPL/ Publiphoto **45 BL** © iStockphoto **47 TR** © Shutterstock **51 BL** © Richard Hamilton Smith/Corbis **54 CR** © Shutterstock **55 TR** © Shutterstock **56 TL** © The Art Archive/Corbis **CL** © Lebrecht Music and Arts Photo Library/Alamy **58 TL** © Bernard Annebicque/Sygma/Corbis **CR** © Shutterstock **BL** © Shutterstock **59 BL** © 2007 Google **62 TR** © Alexandra Kobalenko/Getty Image **63 TR** © Shutterstock **64 TR** © Shutterstock **66 BR** © Shutterstock **67 TR** © Luis Moreira/La Presse Canadienne **BL** © Archives de la Ville de Montréal

Vision 2

69 TL © Shutterstock **TR** © Bill Ross/Corbis **CL** © image100/Corbis **CR** © Shutterstock **70 BR** © akg-images **73 TR** © Shutterstock **74 CL** © iStockphoto **75 TR** 60513764 © 2008 Jupiter Images et ses représentant **76 TR** © Shutterstock **77 TR** © iStockphoto **CL** © Shutterstock **BR** © Damien Meyer/AFP/Getty Images **78 TR** © Keith Dannemiller/Corbis **CR** © Shutterstock **83 CR** © Leslie Foster **84 TR** © akg-images/Cameraphoto **85 BL** © Shutterstock **BC** © Shutterstock **BR** © Nathalie Ricard **86 CR** © Shutterstock **87 TR** © Robert Laberge/Allsport/ Getty Images **CR** © Henry Romero/Reuters/Corbis **BL** © Y. Marcoux/SPL/Publiphoto **88 BR** © Blaine Harrington III/Corbis **90 TR** © Steve Allen/SPL/Publiphoto **91 CL** © Shutterstock **CR** © Shutterstock **98 BR** © Club de Hockey Canadien Inc. **100 CR** © Shutterstock **101 BL** © Dan Lamont/Corbis **103 BR** © Medimage/SPL/Publiphoto **104 BR** © Shutterstock **105 TR** © Shutterstock BL © Daniel J. Cox/Corbis **106 BR** © DLILLC/Corbis **107 BL** © Sam Ogden/SPL /Publiphoto **115 CR** © Hank Morgan/SPL/Publiphoto **BL** © Gilbert Iundt/TempSport/Corbis **119 TR** 37006891 © 2008 Jupiter Images et ses représentants **CR** © Bettmann/Corbis **120 BR** © Jimin Lai/AFP/Getty Images **121 BL** © Niall Benvie/Corbis **127 CR** © Dr Luc Archambault **BR** © iStockphoto **128 BR** © iStockphoto **129 BL** © Kevin Frayer/La Presse Canadienne **130 BR** © George Bernard/SPL/Publiphoto **131 BR** © iStockphoto **132 TL** © akg-images **134 TR** © SPL/Publiphoto **CR** © George Steinmetz/Corbis **BL** © Heals protection Agency/ Corbis **BR** © SPL/Publiphoto **138 TR** © Orban Thierry/Corbis Sygma **CL** © Shutterstock **CR** © Shutterstock **140 BR** © Jeff Lewis/Icon SMI/Corbis **141 CR** © Nathalie Ricard **142 BL** © Reuters/Corbis **143 BR** © iStockphoto **144 CL** © Shutterstock **CR** © Shutterstock **145 TR** © Shutterstock

Vision 3

147 TL © Hisham Ibrahim/Getty Images **TR** © Paul Russell/Corbis **CL** © Sinclair Stammers/SPL/Publiphoto **CR** © Philip James Corwin/Corbis **148 BR** © Louie Psihoyos/Getty Images **154 TR** © iStockphoto **156 BR** © Gilles Paire/Fotolia.com **157 TL** © Shutterstock **TCL** © Shutterstock **TCR** © Shutterstock **TR** © iStockphoto **158 BR** © Shutterstock **164 TR** © Shutterstock **166 CR** © Shutterstock **167 CL** © Tim Ockenden/La Presse Canadienne **CR** © iStockphoto **BL** © iStockphoto **168 BL** © Shutterstock **169 TR** © Bettmann/Corbis **176 TC** © Shutterstock **178 BR** © iStockphoto **179 BL** © Franck Raux/RMN/Art Resource, NY **183 BR** © iStockphoto **BR** © Shutterstock **184 BL** 34638400 © 2008, Jupiter Images et ses représentants **185 CR** © Dinodia Images/Alamy **186 CL** © SPL/ Publiphoto **BL** © iStockphoto **187 TR** © Shutterstock **CL** © iStockphoto **188 TR** © Jorn Tomter/Zefa/Corbis **CL** © Viviane Moos/Corbis **CR** © Shutterstock **189 TR** © Shutterstock **193 BR** © Nathalie Ricard **195 CR** © Shutterstock

Learning and evaluation situations

198 BL © iStockphoto **199 CR** © iStockphoto **200 BL** 37806583 © 2008 Jupiter Images et ses représentants **BC** © Shutterstock **BR** 37834078 © 2008 Jupiter Images et ses représentants **201 CR** © Shutterstock **BL** © Shutterstock **202 BR** © Shutterstock **203 BL** © Shutterstock **BR** © Shutterstock **204 TL** © iStockphoto **BR** © Dr Jeremy Burgess/SPL/Publiphoto **205 BR** © Reuters/Corbis **206 BL** © Shutterstock **BR** © David Papazian/Corbis